Filter Design for System Modeling, State Estimation and Fault Diagnosis

This book analyzes the latest methods in the design of filters for system modeling, state estimation and fault detection with the intention of providing a new perspective of both theoretical and practical aspects.

The book also includes fault diagnosis techniques for unknown but bounded systems, their real applications on modeling and fault diagnosis for lithium battery systems, DC-DC converters and spring damping systems. It proposes new methods based on zonotopic Kalman filtering, a variety of state estimation methods of zonotope and its derived algorithms, a state estimation method based on convex space, set inversion interval observer filtering-based guaranteed fault estimation and a novel interval observer filtering-based fault diagnosis.

The methods presented in this text are more practical than the common probabilistic-based algorithms, since these can be applied in unknown but bounded noisy environments. This book will be an essential read for students, scholars and engineering professionals who are interested in filter design, system modeling, state estimation, fault diagnosis and related fields.

Ziyun Wang is an associate professor at Jiangnan University, China. His research interests include fault detection, state estimation and filtering methods.

Yan Wang is a professor at Jiangnan University, China. Her research interests include fault detection and set-membership filtering methods.

Zhicheng Ji is a professor at Jiangnan University, China. His research interests include state estimation and control theory in practical engineering.

Filter Design for System Modeling, State Estimation and Fault Diagnosis

Ziyun Wang, Yan Wang, Zhicheng Ji

CRC Press
Taylor & Francis Group
Boca Raton London New York

CRC Press is an imprint of the
Taylor & Francis Group, an **informa** business

This work is supported in part by the National Key Research and Development Program of China (2020YFB1710600), the Natural Science Foundation of Jiangsu Province(BK20221533) and the Jiangsu Science and Technology Association Young Science and Technology Talents Lifting Project (TJ-2021-006).

This book is published with financial support from National Key Research and Development Program of China, the Natural Science Foundation of Jiangsu Province, and the Jiangsu Science and Technology Association Young Science and Technology Talents Lifting Project.

First edition published 2023
by CRC Press
6000 Broken Sound Parkway NW, Suite 300, Boca Raton, FL 33487-2742

and by CRC Press
4 Park Square, Milton Park, Abingdon, Oxon, OX14 4RN

CRC Press is an imprint of Taylor & Francis Group, LLC

ISBN: 978-1-032-35512-2 (hbk)
ISBN: 978-1-032-35513-9 (pbk)
ISBN: 978-1-003-32721-9 (ebk)

DOI: 10.1201/b23146

Typeset in Nimbus Roman
by KnowledgeWorks Global Ltd.

Publisher's note: This book has been prepared from camera-ready copy provided by the authors.

Contents

Preface

System modeling and state estimation are two common ways to help engineers analyze the structure and working principles of equipment and industrial activities. By using some mathematical formulas, symbols, diagrams, we can obtain some necessary knowledge about the unknown system. Since the system model is adopted to describe or abstract the real world, essential input and output data are collected. The main objectives of system modeling and state estimation are giving maximum reduction of the system from the collected data. However, some noise term exist in the sampling process that affects the accuracy of modeling. For example, the irregular weak vibration, compared with the gear revolve, is a type of noise and it is hardly eliminated. To deal with the noise disturbance, some probabilistic-based algorithms are studied, such as Kalman filtering. These probabilistic-based algorithms have had widespread application in the past decades, but still have their disadvantages, mainly with dealing with the influence of unknown but bounded noise. In addition, when a fault occurs, some danger is possible if we do not detect it timely. Especially in cases of unknown but bounded noise, we cannot directly suppose the noise term satisfies any probability distribution, which makes the fault detection more difficult.

This book analyzes newer methods in the design of filters for system modeling, state estimation and fault detection, and it is intended to provide a new perspective of theoretical and practical aspects of the subject matter, which includes not only the system modeling and state estimation methods, but also fault diagnosis techniques for unknown but bounded systems and their real applications on modeling and fault diagnosis for lithium battery systems, DC-DC converters and spring damping systems. This book will offer a systematic presentation of all types of filter designs for system modeling, state estimation and fault diagnosis, which encompasses the definition of key concepts, system modeling for unknown but bounded systems, as well as filter design considerations for state estimation. Some novel fault diagnosis methods based on set-membership ideas will also introduced.

This book is intended primarily for an audience of senior undergraduate and graduate students whose major areas of interest are filter design, system modeling, state estimation, fault diagnosis and related fields, as well as engineering professionals from both academia and control engineering.

Ziyun Wang
Jiangnan University
Yan Wang
Jiangnan University
Zhicheng Ji
Jiangnan University

Symbol Description

Symbol	Description
$\mathbb{R}$	Set of real numbers
$\mathbb{R}^n$	Set of n-dimensional real vector
$\mathbf{B}^n$	Unitary box in $\mathbb{R}^n$
A^{T}	Transpose of matrix A
$\hat{A}$	Estimated value of matrix A
A^{-1}	Inverse of matrix A
$det(A)$	Determinant of matrix A
$tr(A)$	Trace of matrix A
$A > 0$	General notation for strictly positive definite matrix A
$A \geqslant 0$	General notation for positive definite matrix A
$A < 0$	General notation for strictly negative definite matrix A
$A \leqslant 0$	General notation for negative definite matrix A
$\boldsymbol{I}_n$	Identity matrix in $\mathbb{R}^{n\times n}$
$\mathbf{0}_n$	Zeros matrix in $\mathbb{R}^{n\times n}$
$\mathrm{diag}(a_1,\ldots,a_n)$	Diagonal matrix of dimension n
$\lvert\cdot\rvert$	Absolute value
$\Vert\cdot\Vert_\infty$	Infinity norm
$\Vert\cdot\Vert_P$	P-norm
$\Vert\cdot\Vert_F$	Frobenius norm
$s.t.$	Subject to
$\in$	Belongs to
$\notin$	Not belongs to
$\subset$	Subset
$\supset$	Contained in
$\cap$	Intersection
$\oplus$	Minkowski sum
$\odot$	Linear mapping
$conv(\cdot)$	Convex hull
max	Maximum value
min	Minimum value
$\emptyset$	Empty set

Chapter 1

Introduction

1.1 System modeling

1.1.1 Background

Automation refers to the use of equipment and systems in place of people to complete certain production tasks or perform affairs management. It includes production control automation and business management automation; these two aspects are interrelated and mutually promote each other, and have been involved in all aspects of our lives [56, 119]. This technology has established a new, perfect and advanced production and lifestyle for humankind. With the continuous development and innovation of science and technology, the application of automatic control technology has become more and more extensive and necessary in the current era.

The basic content of automatic control system research is to establish a mathematical model of the control system, which is called system modeling. The mathematical model is a mathematical description describing the dynamic characteristics of the system.

The transfer function in classical control theory is the mathematical model of the system in the complex domain. The differential equation obtained from the input and output of the system is the most basic form of the system mathematical model, and the transfer function is obtained from the Laplace transform of the differential equation. The transfer function is an external descriptive mathematical model that characterizes the system. It is independent of the input signal and can not only characterize the dynamic performance of the system but also study the impact of changes in system structure or parameters on system performance. Therefore, it is one of the main tools for studying classical control theory.

In classical control theory, for a linear time invariant system described by transfer function, a single variable can be directly connected with the input as the output. In

DOI: 10.1201/b23146-1

fact, the system also contains other independent variables in addition to the output variable. The transfer function is inconvenient to describe the intermediate variables of these contents, so it cannot contain all the information of the system. Obviously, in terms of whether the whole motion states of the system can be fully revealed, the transfer function has its shortcomings in describing a linear constant system.

With the time increasing, researchers prefer to understand the internal operation state of the system, but which classical control theory does not apply. Therefore, the modern control theory came into being and the mathematical model of state space equation is proposed. The state space equation includes a state equation expressing the relationship between input and state and an output equation expressing the relationship between output and state. When the state space equation is used to analyze a system, the dynamic characteristics of the system are described by the first-order differential equations composed of state variables. It can reflect the changes of all independent variables of the system, so as to determine all internal motion states of the system at the same time. In this way, designing a control system is no longer limited to inputs and outputs, which provides a powerful tool for improving system performance. In addition, computer can be used for analysis, design and real-time control, so modern control theory can be widely applied to nonlinear systems, time-varying systems and stochastic processes.

Whether in classical control theory or modern control theory, the establishment of system mathematical model is the basis of all subsequent control design.

1.1.2 Methods of system modeling

Nowadays, there are two methods of system modeling: analytical method and experimental method. Analytical method is to establish the dynamic equations of the system according to the laws of physics and chemistry, also known as white box method. The experimental method uses the system identification method, that is, adding a known signal to the system, recording the output of the system and then approximating it with a mathematical model to obtain the mathematical model of the system, also known as the black box method.

Since most industrial systems are complex and changeable, it is difficult to find physical and chemical laws to establish the system model, and system identification methods are generally used for system modeling.

The system is usually described by a mathematical model representing the input-output relationship of the system, which has its specific structure and parameters. Therefore, system identification includes system structure identification and parameter estimation.

System model structure is the form of system mathematical expression. For SISO linear systems, the model structure is the order of the system. For multivariable linear systems, the model structure is the controllability structure index or observability structure index of the system, and the system order is equal to the sum of the controllability structure index or observability structure index of the system. For the transfer function, the parameters of the system are the coefficients of its numerator denominator polynomial, and the order of the system is the degree of the denominator polynomial. For the state space model, the system parameters are the matrices of the state space model, and the system order is the dimension of the state vector or the dimension of the a matrix, which is equal to the sum of the controllable structure index or observable structure index of the system.

System identification is the mechanism modeling and parameter estimation of the internal structure of the system according to the input and output information of the system. Its principle is as follows: first provide a reference model type $u = M$ (the type and characteristics of this model are known), an input value u, and a more appropriate criterion function $J = l(y, YM)$ (generally, J can choose its absolute error function); then, through various system identification algorithms for calculation and approximation, the model that minimizes the selected absolute error J is the solution of the identification work. The mathematical model of the system obtained by the identification method is widely used in many fields. For example, predict the important parameters of the dynamic and static characteristics of the control system, build a model that can simulate the performance of the system characteristics under the actual working conditions and reliably predict the unknown changes of the system in the future through the measured system input and output, then select and construct the control strategy and means suitable for the system.

In many practical system identification problems, the structure of the system model can generally be determined. More of a problem is the estimation of important parameters in the model that are unknown or difficult to measure directly, that is, parameter estimation. These parameters represent some important properties of the system, which may be time-varying or time invariant. In this way, a model that can simulate the behavior of the real system is established, the future evolution of the system output is predicted with the input and output of the currently measurable system, and the controller is designed. Parameter identification provides a solution for system parameter calculation, which in turn provides model basis for object characterization, analysis, optimization, control and other applications.

System modeling is the basis of all automatic control problems, and system identification is the common method to solve it. First, system modeling is the key to the accurate understanding of specific physical and engineering problems from qualitative to quantitative. Second, the research and analysis of a control system can not only understand the working principle and characteristics of the system qualitatively but also describe the dynamic performance of the system quantitatively. Finally, system modeling is also the basis of system simulation and the cornerstone of subsequent state estimation and fault diagnosis.

1.2 State estimation

State estimation is to estimate the internal state of the dynamic system according to the available observation data and measurement data. The data obtained by measuring the input and output of the system can only reflect the external characteristics of the system, and the dynamic law of the system needs to be described by internal state variables (usually unable to be measured directly). Therefore, state estimation is of great significance for understanding and controlling a system.

The state space equation of the system is the basis of state estimation and includes state equation and output equation. Usually, we have two means to estimate the internal state of a system. One is to deduce the state of the next time through the state equation of the system and combined with the state of the previous time. Another is to obtain the system state with the help of the output equation. Both methods have their own uncertainties. The derivation using the state transition equation will be affected by the process noise. The process noise represents the characteristics of the system state changing with time, but we do not know the exact details of these changes. The method using the output equation will be affected by the measurement noise, which refers to the disturbance of the measured value caused by the external environment or the measuring instrument itself when measuring the output data. Both process noise and measurement noise will cause the inaccuracy of state estimation. Therefore, we generally use a combination of two methods, that is, the state space expression composed of state equation and output equation for state estimation.

Because of the presence of process noise and measurement noise, the system state required to establish the control law is not always available. In this case, the implementation of a state estimator is necessary. This state estimation problem can be solved by different methods, such as Luenberger observer [63], functional observer [74] and set-membership estimation. Owing to its ability to deal with uncertainties and disturbances, the set-membership estimation method [68] has been chosen in this book. This approach has been applied to the problem of state estimation of uncertain systems since the 1960s [3, 11, 80]. The set-membership estimation allows us to obtain a set containing the real system state, which is consistent with disturbances and measurement noise. With the development of robust control theory, the set-membership estimation technique is shown to be suitable for dealing with unknown but bounded uncertainties, disturbances and measurement noise.

State estimation is a popular approach in power system control and monitoring because it minimizes measurement noise level and obtains non-measured state variables. In the literature [83], an extended Kalman filter (EKF) was used to estimate the doubly fed induction generator (DFIG) wind turbines state variables. To estimate all state variables of DFIG wind turbine, it is necessary to develop a model that considers all state variables. So, an 18th-order nonlinear model is proposed for DFIG wind turbines, which is more complete than the models previously developed. Then the EKF method was used to estimate the DFIG state variables based on the proposed model. Simulation studies are done in four cases to verify the ability of the proposed model in the estimation of state variables under noise, wind speed variation and fault condition. The results show that EKF method can well estimate the internal state of DFIG system.

The literature [68] proposed an application of a guaranteed ellipsoidal-based set-membership state estimation technique to estimate the linear position of an octorotor used for radar applications. The size of the ellipsoidal set containing the real state is minimized at each sample time taking into account the measurements performed by the drone's sensors. Three case studies highlight the efficiency of the estimation technique in finding guaranteed bounds for the octorotor's linear position. The computed guaranteed bounds in the linear trajectory of the estimated results are exploited to find the maximum operating frequency of the radar, which are necessary information in radar applications.

1.3 Fault diagnosis

With the rapid development of science and technology, the scale of modern complex engineering systems is becoming larger and larger, such as power system engineering [45], metallurgy and petrochemical industry [97] and power grid system [84]. These modern industries are characterized by large-scale, integration and automation and high dependence on equipment. System modeling and state estimation are important for these systems, and so is fault diagnosis. Once the system fails, it will cause economic losses and endanger the personal safety of employees. Therefore, timely and rapid fault diagnosis of industrial system is very important.

Fault diagnosis refers to a means to evaluate the overall operation and fault status of the system by using various information in the system. Specifically, it can detect whether there is a fault in the system and equipment by using various inspection and test methods, and then further determine the location of the fault and the specific fault type or value in the failed system. The process of fault diagnosis can include three links: fault detection, fault isolation and fault identification.

The research on fault diagnosis technology was first put forward in the doctoral thesis published by Beard scholars in 1971 [9], which has attracted the attention of the academic circle. After that, with the investment of a large number of funds, human and material resources, fault diagnosis technology gradually started and continued research and development. In 1976, Willsky published the first comprehensive article related to fault detection and diagnosis on Automatica [109]. Subsequently, the first academic work in the field of fault detection and diagnosis was published in 1978, which laid a solid theoretical foundation for the development of fault detection and diagnosis technology [40]. Under the research and discussion of scholars for nearly half a century, fault diagnosis technology has shown a good development trend all the way. Various algorithms have been proposed, developed, improved and become mature. At present, fault diagnosis technology is mainly divided into three categories: analytical model-based method [78], knowledge-based method [33] and signal processing method [62].

The fault diagnosis method based on analytical model is not only the earliest research method but also the most in-depth and mature method. This method is based on the accurate mathematical model of the diagnosed object and diagnose the system according to certain mathematical means, which has high diagnosis accuracy. Thus it is not suitable for systems that are difficult to obtain accurate mathematical models, and is sensitive to unknown faults. The fault diagnosis methods based on analytical model mainly include state estimation method, parameter estimation method and equivalent space method. There are some connections between the three methods, but they develop independently at the same time. Both state estimation method and parameter estimation method belong to the estimation field. The accuracy of fault diagnosis depends on the accurate analysis of the system model, that is, by processing a series of actual observation data with interference signals and measurement noise, we can obtain the estimated values of various required states or parameters. The basic principle of equivalent space method is to detect the equivalence of the system mathematical model of the diagnosed object through the actual system input and output values to carry out fault diagnosis, that is, the direct redundancy or instantaneous redundancy of the diagnosed system is given according to the relationship between input and output, so as to detect whether the system is faulty.

The knowledge-based fault diagnosis method does not need an accurate system mathematical model or an in-depth understanding of the specific working mechanism of the system. It is a method to intelligently process the object information and realize fault diagnosis from the perspective of knowledge. With the development of artificial intelligence technology, the application research of knowledge-based fault diagnosis method is deepening and expanding. This method is suitable for the field of nonlinear systems, makes clever use of the characteristics of intelligent knowledge and has good research prospects, but it also depends too much on relevant empirical knowledge. The knowledge-based fault diagnosis methods mainly include the methods based on expert system, neural network and knowledge-based fault tree.

It is often unrealistic to establish an accurate mathematical model for some system processes with dynamic performance. But considering that the system process will produce various signals, the fault diagnosis method based on signal processing is widely studied. Through the measured value of the obtained signal, the signal processing methods such as wavelet transform, spectrum analysis and principal component analysis are used for analysis and processing, and the eigenvalues such as amplitude, variance and frequency of the measurable signal are extracted for fault diagnosis. The commonly used methods based on signal processing include wavelet analysis, spectrum analysis, information fusion and so on. In reality, some systems are very complex and have strong nonlinearity, so it is difficult to obtain their accurate mathematical model. The fault diagnosis method based on signal processing just solves this problem effectively. It is widely used in large-scale production and chemical process where mathematical model cannot be established because of good real-time performance and simple implementation. But at the same time, this method is ineffective for early potential faults, because it is effective only when the fault occurs to a certain extent and affects the external characteristics, which is difficult to locate the fault in most cases. Therefore, this method is often combined with other fault diagnosis methods to improve the performance of fault diagnosis.

This book will use the method based on analytical model and combine with filter theory for fault diagnosis. Filtering refers to filtering the system noise according to the observed signal and algorithm criteria in a system disturbed by noise, which is an effective measure to suppress system interference. The idea of filtering is integrated into fault diagnosis, which can effectively filter most of the noise interference of the system, improve the accuracy of fault diagnosis and improve the practicability and reliability of the algorithm. The fault diagnosis method based on filter is an important method in the fault diagnosis method based on analytical model.

1.4 Summary of filtering design methods

1.4.1 Traditional filtering design methods

Generally, system identification, state estimation and fault diagnosis are carried out according to the available input, output and measurable observation data of the system, which are essential and will be affected by noise. The noise will cause the inaccuracy of data, so it is necessary to filter the data to obtain the true data.

The first proposed filtering theory is Wiener filtering. Wiener, an American scientist known as the pioneer of random process and noise signal processing, put forward Wiener filtering theory in 1942 [114]. This theory realizes the filter design by analyzing the power spectrum, but it requires too much calculation of Wiener equation, uses infinite historical data, cannot process real-time data, and is not suitable for the filtering of multivariable, time-varying and non-stationary random signals. Therefore, these limitations make Wiener filtering very limited in practical application.

In order to solve the above problems, R.E. Kalman and R.S. Bucy proposed a filter recursive estimation calculation method based on the system state equation, measurement equation and relevant statistical characteristics in the 1960s [47]. The required signal is estimated from the measurement data through the algorithm called Kalman filter. Kalman filter provides an optimal solution for linear Gaussian problem and has been widely used in target tracking [29], robot control and fault diagnosis [23] in recent years.

Whether Wiener filter or Kalman filter, these methods are only suitable for linear systems, and need to have sufficient knowledge of the estimated process. For nonlinear systems or complex estimation problems with incomplete understanding of dynamic system characteristics, further research is needed. Some approximate calculation methods can be used in engineering. The common ones are the extended Kalman filtering [111] and unscented Kalman filtering [91], and adaptive filter [125] that can automatically modify parameters according to the historical knowledge of the filtering process, which extend their application range.

Particle filter realizes recursive Bayesian estimation based on Nonparametric Monte Carlo simulation method, and describes the posterior probability distribution of system state in the form of samples [2, 54]. The characteristic of particle filter is that it needs to sample a large number of particles for estimation, which makes the algorithm have the problems of large amount of calculation and poor real-time performance. However, because particle filter is generally suitable for dealing with nonlinear and non-Gaussian filtering problems, it has attracted the attention of experts and scholars in many disciplines, and an improved algorithm of interdisciplinary fusion is proposed.

The above traditional filter design method requires that the process disturbance and measurement noise of the system meet the specific distribution requirements. However, the prior knowledge of the probability distribution of noise and interference in many practical engineering fields is generally difficult to obtain in practical systems, and even some noise itself does not have random characteristics, which is difficult to be described by statistical laws [1]. If these methods are applied to solve the system modeling, state estimation and fault diagnosis problem of the actual system, they may lead to inaccuracy of system modeling, state estimation and missing judgment or misjudgment of fault diagnosis. Therefore, considering the existence of the above problems, these methods have some limitations in practical application.

1.4.2 Non-probabilistic filtering design method

In view of the limitations of traditional filtering design methods, set-membership filtering method with unknown but bounded (UBB) noises is applied and has attracted the attention of experts and scholars.

For ease of understanding, take the state estimation of the following discrete-time linear system as an example:

$$\begin{cases} \boldsymbol{x}_{k+1} = A\boldsymbol{x}_k + B\boldsymbol{u}_k + D_1\boldsymbol{w}_k, \\ \boldsymbol{y}_k = C\boldsymbol{x}_k + D_2\boldsymbol{v}_k, \end{cases} \tag{1.1}$$

where $\boldsymbol{x}_k \in \mathbb{R}^n$, $\boldsymbol{u}_k \in \mathbb{R}^r$ and $\boldsymbol{y}_k \in \mathbb{R}^p$ are the state, input and output vectors, respectively. A, B, D_1, C, D_2 are the parameter matrices with appropriate dimensions. $\boldsymbol{w}_k \in \mathbb{R}^w$ and $\boldsymbol{v}_k \in \mathbb{R}^v$ are the unknown but bounded disturbance and measurement noise. It is assumed that the initial state is also unknown but bounded. The set-membership filtering estimator calculates a feasible set containing all the possible system states and consistent with disturbance, uncertainty and the measurement noise at each sampling time. The problem of set-membership estimation is that the complexity of the feasible set increases over time. To overcome this problem, the feasible set can be approximately described by regular spatial geometry [13]. According to different envelope methods, the spatial geometry commonly used to approximately describe feasible sets are ellipsoid [105, 106], cube [104], interval [76, 104], polytope [53], parallelotope [53] and zonotope [4, 22], etc. According to the geometry of the set, the set-membership filtering algorithms mainly include ellipsoidal algorithm, cubic algorithm , interval algorithm, polytopic algorithm parallelotopic algorithm and zonotopic algorithm, etc.

For example, it is assumed that the uncertainties and the initial state in the system (1.1) are bounded by ellipsoids, which means $\boldsymbol{w}_k \in \mathcal{W}, \boldsymbol{v}_k \in \mathcal{V}$ and $\boldsymbol{x}_0 \in \mathcal{X}_0$. Considering the mathematical model and these assumptions, the set-membership estimation computes the ellipsoidal set containing all possible values of the unknown state $\boldsymbol{x}_k$. Similar to the Kalman filter, the set-membership estimation includes three basis steps: prediction, measurement and correction.

Assuming the state set $\hat{\mathcal{X}}_{k-1}$ that contains all the possible values of the system is known, the state set $\hat{\mathcal{X}}_k$ can be estimated using the following algorithm.

1. Prediction step: given system (1.1), use the state equation to compute a predicted ellipsoid $\bar{\mathcal{X}}_k = A\hat{\mathcal{X}}_{k-1} \oplus \mathcal{W}$ that offers a bound for the uncertain trajectory of the system.

2. Measurement step: compute the measurement consistent state set $\mathcal{X}_{\boldsymbol{y}_k}$ using the measurement $\boldsymbol{y}_k$, that is, $\mathcal{X}_{\boldsymbol{y}_k} = \{\boldsymbol{x} \in \mathbb{R}^n : \boldsymbol{y}_k - C\boldsymbol{x} \in \mathcal{V}\}$.

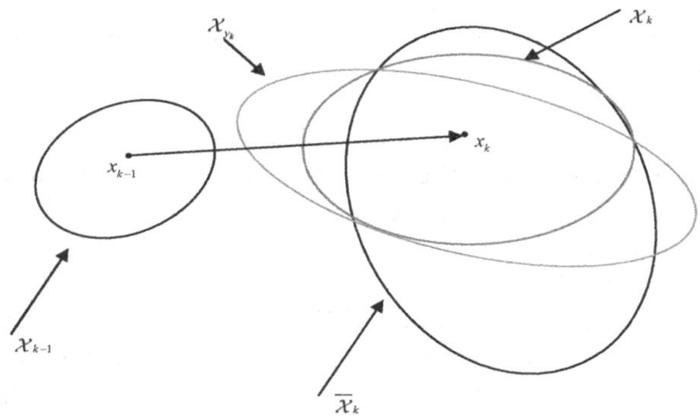

Figure 1.1: The process of set-membership filtering algorithm.

3. Correction step: compute the state estimation set $\hat{\mathcal{X}}_k$ at time k by the intersection between $\mathcal{X}_{\mathbf{y}_k}$ and $\bar{\mathcal{X}}_k$.

The general case of the proposed set-membership filtering algorithm is illustrated in Fig. 1.1, where x_{k-1}, x_k represent the true value of the system at time $k-1$ and time k, respectively.

Compared with traditional filtering methods, set-membership filtering method has the following advantages:

1. UBB noise assumption is more practical than random noise assumption in many practical cases, such as measurement error introduced by analog-to-digital converter or sensor, rounding error of machine number and modeling error, which can be regarded as bounded error.

2. UBB noise hypothesis requires less priori knowledge. It only requires the upper and lower bounds of the uncertainty to be known regardless of the distribution within the bound. Therefore, the set-membership filtering method under UBB noise assumption is more robust than that under random noise assumption.

3. Because the modeling uncertainty of the system can also be regarded as a bounded noise when dealing with the parameter uncertain system, only the equivalent bounded noise characteristics of the uncertain system can be obtained. Therefore, the set-membership filtering method can conveniently deal with the situation that the system has unmodeled dynamics.

Ellipsoidal set-membership filtering algorithm is a set-membership filtering algorithm using ellipsoid as an approximate feasible set. The advantages of ellipsoidal set-membership filtering algorithm include:

1. Its recursive form and calculation process is relatively simple.
2. The estimation result can be expressed explicitly and its boundary is smooth and differentiable.

These advantages make the ellipsoidal set-membership filtering algorithm become an important type of set-membership filtering algorithm, which has been widely studied by scholars. Although the ellipsoidal filter method is simple to calculate and high efficiency, the solution process is very conservative, and the enclosed feasible parameter set area is often large [93, 120].

Schweppe was an early scholar who studied the ellipsoidal set-membership estimation algorithm. In 1968, Schweppe used ellipsoid approximation to describe the state feasible set and gave the ellipsoidal set-membership filtering algorithm [34]. In 1982, Fogel and Huang introduced ellipsoid into the approximate description of parametric feasible set and proposed two recursive external ellipsoid algorithms (FH algorithm for short) [35], one is to minimize ellipsoid volume, the another is to minimize the ellipsoid trace algorithm. In order to deal with the complexity of FH algorithm, Deller gives an algorithm implementation of systolic structure, which effectively improves the operation speed [30]. Belforte and Bona proposed a modified FH algorithm [11] (referred to as MFH algorithm for short) based on FH algorithm. Compared with FH algorithm, MFH algorithm can obtain ellipsoid with smaller volume. Pronzato proved that MFH algorithm is mathematically equivalent to minimum volume ellipsoid EPC algorithm with parallel cuts. Cheung derived the so-called optimal volume ellipsoid algorithm, which is actually equivalent to EPC algorithm by geometric method, and gave the convergence conclusion [24]. Then, the ellipsoid set-membership filter estimation method has been developed rapidly, achieved fruitful research results, and solved a series of practical problems in system identification, state estimation and fault diagnosis.

In the literature [126], an ellipsoid bounding set-membership identification algorithm was proposed to propagate bounded uncertainties rigorously and the guaranteed feasible set of fault parameters enveloping true parameter values is given. Fault caused by abrupt parameter variations can be detected and isolated on-line by consistency check between predicted and observed parameter sets obtained in the identification procedure. The proposed approach provides the improved robustness with its ability to distinguish real faults from model uncertainties, which comes with the inherent guaranteed robustness of the set-membership framework.

The literature [79] used the ellipsoid to wrap the real parameter values, and judged whether there is a fault by detecting whether the ellipsoid and the measuring strip overlap. After the fault occurs, the ellipsoid is expanded globally and updated with the measuring strip to complete the fault isolation. The numerical simulation analysis results demonstrate the effectiveness of the proposed algorithm.

Milanese and Belforte proposed the cubic set-membership estimation method for linear systems [70]. They transformed the cubic external calculation problem of feasible sets into solving several linear programming problems. Later, Milanese and Vicino introduced the geometric programming method into set-membership identification, and gave a cubic set-membership estimation algorithm for a class of nonlinear systems [71]. Because the ellipsoid and cube of feasible sets usually contain a large number of incompatible elements, they are relatively conservative in practice. In order to accurately describe feasible sets, approximate or exact polytope algorithms are given [88]. But then it was found that the improvement of accuracy brought a huge computational burden. In order to make a better compromise between accuracy and computational complexity, Vicino and Chisci proposed the minimum volume accumulation parallelotope algorithm [95]. The simulation results show that the parallelotope algorithm can obtain a satisfactory approximate description effect at a low computational cost.

The literature [14] proposed a robust set-membership fault detection method based on the use of polytopes to bound the parameter uncertainty set. The algorithm is able to handle systems with bounded parameter variation between samples. It is shown that consistency checks indicating faults can be performed in a natural manner with a polytopic description of the feasible parameter set.

Jaulin and Walter were the first scholars to apply interval analysis to set-membership estimation. In 1993, Jaulin and Walter proposed a set inversion via interval analysis (SIVIA) algorithm, which can deal with the set-membership identification problem of nonlinear systems [43]. After 2000, set-membership filtering algorithms based on interval analysis have also been published [49, 75]. Due to the need for dichotomy, SIVIA is not suitable for processing high-dimensional data. To solve this problem, literatures [42, 44] proposed an improved algorithm that can overcome the difficulty of dimensionality disaster. The set-membership filtering algorithm based on interval estimation can deal with the set-membership estimation problem of a large class of nonlinear systems, and the effect is very remarkable. In addition, one of its advantages is that it can ensure the characteristics of global optimization, which most algorithms do not have.

In literature [129], an interval observer for part of the measurable output is proposed and an unknown input reconstruction method based on the interval observer was developed. Finally, an observer-based state feedback and unknown input controller is designed and the closed-loop system stability is analyzed and simulation results are provided to illustrate the effectiveness of the proposed method.

Zonotope is the affine transformation of the unit hypercube. In 1998, Kunn first pointed out in his article that the use of zonotope can effectively control the warping effect [50]. Alamo et al. [4] and Combastel [26] gave zonotope set-member filtering algorithms for nonlinear discrete-time systems and nonlinear continuous time systems respectively. Because the time-varying property of system parameters can be accurately expressed by zonotope without approximation, Bravo et al. [15] also used this excellent property to give a zonotope set identification algorithm for time-varying parameter discrete-time systems based on the research work of [4]. In the measurement updating process of the zonotope set-membership estimation algorithm, it is necessary to calculate the intersection of the zonotope and the strip. The external calculation methods were proposed in literatures [4, 15]. To overcome the flaw of external calculation methods, the literature [26] proposed the singular value decomposition method to calculate the zonotope containing the intersection of the zonotope and the strip. After that, the zonotope set-membership filtering algorithms have been developed rapidly [22, 55].

The literature [22] proposed an improved zonotopic method to set-membership state estimation for nonlinear discrete-time systems with a bounded description of noise and parameters. According to the previous method, a zonotope containing the uncertain trajectory of the system is computed using interval arithmetic in the time update. In the observation update, an improved intersection algorithm of zonotope and strip is used to obtain a zonotope bounding the intersection of the uncertain trajectory with the region of the state space consistent with the observed output. This obtained zonotope is used to describe the feasible solution set of state. Simulation experiments proved that the zonotope obtained by the improved algorithm can be tighter than the one obtained by the previous algorithm.

The literature [103] designed a zonotopic fault estimation filter based on the analysis of fault detectability indexes. To ensure estimation accuracy, the filter gain in the zonotopic fault estimation filter is optimized through the zonotope minimization. The designed zonotopic filter can not only estimate fault magnitudes, but also provide fault estimation results in an interval, i.e., the upper and lower bounds of fault magnitudes. Moreover, the proposed fault estimation filter has a non-singular structure and hence it is easy to implement. Finally, simulation results are provided to illustrate the effectiveness of the proposed method.

The literature [55] studied zonotopic fault detection observer design for a class of linear parameter-varying descriptor systems. The disturbance and measurement noise are unknown but can be bounded by zonotopes. A zonotopic method is presented to estimate the envelope of residual. To attenuate the effect of disturbance and measurement noise, a zonotopic size optimization criterion is implemented based on P-radius. To improve the fault detection performance, a finite frequency is introduced based on the generalized Kalman-Yakubovich-Popov lemma. Simulation examples of a mass spring and a direct current motor are utilized to demonstrate the performance of the proposed method.

Among them, the ellipsoidal and parallelotopic set-membership filtering algorithm are generally recursive algorithm, the calculation process is relatively simple, but the accuracy is relatively low. The cubic set-membership filtering algorithm is usually a batch algorithm with moderate accuracy, but it has the problem of large amount of computation. The polytopic set-membership filtering algorithm has the highest accuracy, but the calculation process is more complex and requires large computer storage capacity. The calculation process of the zonotopic set-membership filtering algorithm is slightly complex, but the accuracy is improved. Compared with the above algorithms, the interval set-membership filtering algorithm has different ideas in the approximate description of feasible sets. The result of the algorithm is not a single geometry, but the union of multiple geometry, and the union can approach the feasible set with arbitrary accuracy. However, due to the defects of interval analysis, the convergence speed of the algorithm may not be good.

Because the set-membership filtering method has the above advantages, as soon as the set-membership filtering method was proposed, it has been widely used in system modeling [127], state estimation [8] and fault diagnosis [77].

1.5 Motivation and objective

Based on the above literature analysis, the filtering theories and methods have been widely developed in recent decades. The application of filtering theory in system identification, state estimation and fault diagnosis has also been highly valued by scholars, and more innovative research results have been obtained.

As one of the important application fields of filtering theory, state estimation can be carried out by designing appropriate filters, which can be successfully applied to the research of state estimation in electric power, chemical industry, machinery and other industries, so as to obtain the real state of real-time operation of the system. At present, for the problems of system identification, state estimation and fault diagnosis, although the research based on filter design method has achieved a lot of research results, its theoretical research is far from perfect, and a lot of research work still needs to be done. Moreover, most traditional filter designs must assume that the system noise satisfies a certain probability distribution; however, as the complexity of the system increases and the operating environment of the system continues to change, it is increasingly difficult to obtain system noise that satisfies the requirements of the probability distribution assumptions. In order to solve this problem, set-membership filter estimation method came into being, and effectively widened the research innovation and practical application field based on filter design methods.

Nowadays, the set-membership filtering method based on unknown but bounded noise has been applied to system identification, state estimation, fault diagnosis and other fields, but the theoretical research still has room for further improvement and expansion. For example, for the system with unknown noise distribution parameters to be identified, the existing methods only focus on the constant system parameters, and then the actual system is complex and changeable, and its system parameters are constantly changing. For another example, the existing fault diagnosis often focuses on the sudden fault type, and there is less research on the slow fault with the continuous increase of fault amplitude but its influence cannot be ignored. However, the slow change fault is the root that affects the service life of industrial equipment and brings uncontrollable production safety.

To sum up, we must continue to research and explore in this field in order to improve the theoretical system of system modeling, state estimation and fault diagnosis methods based on filter design. On the basis of the research results of existing scholars, it is of great practical significance to continue to carry out learning and research and apply it to the field of practical engineering.

1.6 Outlines

Although some results have been achieved in the research on the theory, method and application of set-membership filtering, there is still a lot of work to be done. Based on analyzing the existing results, further research work is carried out in this book. The research work here focuses on the improvement of existing methods, the proposal of new methods, and the application of set-membership filtering theory and methods in system modeling, state estimation and fault diagnosis.

The following contents and arrangements of this book are as follows: In Chapter 2, for the traditional time-varying parameter estimation method that uses ellipsoids and zonotopes as the feasible parameter set is highly conservative and complexity, a time-varying parameter system estimation method based on zonotope-ellipsoid double filtering is proposed. In Chapter 3, the zonotope and its derived innovative algorithm are proposed for different practical application scenarios. In Chapter 4, based on the state estimation of convex space structure, the state estimation methods of hyperparallel space filtering and axisymmetric box space filtering for nonlinear systems are proposed respectively. In Chapter 5, a guaranteed fault diagnosis algorithm based on interval set inversion observer filtering and a novel interval observer filtering methods are proposed for linear discrete-time systems first. Second, an orthometric hyperparallel spatial directional expansion filtering based fault diagnosis method is proposed for discrete systems. In Chapter 6, three fault diagnosis methods based on zonotopic Kalman filtering are proposed. Finally, in Chapter 7, the research contents of this book are summarized, and the future development direction of system modeling, state estimation, fault diagnosis methods based on set-membership filtering method is prospected.

Chapter 2

Parameter estimation algorithm based on zonotope-ellipsoid double filtering

2.1 Problem description

2.1.1 Model parameterization

Consider the following linear discrete-time system:

$$y(k) = \Psi_k^{\mathrm{T}} \theta(k) + e(k), \tag{2.1}$$

$$\theta(k) = \theta(k-1) + w(k), \tag{2.2}$$

where Ψ_k is a regression vector consisting of input and output signals, $\theta(k)$ is the parameter to be identified, $e(k)$ is unknown but bounded noise that satisfies $|e(k)| \leqslant \sigma$ and $w(k)$ is an unknown system parameter change vector and satisfies $\|w(k)\|_\infty \leqslant \gamma$. This study considers the bounded parameter changes, i.e., γ satisfies $0 < \gamma < \infty$.

The feasible parameter set $\mathcal{S}(k)$ describing the compatible measured value, output value and bounded noise at time k is defined as

$$\mathcal{S}_k = \left\{ \theta_k : \left| y(k) - \Psi_k^{\mathrm{T}} \theta_k \right| \leqslant \sigma \right\}, \tag{2.3}$$

where $k = 1, \ldots, N$. The feasible parameter set satisfies a constraint strip $\mathcal{S}(k)$ expressed by Eq. (2.3). The intersection of these constraint strips constitutes the

DOI: 10.1201/b23146-2

feasible parameter set for the time-varying system at time k as follows:

$$\Theta_k = \Theta_{k-1} \cap \mathcal{S}_k = \left(\bigcap_{i=1}^{k-1} \mathcal{S}(i) \right) \cap \Theta_0,$$

where Θ_0 is the initial feasible parameter set. Assume that $\{y(k), \Psi_k\}$ are known at $k = 1, 2, \ldots, N$, and the system defined by Eq. (2.1) is not faulty, and the system satisfies: (1) $\theta = \theta_N$; (2) $\theta_0 = \Theta_0$; (3) $\left| y(k) - \Psi_k^{\mathrm{T}} \theta \right| \leqslant \sigma, k = 1, 2, \ldots, N$; (4) $|\theta_k - \theta_{k-1}| \leqslant \gamma, k = 1, 2, \ldots, N$.

Although an accurate description can accurately obtain the parameter change of the time-varying system, the computation of the solution set of feasible parameters is too complex. In the process of using approximate parameter feasible solution sets to identify time-varying parameters, it is often necessary to use a regular geometric space structure to expand the feasible parameter solution set at each moment. Considering the advantages of ellipsoids and zonotopes, which are graphically regular and easy to calculate, this study uses ellipsoids and zonotopes as approximate parameter feasible sets and finally uses ellipsoids to describe the feasible parameter set.

2.1.2 Symbol definitions

Given two sets X and Y, the Minkowski sum can be expressed as

$$X \oplus Y = \{x + y : x \in X, y \in Y\}. \tag{2.4}$$

A unit interval is defined as $\mathbf{B} = [-1 \quad 1]$. A unit box is composed of multiple unit intervals. For example, m unit intervals can form an m-order box $\mathbf{B}^m$. Assume a box $\mathbf{Q} = \left[\, [\, a_1 \quad b_1 \,] \quad \cdots \quad [\, a_n \quad b_n \,] \, \right]^{\mathrm{T}}$; then, its center and width can be expressed, respectively, as

$$\mathrm{mid}(\mathbf{Q}) = \left[\frac{a_1 + b_1}{2}, \cdots, \frac{a_n + b_n}{2} \right]^{\mathrm{T}}, \tag{2.5}$$

$$\mathrm{diam}(\mathbf{Q}) = \left[\frac{b_1 - a_1}{2}, \cdots, \frac{b_n - a_n}{2} \right]^{\mathrm{T}}. \tag{2.6}$$

An m-order zonotope is defined as

$$p \oplus H\mathbf{B}^m = \{p + H\Delta : \Delta \in \mathbf{B}^m\}, \tag{2.7}$$

where $p \in \mathbb{R}^n$ is the center of the zonotope, and $H \in \mathbb{R}^{n \times m}$ is the shape matrix of the zonotopes.

An n-order ellipsoid $E(c, P)$ is defined as

$$\{x \in \mathbb{R}^n : (x - c)^{\mathrm{T}} P^{-1} (x - c) \leqslant 1\}, \tag{2.8}$$

where $c \in \mathbb{R}^n$ is the center of the ellipsoid, and $P^{-1} \in \mathbb{R}^{n \times n}$ is the symmetric and positive definite envelope matrix of the ellipsoid.

Lemma 2.1
The Minkowski sum of the zonotope can decompose the zonotope into the following forms: $p \oplus H\mathbf{B}^m = p \oplus H_1\mathbf{B} \oplus H_2\mathbf{B} \oplus \cdots H_m\mathbf{B}$, $H_i\mathbf{B} = \{H_ib : b \in \mathbf{B}\} = \{H_ib : b \in [-1 \quad 1]\}$ *is a part of the sum, where* H_i *denotes the ith column of* H *[4].*

The Minkowski sum for time-varying parameters $|\theta_k - \theta_{k-1}| \leqslant \gamma$ in a bounded time-varying parameter system is of the form

$$\theta_{k+1} \in \theta_k \oplus \Gamma\mathbf{B}^{n_\theta} = \{\theta_k + \Gamma\Delta : \Delta \in \mathbf{B}^{n_\theta}\}, \tag{2.9}$$

where n_θ is the dimension of the parameter vector, Γ is a diagonal matrix and $\Gamma_{i,i} = \gamma_i$. The feasible parameter set can be obtained through the following iterations:

$$FSS_{k+1} = (FSS_k \cap \Theta_k) \oplus \Gamma\mathbf{B}^{n_\theta}. \tag{2.10}$$

Assuming that the kth zonotope in Eq. (2.9) is known, according to Lemma 2.1, the feasible set of parameters in the $k+1$ step must be satisfied:

$$FSS_{k+1} \subseteq p \oplus H\mathbf{B}^m \oplus \Gamma\mathbf{B}^{n_\theta} = p \oplus [H \quad \Gamma]\,\mathbf{B}^{m+n_\theta}. \tag{2.11}$$

2.2 Main results

2.2.1 Ellipsoid-filtering-based estimation algorithm

2.2.1.1 Prediction step

Assuming that the time-varying parameter changes within the ellipsoid $E(0,\Gamma)$, the parameter changes can be added directly to the $k-1$ parameter feasible set Θ_{k-1}. However, $\Theta_{k,k-1}$ is often an ellipsoid, and hence, it needs to be approximated as an ellipsoid.

$$\Theta_{k,k-1} \subseteq E(c_{k,k-1}, P_{k,k-1}). \tag{2.12}$$

The choice of $E(c_{k,k-1}, P_{k,k-1})$ can be in the following form by minimizing the size of the ellipsoid:

$$c_{k,k-1} = c_{k-1}, P_{k,k-1} = \frac{P_{k-1}}{1-\beta_k} + \frac{\Gamma}{\beta_k}. \tag{2.13}$$

Among them, $0 \leqslant \beta_k < 1$ and the choice of β directly affects the shape of the ellipsoid. Let $tr(A)$ be the trace of the matrix A. If the time-invariant system is estimated, then $\beta_k = 0$, $P_{k,k-1} = P_{k-1}$; otherwise, usually, through the minimum trace or minimum volume to select the optimal ellipsoid, this study uses the minimum trace method to calculate β_k to obtain the prediction step ellipsoid.

$$\beta_k = \frac{\sqrt{tr(\Gamma)}}{\sqrt{tr(P_{k-1})} + \sqrt{tr(\Gamma)}}. \tag{2.14}$$

2.2.1.2 Update step

In the update step, $E(c_k, P_k)$ is obtained by fusing the finite model error $e(k)$ into the prediction set. This is achieved by intersecting $E(c_{k,k-1}, P_{k,k-1})$ with the S_k defined in Eq. (2.3) [66]. This indicates that the exact S_k is usually a polyhedron.

$$\mathcal{S}_k = \bigcap_{i=1}^{n} \mathcal{S}_{k,i} = \bigcap_{i=1}^{n} \left\{\theta_k : \left|y_{k,i} - \psi_{k,i}^{\mathrm{T}}\theta_k\right| \leqslant \sigma_{k,i}\right\}, \tag{2.15}$$

where $\mathcal{S}_{k,i}$ is the constrained region of the hyperplane $\left\{\theta_k : \left|y_{k,i} - \psi_{k,i}^{\mathrm{T}}\theta_k\right| = \sigma_{k,i}\right\}$, $y_{k,i}$ is the ith element of y_k and $\psi_{k,i}$ is the ith column of Ψ_k.

The update step ellipsoid is initialized as

$$c_k^0 = c_{k,k-1}, P_k^0 = P_{k,k-1}. \tag{2.16}$$

At iteration $i, i = 1, \ldots, n$, the external ellipsoid is defined as follows:

$$E(c_k^i, P_k^i) \supseteq E(c_k^{i-1}, P_k^{i-1}) \cap \mathcal{S}_{k,i}. \tag{2.17}$$

First, the distance from the center point of the previous ellipsoid c_k^{i-1} to the hyperplane $\mathcal{S}_{k,i}$ is calculated as follows:

$$d_{k,i}^{+} = \frac{y_{k,i} - \psi_{k,i}^{\mathrm{T}} c_k^{i-1} + \sigma_{k,i}}{\sqrt{\psi_{k,i}^{\mathrm{T}} P_k^i \psi_{k,i}}}, \tag{2.18}$$

$$d_{k,i}^{-} = \frac{y_{k,i} - \psi_{k,i}^{\mathrm{T}} c_k^{i-1} - \sigma_{k,i}}{\sqrt{\psi_{k,i}^{\mathrm{T}} P_k^i \psi_{k,i}}}. \tag{2.19}$$

If $d_{k,i}^{-} > 1$ or $d_{k,i}^{+} < -1$, then the feasible parameter set is empty $E(c_k^{i-1}, P_k^{i-1}) \cap \mathcal{S}_{k,i} = \varnothing$. The iteration is stopped and the following is set:

$$c_k = c_{k,k-1}, P_k = P_{k,k-1}. \tag{2.20}$$

Otherwise, the following distance is normalized:

$$d_{k,i}^{-} = \max(d_{k,i}^{-}, -1), \; d_{k,i}^{+} = \min(d_{k,i}^{+}, 1). \tag{2.21}$$

If $d_{k,i}^{-} d_{k,i}^{+} \leqslant -1/n_\theta$, then $c_k^i = c_k^{i-1}, P_k^i = P_k^{i-1}$; else,

$$c_k^i = c_k^{i-1} + \lambda_i \frac{\eta_k^i \psi_{k,i} e_i}{a_k^2}, \tag{2.22}$$

$$P_k^i = \left(1 + \lambda_i - \frac{\lambda_i e_i^2}{a_i^2 + \lambda_i g_i}\right) \eta_k^i, \tag{2.23}$$

where

$$
\begin{aligned}
\eta_k^i &= P_k^{i-1} - \frac{\lambda_i}{a_i^2+\lambda_i g_i} P_k^{i-1}\psi_{k,i}\psi_{k,i-1}^{\mathrm{T}} P_k^{i-1}, && (2.24)\\
e_i &= \sqrt{g_i}\frac{d_{k,i}^{+}+d_{k,i}^{-}}{2}, && (2.25)\\
a_i &= \sqrt{g_i}\frac{d_{k,i}^{+}-d_{k,i}^{-}}{2}, && (2.26)\\
g_i &= \psi_{k,i}^{\mathrm{T}} P_k^{i-1}\psi_{k,i-1}, && (2.27)
\end{aligned}
$$

and λ_i are the positive roots of the following equation:

$$(n_\theta-1)g_i^2\lambda_i^2+((2n_\theta-1)a_i^2-g_i+e_i^2)g_i\lambda_i+(n_\theta(a_i^2-e_i^2)-g_i)a_i^2=0. \quad (2.28)$$

Finally, the preliminary ellipsoid estimation results are obtained as follows:

$$c_k = c_k^n, P_k = P_k^n. \quad (2.29)$$

2.2.2 Zonotopic dimensional reduction filtering

Lemma 2.2

Suppose there is a zonotope $\mathcal{X} = \widehat{\theta} \oplus H\mathbf{B}^r = \widehat{\theta} \oplus [H_1H_2\cdots H_r]\mathbf{B}^r$ and the constraint strip satisfied by the feasible parameter set at a certain time obtained by $\mathcal{S} = \{x : |\psi^{\mathrm{T}}x - y| \leqslant \sigma\}$. Then, we can obtain the fully symmetric polysome with the smallest volume at the next moment that satisfies [3]:

$$\mathcal{X} \cap \mathcal{S} \subseteq v(j) \oplus T(j)\mathbf{B}^r, \quad (2.30)$$

where

$$v(j) = \begin{cases} \widehat{\theta} + \left(\frac{y-\psi^{\mathrm{T}}\widehat{\theta}}{\psi^{\mathrm{T}}H_j}\right)H_j, & \text{if } 1 \leqslant j \leqslant r \text{ and } \psi^{\mathrm{T}}H_j \neq 0 \\ \widehat{\theta}, & \text{otherwise} \end{cases} \quad (2.31)$$

$$T(j) = \begin{cases} \left[T_1^j\ T_2^j\ \ldots\ T_r^j\right], & \text{if } 1 \leqslant j \leqslant r \text{ and } \psi^{\mathrm{T}}H_j \neq 0 \\ H, & \text{otherwise} \end{cases} \quad (2.32)$$

$$T_i^j = \begin{cases} H_i - \left(\frac{\psi^{\mathrm{T}}H_i}{\psi^{\mathrm{T}}H_j}\right)H_j, & \text{if } i \neq j \\ \left(\frac{\sigma}{\psi^{\mathrm{T}}H_j}\right)H_j, & \text{if } i = j \end{cases} \quad (2.33)$$

According to Lemma 2.2, the position of the constraint strip $\mathcal{S}$ in Eq. (2.30) is determined by the parameter y. Different y values will result in different intersections of the zonotope and the constraint strip. For example, when the parameters of the zonotope and constraint strip are as follows:

$$\widehat{\theta} = \begin{bmatrix} 0 \\ 0 \end{bmatrix}, \quad H = \begin{bmatrix} 4 & 2 & 1 \\ 1 & 2 & 3 \end{bmatrix},$$

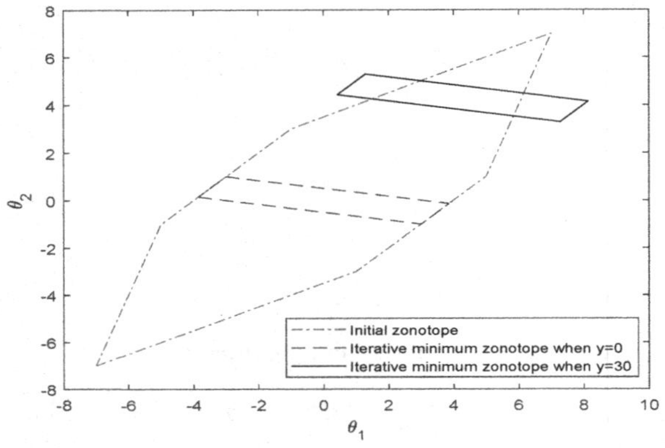

Figure 2.1: Intersection of constraint strips and zonotope at different y values.

$$c = \begin{bmatrix} 1 \\ 6 \end{bmatrix}, \quad y_1 = 0, \quad y_2 = 30, \quad \sigma = 3,$$

the intersection of the constraint strip and the zonotope under the action of y_1 and y_2 is shown in Fig. 2.1.

It can be observed from Fig. 2.1 that different positional parameters of the constraint strip will obtain the zonotope with the minimum volume at different positions. When $y = 30$, it can be observed that the zonotope obtained by iteration is not the zonotope with the smallest volume containing the feasible parameter set. Therefore, updating the zonotope only according to Lemma 2.2 does not guarantee that the volume of the zonotope obtained at each step is the smallest. This will lead to a reduction in the contraction efficiency of the feasible parameter set. In addition, when identifying time-varying parameters, the dimension of the zonotopic shape matrix increases with the number of iterations, which increases the computational complexity.

Considering that the dimension of the zonotopic shape matrix H continues to increase with the identification time, the calculation complexity also increases. Under the premise of reducing the conservativeness and accuracy of the calculation, in this study, singular value decomposition is used to reduce the dimension of the H matrix to obtain a lower-dimensional minimum zonotope.

Theorem 2.1

Suppose there are two zonotopes $\mathcal{X} = \widehat{\theta} \oplus H\mathbf{B}^r$ and $\mathcal{P} = \widehat{\theta} \oplus G\mathbf{B}^m$. If $\mathcal{P}$ is a low-dimensional zonotope wrapped around $\mathcal{X}$, then $\mathcal{X} \subseteq \mathcal{P}$, $m \leqslant r$ is established.

Proof 2.1 The singular value decomposition of the matrix H can be obtained $H = U\Sigma V^{\mathrm{T}}$. Assume that $G = UD$, where D is a diagonal matrix with a diagonal value of

d_i. For $\mathcal{X} \subseteq \mathcal{P}$ to be established, the following box inclusion relationship must exist:

$$\Sigma V^{\mathrm{T}} \mathbf{B}^r \subseteq D\mathbf{B}^m. \tag{2.34}$$

Now, $d_i = \|\sigma_i V_i\|_1$ is selected, where V_i is the column vector of the matrix V, and σ_i is the diagonal element of the diagonal matrix Σ. For the singular value matrix Σ, $\sigma_i V_i < \|\sigma_i V_i\|_1$ is established and can be obtained at the same time

$$\begin{aligned}
&\min(size(\Sigma)) \leqslant \min(size(V)),\\
&\min(size(V)) = r,\\
&\min(size(\Sigma)) = \min(size(D)) = m,
\end{aligned}$$

then, $m \leqslant r$.

Consider a zonotope

$$\mathcal{X} = \widehat{\theta} \oplus H\mathbf{B}^r = \widehat{\theta} \oplus [H_1\ H_2\ \cdots\ H_r]\,\mathbf{B}^r$$

and the constraint strip satisfied by the feasible parameter set at a certain time obtained by $\mathcal{S} = \{x : |\psi^{\mathrm{T}} x - y| \leqslant \sigma\}$. Using the zonotopic transformation method in [21], for the integer j and $0 \leqslant j \leqslant r$, we have

$$\mathcal{X} \cap \mathcal{S} \subseteq v(j^*) \oplus \tilde{T}(j^*)\mathbf{B}^r \tag{2.35}$$

where

$$\begin{aligned}
j^* &= \arg \min_{0 \leqslant j \leqslant r} 2^{n_\theta} \sqrt{\det\left(T(j)T(j)^{\mathrm{T}}\right)}\\
&= \arg \min_{0 \leqslant j \leqslant r} \det\left(T(j)T(j)^{\mathrm{T}}\right).
\end{aligned} \tag{2.36}$$

$v(j)$, $T(j)$ are iteratively obtained the center point and shape matrix of a zonotope.

Remark 2.1 From Eqs. (2.30)–(2.33), the smallest zonotope is calculated, then the full symmetry polysome $\mathcal{X}$ is expanded, and singular value decomposition $[T(j^*)\ \Gamma] = U\Sigma V^{\mathrm{T}}$ is performed. According to Theorem 2.1, $\tilde{T}(j^*) = UD$. Thus, the zonotope at this time $v(j^*) \oplus \tilde{T}(j^*)\mathbf{B}^r$ is the smallest zonotope with the lowest dimension. ▮

2.2.3 Discretization of zonotope into constraint strips

After the dimensionality reduction filter estimation of the zonotope, the column dimensions of the zonotopic shape matrix always remain the same as the dimension of the parameter to be estimated. The following will take the two-dimensional parameter as an example to introduce the method of discretizing the zonotope into Eq. (4.62).

The coordinates of the vertices of the zonotope are set as $(\bar{\theta}_{1,i}, \bar{\theta}_{2,i})$, $i \in \{1,2,3,4\}$. We only take the line segment l_{12} and the line segment l_{34} as examples to

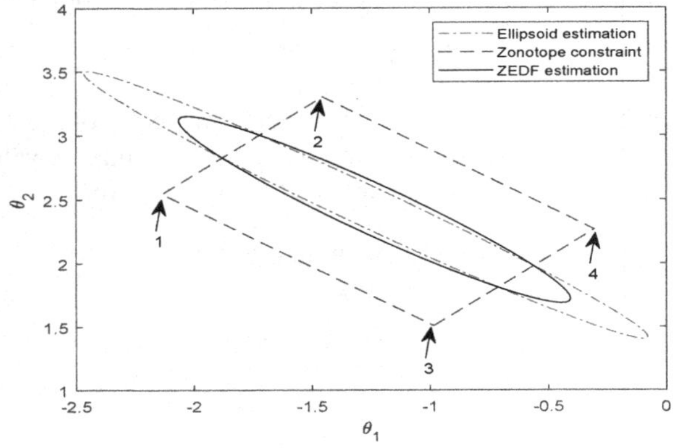

Figure 2.2: Estimation algorithm based on zonotope-ellipsoid double filtering for time-varying parameters.

introduce the discrete method. Similarly, the line segment l_{14} and the line segment l_{23} can also be discretized. Suppose the equations of line segment l_{12} and line segment l_{34} are, respectively,

$$a_{12}\theta_1 + b_{12}\theta_2 = f_{12}, \tag{2.37}$$

$$a_{34}\theta_1 + b_{34}\theta_2 = f_{34}. \tag{2.38}$$

It is known that the slopes of the symmetrical line segments of the zonotope about the center point are the same. Then, $a_{12} = a_{34}, b_{12} = b_{34}$, where

$$a_{12} = a_{34} = \bar{\theta}_{2,1} - \bar{\theta}_{2,2}, \tag{2.39}$$

$$b_{12} = b_{34} = \bar{\theta}_{1,2} - \bar{\theta}_{1,1}, \tag{2.40}$$

$$f_{12} = \bar{\theta}_{1,1}\bar{\theta}_{2,2} - \bar{\theta}_{1,2}\bar{\theta}_{2,1}, \tag{2.41}$$

$$f_{34} = \bar{\theta}_{1,3}\bar{\theta}_{2,4} - \bar{\theta}_{1,4}\bar{\theta}_{2,3}. \tag{2.42}$$

Then, it is converted into Eq. (2.3). The constraint strip equation is

$$\psi^{\mathrm{T}} = [a_{12}, b_{12}], y = -\frac{f_{12} + f_{34}}{2}, \tag{2.43}$$

$$\sigma = \begin{cases} \frac{f_{12} - f_{34}}{2}, & f_{12} \geqslant f_{34} \\ \frac{f_{34} - f_{12}}{2}, & f_{34} > f_{12} \end{cases} \tag{2.44}$$

It can be observed from Fig. 2.2 that the ellipse obtained after transforming the zonotope into constraint strips based on zonotope-ellipsoid double filtering is smaller than the feasible parameter set of the ellipsoid obtained in the initial iteration.

In summary, the algorithm steps of the ZEDF-based estimation algorithm for time-varying parameter systems proposed in this chapter are summarized as follows:

1. Define the number of estimation steps as L, and define the initial zonotope $\mathcal{X}_0 = \widehat{\theta}_0 \oplus H_0\mathbf{B}^l$, ellipsoid $E_0(\widehat{\theta}_0, P_0)$ and initial parameter feasible set Θ_0; set $k = 1$;

2. According to the constraint strips $\mathcal{S}(k)$ satisfied by the feasible parameter set at time k, an updated feasible parameter set is obtained according to the ellipsoid algorithm. Simultaneously, in the iterative process of the zonotope, it is calculated that all the zonotopes after the constraint strips intersect with the zonotope;

3. Calculate the smallest fully symmetric multibody $v(j^*) \oplus T(j^*)\mathbf{B}^r$;

4. Use parameter changes to expand the shape matrix of the zonotope to obtain $[T(j^*) \quad \Gamma]$;

5. Perform dimensionality reduction on the updated matrix, and calculate the shape matrix of the zonotope after dimensionality reduction;

6. Construct the zonotope $v(j^*) \oplus \tilde{T}(j^*)\mathbf{B}^l$ after dimensionality reduction, and discretize the zonotope into the constraint strips;

7. Update the ellipsoid for the second time using the discrete constraint strips, and record the center point of the ellipsoid after the second update as the estimated point of the time-varying parameter identification at this moment;

8. Set $k = k+1$, return to Step 2; when $k = L$, the algorithm ends.

2.3 Numerical examples

Example 1: Consider the following time-varying model:

$$y(k) = \theta_1 u(k) + \theta_2 u(k-1) + e(k), \tag{2.45}$$

where $k = 1, \ldots, 100$, $\theta_1 = -2 \times \sin(\pi \times k \times 0.02)$, $\theta_2 = 4 \times \sin(\pi \times k \times 0.02)$. $u(k)$ is evenly distributed in the interval $[-1,\ 1]$ and $e(k)$ is evenly distributed in the interval $[-0.1,\ 0.1]$.

The initial ellipsoid is set as $E_0(\widehat{\theta}_0, P_0)$ and the zonotope is set as $\mathcal{X}_0 = \widehat{\theta}_0 \oplus H_0\mathbf{B}^l$, where $\widehat{\theta}_0 = \begin{bmatrix} 0 \\ 0 \end{bmatrix}$, $H_0 = P_0 = \begin{bmatrix} 2 & 0 \\ 0 & 2 \end{bmatrix}$, $l = 2$. Moreover, compare the algorithm with the "minimizing the segments of the zonotope method" (MSZM) described in [4] and the ellipsoid filtering method (EFM) described in [126]. The simulation results are shown in Figs. 2.3 and 2.4.

As can be observed from Figs. 2.3 and 2.4, EFM and MSZM as well as the ZEDF-based estimation algorithm for time-varying parameter systems proposed in this chapter can follow the true value changes of the parameters well. It can be observed that the upper and lower bounds of ZEDF are more compact and closer to the true value during the estimation process, which reduces the conservativeness of the

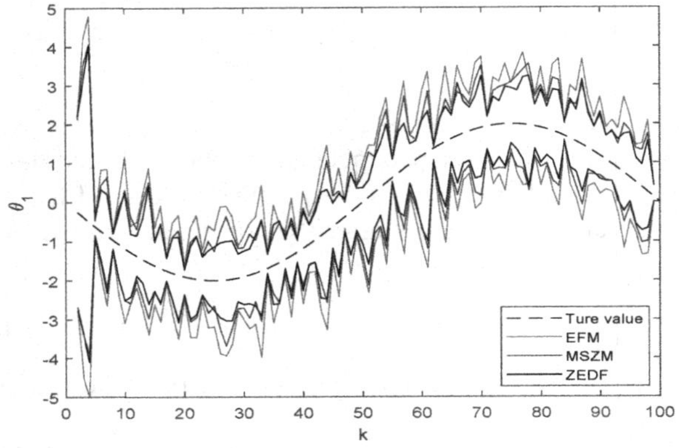

Figure 2.3: Comparison of time-varying parameter θ_1.

time-varying parameter estimation, reflects the effectiveness of the proposed algorithm in estimating time-varying parameters and improves the algorithm estimation accuracy. Furthermore, as can be observed from Figs. 2.3 and 2.4, when the method proposed in this chapter tracks the change of parameters, the estimated upper and lower bounds are relatively constant with respect to the true value, and the fluctuation is small, which reflects the stability of the estimation process.

Figs. 2.5 and 2.6 show the two-dimensional variable parameter feasible set consisting of the parameters θ_1 and θ_2 when the number of iterations is an integer multiple of 5. Here, the X-axis is the number of iterations k, the Y-axis is the time-varying

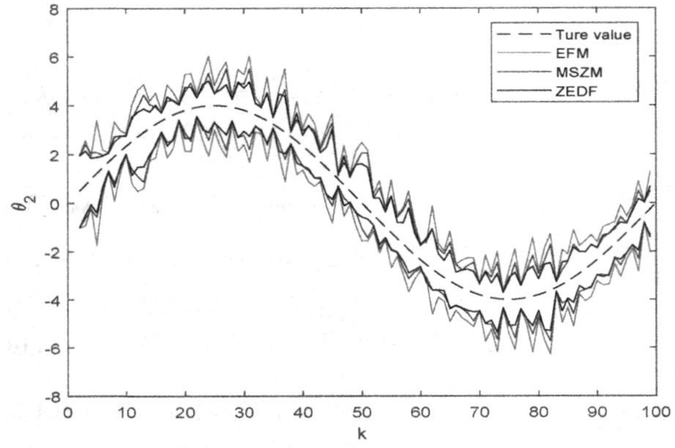

Figure 2.4: Comparison of time-varying parameter θ_2.

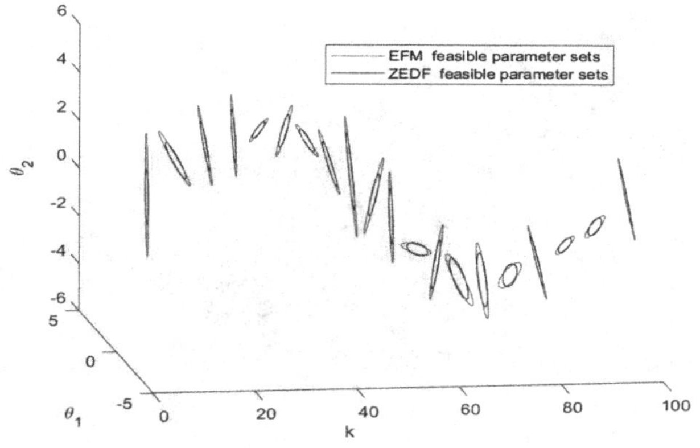

Figure 2.5: Changes in the feasible parameter set of time-varying parameters θ_1 and θ_2.

parameter θ_1 and the Z-axis is the time-varying parameter θ_2. The black solid line elliptical wrapped area is the feasible parameter set obtained using ZEDF, and the red solid line elliptical wrapped area is the feasible parameter set obtained using EFM. As can be observed from Figs. 2.5 and 2.6, the area of the solid black ellipse is always smaller than that of the red solid ellipse. When the true value of the parameter is included, the conservativeness of the algorithm is considerably reduced and the estimation accuracy is improved.

Example 2: A typical kinematic model of an outdoor tracked mobile robot is expressed as follows [46, 126], the physical drawing of tracked mobile robot is shown

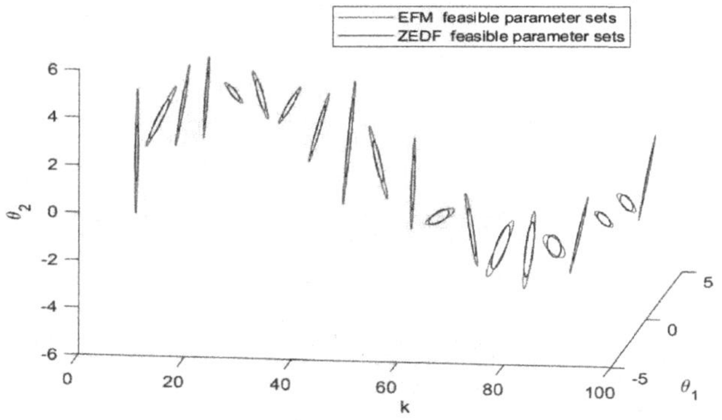

Figure 2.6: Changes in the feasible parameter set in different perspectives.

Figure 2.7: Physical drawing of tracked mobile robot.

in Fig. 2.7 and the planar motion of the tracked mobile robot is given in Fig. 2.8 [17]:

$$\begin{pmatrix} x_{k+1} \\ y_{k+1} \\ \psi_{k+1} \end{pmatrix} = \begin{pmatrix} x_k + s_{1,k} r \left(\omega_{1,k} + \omega_{2,k}\right) \cos \psi_k \Delta T/2 \\ y_k + s_{1,k} r \left(\omega_{1,k} + \omega_{2,k}\right) \sin \psi_k \Delta T/2 \\ \psi_k + s_{2,k} r \left(-\omega_{1,k} + \omega_{2,k}\right) \Delta T/b \end{pmatrix} \tag{2.46}$$

where $(x_k, y_k)^{\mathrm{T}}$ and ψ_k denote the position and orientation of the robot, respectively, and they can be measured using sensors; $w_{1,k}$ and $w_{2,k}$ are the rotation angular speeds of the left and right driving wheels, respectively; r is the wheel radius; b is the wheel distance; ΔT is the time interval and $s_{1,k}$ and $s_{2,k}$ are the slipping

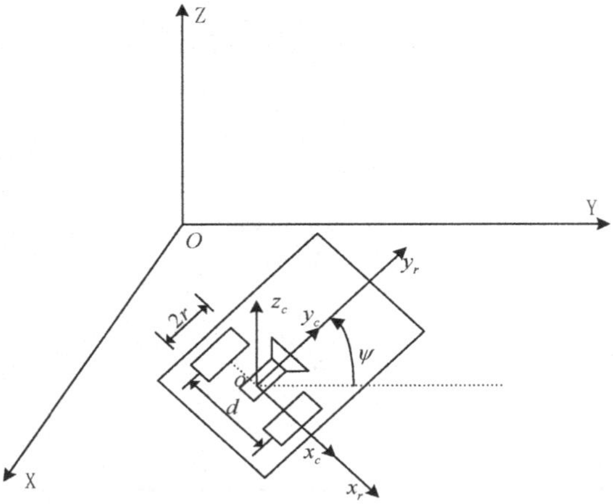

Figure 2.8: Planar motion of tracked mobile robot.

Table 2.1: Parameters of mobile robot dynamics system

Parameter	r/m	b/m	$\omega_{1,k}$/(rad/s)	$\omega_{2,k}$/(rad/s)	ΔT/s	$\sigma_{k,j}(j=1,2,3)$
Value	0.1	0.35	1	7	0.01	0.02

ratios of the robot. The time-varying parameters are set as $\theta_{1,k} = s_{1,k} r \sin(2\pi k \Delta T)$, $\theta_{2,k} = s_{1,k} r/b \sin(\pi k \Delta T)$.

Eq. (2.46) is rewritten to the form of Eq. (2.1), which is linear in parameters

$$\begin{pmatrix} x_{k+1} - x_k \\ y_{k+1} - y_k \\ \psi_{k+1} - \psi_k \end{pmatrix} = \begin{pmatrix} v_1 \cos \psi_k \Delta T & 0 \\ v_1 \sin \psi_k \Delta T & 0 \\ 0 & v_2 \Delta T \end{pmatrix} \theta_k + e_k, \tag{2.47}$$

where $v_1 = (\omega_{1,k} + \omega_{2,k})/2$, $v_2 = \omega_{2,k} - \omega_{1,k}$. $\theta_k = [\theta_{1,k},\ \theta_{2,k}]^{\mathrm{T}}$ is the parameter to be identified, and $e(k)$ is unknown but bounded noise. The parameters of the mobile robot dynamics system in the simulation are shown in Table 2.1. The initial position of the robot is $(0 \quad 0 \quad 0)^{\mathrm{T}}$, the initial ellipsoid is $E_0(\widehat{\theta}_0, P_0)$ and the zonotope is $\mathcal{X}_0 = \widehat{\theta}_0 \oplus H_0 \mathbf{B}^l$, where $\widehat{\theta}_0 = \begin{bmatrix} 0 \\ 0 \end{bmatrix}$, $H_0 = P_0 = \begin{bmatrix} 0.1 & 0 \\ 0 & 0.1 \end{bmatrix}$, $l = 2$. The simulation results are shown in Figs. 2.9 and 2.10.

It can be observed from Figs. 2.9 and 2.10 that the upper and lower bounds of the MSZM identification results are always maintained to have a relatively large difference and are highly conservative. The upper and lower bounds of ZEDF are, in most cases, within them, indicating that the upper and lower limits of the algorithm

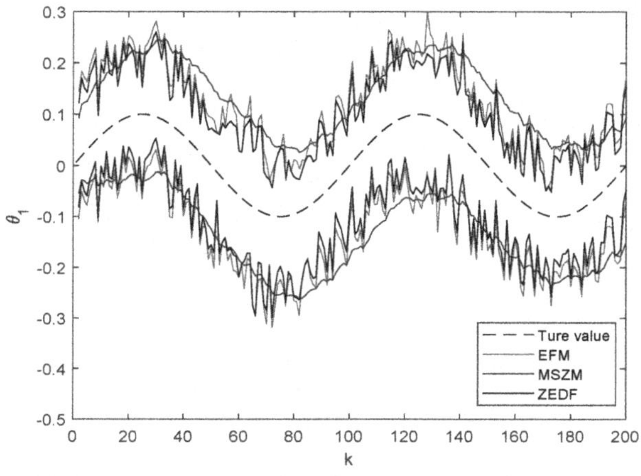

Figure 2.9: Comparison of time-varying parameter θ_1.

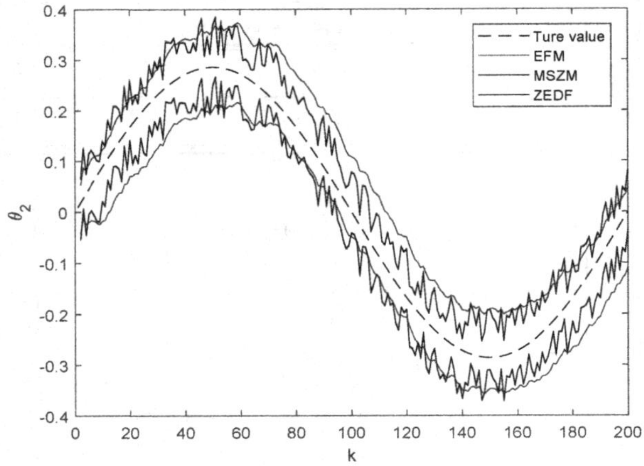

Figure 2.10: Comparison of time-varying parameter θ_2.

are closer to the true value. Compared with EFM, although the two estimation effects are similar, this method is always inside, which reflects the effectiveness of this algorithm.

Fig. 2.11 shows the change of the two-dimensional parameter feasible set consisting of time-varying parameters θ_1 and θ_2, where the black solid ellipse is the parameter feasible set of ZEDF, and the red solid ellipsoid is the parameter feasible set

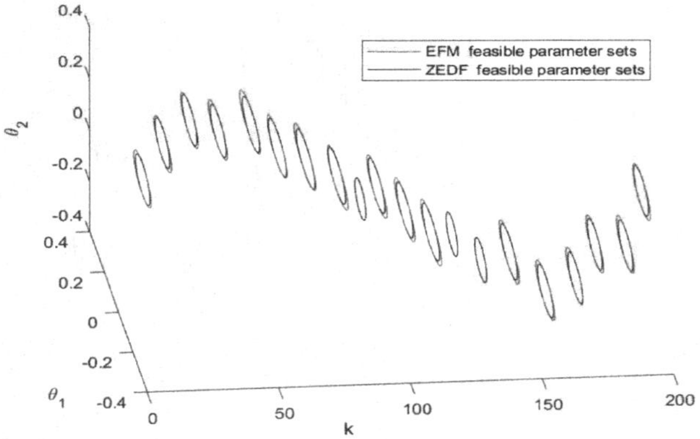

Figure 2.11: Changes in the feasible parameter set of time-varying parameters θ_1 and θ_2.

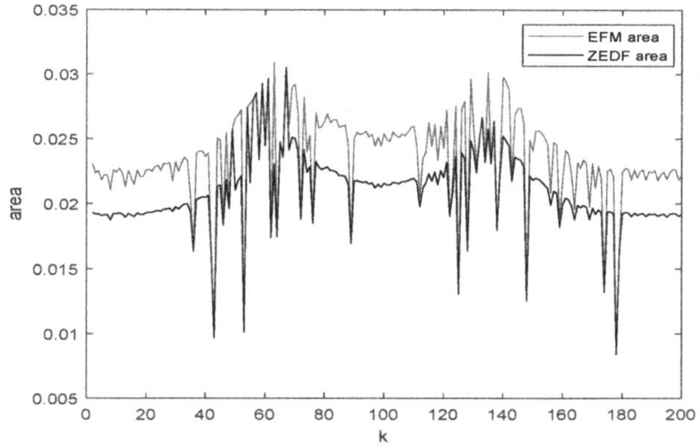

Figure 2.12: Comparison of parameter feasible set area.

of EFM. The X-axis is the number of iterations k, the Y-axis is the time-varying parameter θ_1 and the Z-axis is the time-varying parameter θ_2. To show the change of the parameter feasible set, Fig. 2.11 shows the parameter feasible set when the number of iterations is an integer multiple of 10. Fig. 2.11 shows the change of the feasible parameter set at different times and the size of the feasible set of the two algorithms. To help explain the change in the size of the feasible parameter set, Fig. 2.12 shows the comparison of the ellipse areas of the two algorithms. It can be observed that the area of ZEDF is always small, indicating that EFM is more conservative, confirming the superiority of the method proposed in this chapter.

2.4 Conclusions

This chapter proposes a ZEDF-based estimation algorithm for time-varying parameter systems. By using ellipsoids to describe the feasible parameter set, an ellipsoid estimation is first used to obtain an updated parameter feasible set. In the analysis of the intersecting relationship between the zonotope and the constraint strip, singular value decomposition is used to reduce the shape matrix dimension of the zonotope. In each step of the construction of a zonotope, it is ensured that the dimension of the zonotope is consistent with the dimension of the time-varying parameters. The zonotope is discretized into constraint strips and the ellipsoids are updated again. Compared with the existing methods, the proposed algorithm improves the operation efficiency of the identification process; simultaneously, it effectively improves the identification accuracy and reduces the conservativeness of the identification results. The identification algorithm proposed in this chapter can be further applied to solve the modeling and fault diagnosis problems of industrial processes.

Chapter 3

State estimation based on zonotope

3.1 Set-membership filtering based bi-directional DC-DC converter state estimation algorithm for lithium battery formation

3.1.1 Problem description

Bi-directional DC-DC converter is a two-quadrant operating converter, through which the electric power can flow in either direction [90]. The partial battery formation procedure applying a non-isolated bi-directional DC-DC converter is shown in Fig. 3.1, where i_L is the inductive current and u_C is the capacitive voltage, E, L, C, R and V_o are constant current voltage source, inductance, capacitance, load resistance and output voltage, respectively.

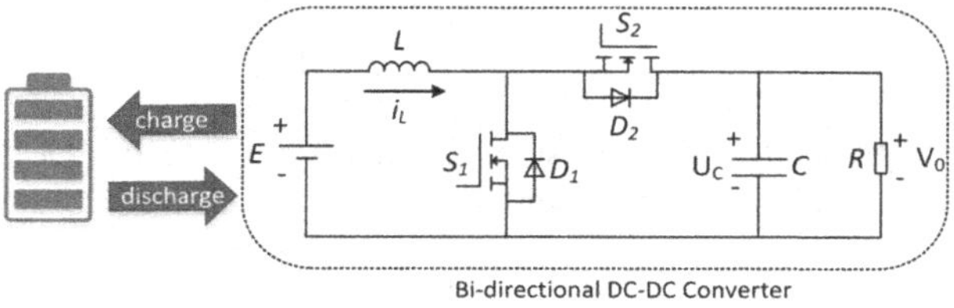

Figure 3.1: Battery formation by bi-directional DC-DC converter.

DOI: 10.1201/b23146-3

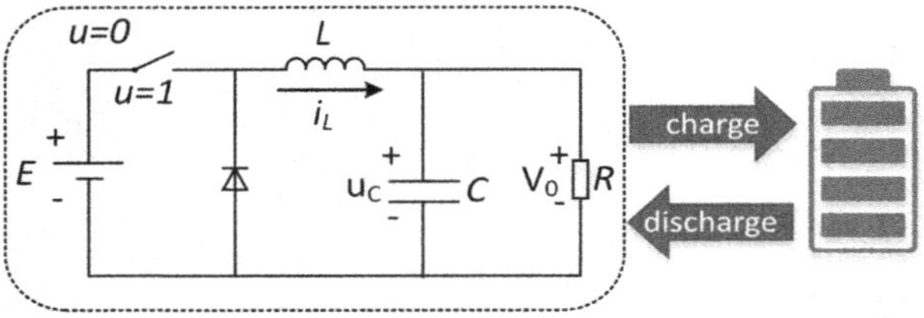

Figure 3.2: Buck mode of converter.

In Fig. 3.1, on condition that S_1 is open while S_2 is closed periodically, it works in buck mode since it reduces the output voltage; on condition that S_1 is closed periodically while S_2 is open, it works in boost mode since it increases its output voltage [31,98].

3.1.1.1 Buck mode

In this section, the buck equivalent converter in [31] is quoted, as shown in Fig. 3.2. Whether $u = 1$ or $u = 0$ means MOSFET S_2 conducts or not.

Let x_1 be i_L, x_2 be u_C, the expression of buck converter in CCM (continuous conduction mode) [4] is obtained by Lagrange modeling:

$$\begin{bmatrix} \dot{x_1} \\ \dot{x_2} \end{bmatrix} = A \begin{bmatrix} x_1 \\ x_2 \end{bmatrix} + B, \tag{3.1}$$

where

$$A = \begin{cases} \begin{bmatrix} 0 & -\frac{1}{L} \\ \frac{1}{C} & -\frac{1}{RC} \end{bmatrix}, & \text{if } u = 1, \\ \begin{bmatrix} 0 & 0 \\ 0 & -\frac{1}{RC} \end{bmatrix}, & \text{if } u = 0, \end{cases}$$

$$B = \begin{cases} \begin{bmatrix} \frac{E}{L} \\ 0 \end{bmatrix}, & \text{if } u = 1, \\ 0, & \text{if } u = 0. \end{cases}$$

When $u = 1$, we construct subsystem Σ_1 by discretizing Eq. (3.1) and impose process noise w_k as well as measurement noise v_k:

$$\Sigma_1 : \begin{cases} x(k+1) = \begin{bmatrix} 1 & -\frac{T}{L} \\ \frac{T}{C} & 1 - \frac{T}{RC} \end{bmatrix} x(k) + \begin{bmatrix} \frac{ET}{L} \\ 0 \end{bmatrix} + w_k, \\ y(k) = \begin{bmatrix} 1 & 0 \end{bmatrix} x(k) + v_k, \end{cases}$$

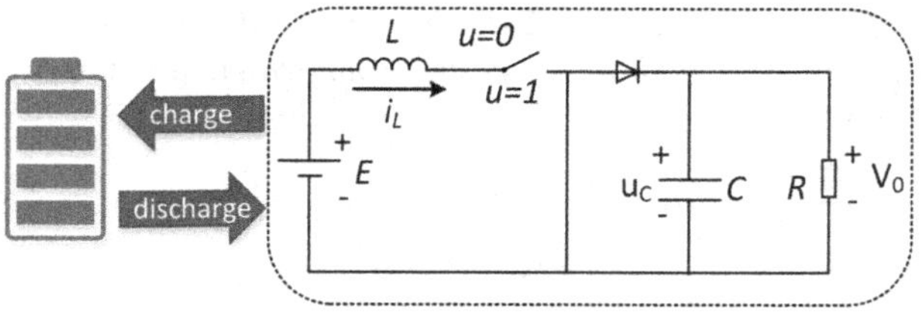

Figure 3.3: Boost mode of converter.

where T is sampling time, $y(k)$ is measurement value at time k. When $u = 0$, construct subsystem Σ_2:

$$\Sigma_2 : \begin{cases} x(k+1) = \begin{bmatrix} 1 & 0 \\ 0 & 1 - \frac{T}{RC} \end{bmatrix} x(k) + w_k, \\ y(k) = \begin{bmatrix} 1 & 0 \end{bmatrix} x(k) + v_k. \end{cases}$$

3.1.1.2 *Boost mode*

The boost equivalent converter is shown in Fig. 3.3. Whether $u = 1$ or $u = 0$ means MOSFET S_1 conducts or not.

Under the circumstance of CCM, the parameter of boost converter in Eq. (3.1) is obtained by Lagrange modeling:

$$A = \begin{cases} \begin{bmatrix} 0 & 0 \\ 0 & -\frac{1}{RC} \end{bmatrix}, & \text{if } u = 1, \\ \begin{bmatrix} 0 & -\frac{1}{L} \\ \frac{1}{C} & -\frac{1}{RC} \end{bmatrix}, & \text{if } u = 0, \end{cases}$$

$$B = \begin{bmatrix} \frac{E}{L} \\ 0 \end{bmatrix}.$$

When $u = 1$, construct subsystem Σ_3:

$$\Sigma_3 : \begin{cases} x(k+1) = \begin{bmatrix} 1 & 0 \\ 0 & 1 - \frac{T}{RC} \end{bmatrix} x(k) + \begin{bmatrix} \frac{ET}{L} \\ 0 \end{bmatrix} + w_k, \\ y(k) = \begin{bmatrix} 1 & 0 \end{bmatrix} x(k) + v_k. \end{cases}$$

When $u = 0$, construct subsystem Σ_4:

$$\Sigma_4 : \begin{cases} x(k+1) = \begin{bmatrix} 1 & -\frac{T}{L} \\ \frac{T}{C} & 1 - \frac{T}{RC} \end{bmatrix} x(k) + \begin{bmatrix} \frac{ET}{L} \\ 0 \end{bmatrix} + w_k, \\ y(k) = \begin{bmatrix} 1 & 0 \end{bmatrix} x(k) + v_k. \end{cases}$$

Assumption 1 *For subsystems $\Sigma_1 \sim \Sigma_4$, it is assumed that the process noise w_k and measurement noise v_k are both unknown but bounded (UBB) noise and satisfy the following equations:*

$$\|w_k\|_\infty \leqslant \varepsilon_w, \tag{3.2}$$

$$\|v_k\|_\infty \leqslant \varepsilon_v, \tag{3.3}$$

where ε_w and ε_v are bounds of w_k and v_k, respectively.

The objective of this section is to estimate the current of inductance and voltage of capacitance under the disturbance of unknown process noise and measurement noise by proposing an improved zonotopic set-membership identification algorithm, which is adapted in bi-directional DC-DC converter.

3.1.2 Preliminaries

3.1.2.1 Definition of strip and zonotope

Definition 3.1 The ith observation strip is defined as [85]

$$S_i = \{\theta \in \mathbb{R}^n \mid \|p_i\theta - c_i\|_\infty \leqslant 1\},\ i = 1, 2, \ldots, m, \tag{3.4}$$

where $p_i \in \mathbb{R}^{1\times n}$ is strip parameter, θ is state, $c_i \in \mathbb{R}^{1\times 1}$ is central point of strip S_i.

Definition 3.2 The zonotope $\mathcal{Z}(\theta_c, H)$ is defined as [4]

$$\mathcal{Z}(\theta_c, H) = \{\theta : \theta = \theta_c + Hz, z \in \mathrm{B}^n\} \triangleq \theta_c \oplus H\mathrm{B}^n, \tag{3.5}$$

where θ and θ_c are feasible state and center of zonotope respectively, H is shape-defining matrix, B^n is a vector composed of n unitary intervals.

3.1.2.2 Properties

Property 1 *Zonotope satisfies the following equations [4]:*

$$\mathcal{L}\mathcal{Z}(\theta, H) + k = \mathcal{Z}(\mathcal{L}\theta + k, \mathcal{L}H), \tag{3.6}$$

$$\mathcal{Z}(\theta_1, H_1) \oplus \mathcal{Z}(\theta_2, H_2) = \mathcal{Z}(\theta_1 + \theta_2, [H_1\ H_2]), \tag{3.7}$$

where $\mathcal{L}$ is multiplicative factor, it can be a constant or matrix.

3.1.3 Main results

The iterative process of set-membership identification includes two parts: prediction and update.

3.1.3.1 Prediction

Assume that there exists a feasible parameter set at time k, denoted as $X_{k|k}$, which contains the true state value $x_{k|k} = [i_L \ u_C]^T$, that is, $x_{k|k} \in X_{k|k}$. Use a zonotope to wrap it up, $X_{k|k} \subseteq \mathcal{Z}_1(\theta_{c1}, H_1)$. According to the state equation in subsystems $\Sigma_1 \sim \Sigma_4$, it is required that linear zonotopic affine transformation is made combining with Property 1. The results are exhibited as follows:

Theorem 3.1

Let two zonotopes $\mathcal{Z}_1(\theta_{c1}, H_1)$ and $\mathcal{Z}_2(\theta_{c2}, H_2)$ contain $X_{k|k}$ and process noise w_k, respectively. $X_{k|k} \subseteq \mathcal{Z}_1(\theta_{c1}, H_1), w_k \subseteq \mathcal{Z}_2(0, H_2)$. An expansive zonotope $\mathcal{Z}(\theta_{k+1|k}, H_{k+1|k})$ is obtained after prediction, where

$$\theta_{k+1|k} = A_k\theta_{c1} + B_k u_k, \tag{3.8}$$

$$H_{k+1|k} = [A_k H_1 \quad H_2]. \tag{3.9}$$

Proof 3.1 Since $x(k|k) \in X_{k|k} \subseteq \mathcal{Z}_1(\theta_{c1}, H_1)$, by taking zonotope $\mathcal{Z}_1(\theta_{c1}, H_1)$ into the state equation of $\Sigma_1 \sim \Sigma_4$, we have

$$\begin{aligned}\mathcal{Z}(\theta_{k+1|k}, H_{k+1|k}) &= A_k\mathcal{Z}_1(\theta_{c1}, H_1) + B_k u_k \oplus \mathcal{Z}_2(\theta_{c2}, H_2) \\ &= \mathcal{Z}_1(A_k\theta_{c1} + B_k u_k, A_k H_1) \oplus \mathcal{Z}_2(\theta_{c2}, H_2) \\ &= \mathcal{Z}_3(A_k\theta_{c1} + B_k u_k, [A_k H_1 \quad H_2]),\end{aligned}$$

thus,

$$\begin{aligned}\theta_{k+1|k} &= A_k\theta_{c1} + B_k u_k, \\ H_{k+1|k} &= [A_k H_1 \quad H_2].\end{aligned}$$

3.1.3.2 Update

Consider that the dimension of measurement value y_k is m, which means strip space can be constructed by intersecting strip $S_1, S_2, ..., S_m,$. Denote SS_k as the strip space at time k, thus, $SS_k = \bigcap_{i=1}^{m} S_i$. Since SS_k is an irregular polyhedron with complex shape, it can be tackled in the way that intersecting the zonotope with only one strip at a time, that is,

$$\begin{aligned}&\mathcal{Z}(\theta_{k+1|k}, H_{k+1|k}) \cap SS_{k+1} = \\ &\{\mathcal{Z}(\theta_{k+1|k}, H_{k+1|k}) \cap S_1\} \cap \cdots \cap S_m\}.\end{aligned} \tag{3.10}$$

Since the designed measurement value y_k of bi-directional DC-DC converter is one-dimensional, Eq. (3.10) can be simplified to $\mathcal{Z}(\theta_{k+1|k}, H_{k+1|k}) \cap S_{k+1}$, where S_{k+1} represents the strip band obtained at the next moment.

The strip can be written as $S_{k+1} = \{\theta \in \mathbb{R}^n \mid \|c^T\theta - d\|_\infty \leqslant \sigma\}$, then according to [4], there exists

$$\mathcal{Z}(\hat{\theta}_{k+1|k}, \hat{H}_{k+1|k}) \supseteq \{\mathcal{Z}(\theta_{k+1|k}, H_{k+1|k}) \cap S_{k+1}\},$$

where

$$\hat{\theta}_{k+1|k} = \theta_{k+1|k} + \lambda(d - c^{\mathrm{T}}\theta), \tag{3.11}$$

$$\hat{H}_{k+1|k} = [(I - \lambda c^{\mathrm{T}})H_{k+1|k} \quad \sigma\lambda], \tag{3.12}$$

$$\lambda = \frac{H_{k+1|k}H_{k+1|k}^{\mathrm{T}}c}{c^{\mathrm{T}}H_{k+1|k}H_{k+1|k}^{\mathrm{T}}c + \sigma^2}. \tag{3.13}$$

From Eq. (3.12), the shape-defining matrix $\hat{H}_{k+1|k}$ has more dimensions than $H_{k+1|k}$, which suggests $\hat{H}_{k+1|k}$ be more complex with increasing circulation iteration time. Denote the zonotope after singular value decomposition as $\mathcal{Z}(\hat{\theta}_{k+1|k}, \hat{H}_{k+1|k}) \downarrow$. Furthermore, the strip S_{k+1} can be reused to tighten $\mathcal{Z}(\hat{\theta}_{k+1|k}, \hat{H}_{k+1|k}) \downarrow$, finally we get $\mathcal{Z}(\theta_{k+1|k+1}, H_{k+1|k+1})$ as follows:

$$\hat{H}_{k+1|k+1} = [h_1, h_2, ..., h_n], \tag{3.14}$$

$$p_0 = \frac{c^{\mathrm{T}}}{\sigma}, \tag{3.15}$$

$$c_0 = \frac{d}{\sigma}, \tag{3.16}$$

$$\varepsilon_0^+ = (p_0\hat{\theta}_{k+1|k} - c_0) + \sum_{i=1}^{n} |p_0 h_i|, \tag{3.17}$$

$$\varepsilon_0^- = (p_0\hat{\theta}_{k+1|k} - c_0) - \sum_{i=1}^{n} |p_0 h_i|. \tag{3.18}$$

Take $n = 2$ for example, the tightening procedure can be summarized as in Fig. 3.4.

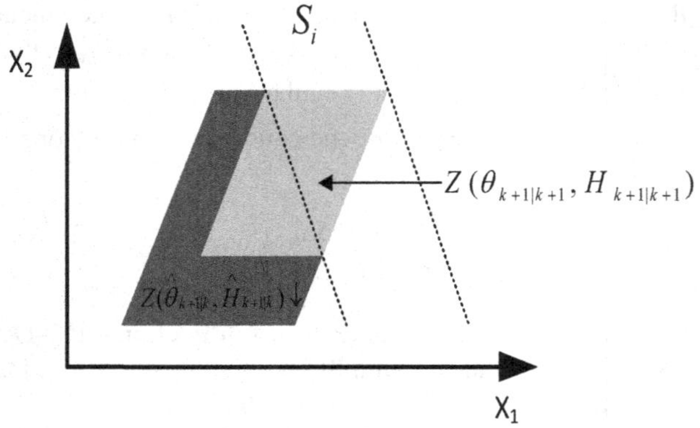

Figure 3.4: Diagram of tightening process.

When $\varepsilon_0^+ < -1$ or $\varepsilon_0^- > 1$, it means that the intersection is empty, that is, a fault occurs and the current cycle exits; otherwise, continue the iteration steps. For $i = 1, 2, ..., n$, we have:

$$r_0^+ = \min(1, \varepsilon_0^+), \tag{3.19}$$

$$r_0^- = \min(1, -\varepsilon_0^-), \tag{3.20}$$

$$\bar{p}_0 = \frac{2}{r_0^+ + r_0^-} p_0, \tag{3.21}$$

$$r_i^+ = \begin{cases} 1, & \text{if } p_0 h_i = 0, \\ \min\left(1, \dfrac{1-\varepsilon_0^-}{|p_0 h_i|} - 1\right), & \text{if } p_0 h_i \neq 0, \end{cases} \tag{3.22}$$

$$r_i^- = \begin{cases} 1, & \text{if } p_0 h_i = 0, \\ \min\left(1, \dfrac{1+\varepsilon_0^+}{|p_0 h_i|} - 1\right), & \text{if } p_0 h_i \neq 0, \end{cases} \tag{3.23}$$

and $\mathcal{Z}(\theta_{k+1|k+1}, H_{k+1|k+1})$ is simply obtained by

$$\begin{cases} \bar{h}_i = \dfrac{r_i^+ + r_i^-}{2} h_i, \\ \theta_{k+1|k+1} = \hat{\theta}_{k+1|k} + \displaystyle\sum_{i=1}^{n} \dfrac{r_i^+ - r_i^-}{2} h_i. \end{cases} \tag{3.24}$$

3.1.4 Simulation results

Based on the design of lithium battery formation estimator with disturbance of unknown but bounded noise, the state parameters of buck and boost converters are identified respectively. Set the circuit parameters as follows: constant current voltage source $E = 20$ V, inductance $L = 4$ mH, capacitor $C = 40\mu$F, load resistance $R = 30\,\Omega$ and sampling time $T = 0.1$ ms. The simulation runs under the condition of CCM, where ε_w and ε_v are both 0.01.

3.1.4.1 Simulation of buck mode

Construct subsystems Σ_1 and Σ_2 with circuit parameters given above, set switching rate of u be 0.5. The results are shown in Figs. 3.5–3.7.

Figs. 3.5 and 3.6 depict the respective trajectory of state i_L and u_C, as well as boundary of ESMI and ZSMI. A conclusion is drawn that both ZSMI and ESMI contain the true value of the state, but the boundary of the former is more compact than that of the latter. Fig. 3.7 depicts the comprehensive trajectory of state i_L and u_C, as well as the feasible parameter set of ESMI and ZSMI, and the dotted line represents this trajectory. It illustrates that both sets of ESMI and ZSMI can surround the true state value under the disturbance of unknown but bounded noises, but the feasible set of the ZSMI is smaller than that of the ESMI.

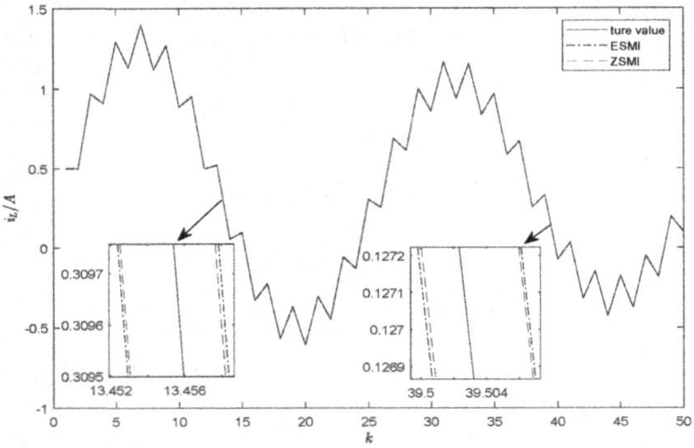

Figure 3.5: Comparison between ESMI and ZSMI methods on inductive current estimation in buck mode.

3.1.4.2 Simulation of boost mode

The parameter setting is the same as previous buck mode. The results are shown in Figs. 3.8–3.10.

Similar to buck mode, Figs. 3.8–3.10 point out that the proposed ZSMI algorithm has less conservativeness than ESMI algorithm, on the premise of estimating the inductive current and capacitive voltage correctly with the disturbance of unknown noises.

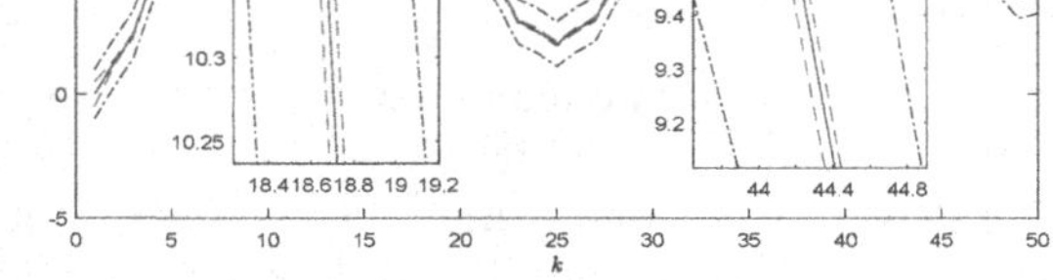

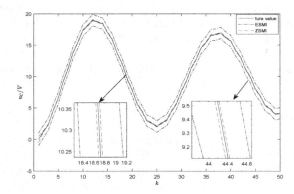

Figure 3.6: Comparison between ESMI and ZSMI methods on capacitive voltage estimation in buck mode.

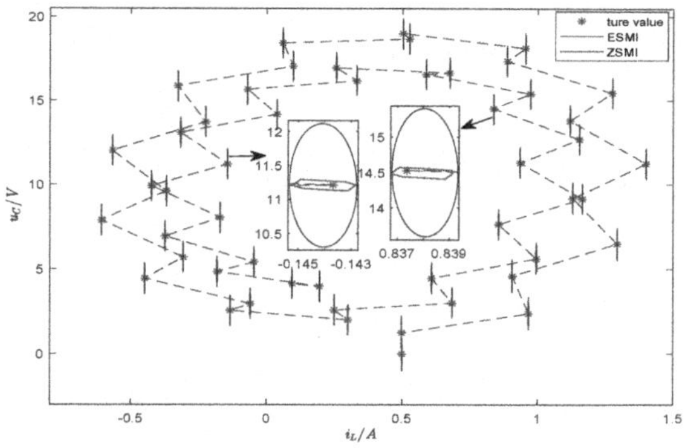

Figure 3.7: Comparison on comprehensive trajectories of inductive current and capacitive voltage in buck mode.

3.1.5 Conclusion

Aiming at the inductive current and capacitive voltage state estimation problem of bi-directional DC-DC converter under the interference of unknown noises, this section presents a set-membership filtering based bi-directional DC-DC converter estimation algorithm for lithium battery formation. It is turned out that the proposed ZSMI tracks the trajectories of true state value i_L and u_C, and estimates them accurately with less conservativeness.

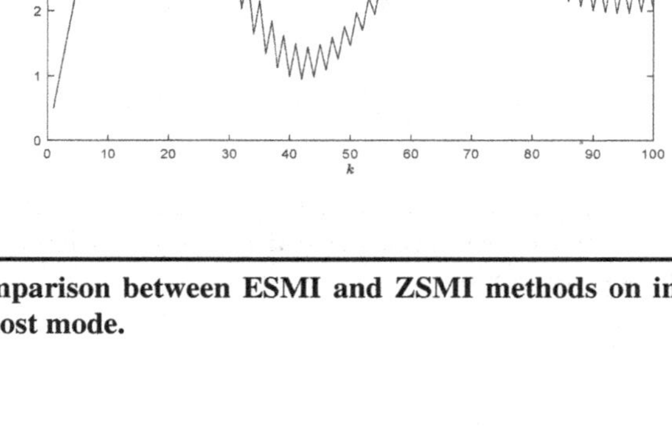

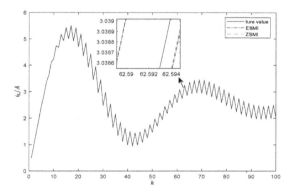

Figure 3.8: Comparison between ESMI and ZSMI methods on inductive current estimation in boost mode.

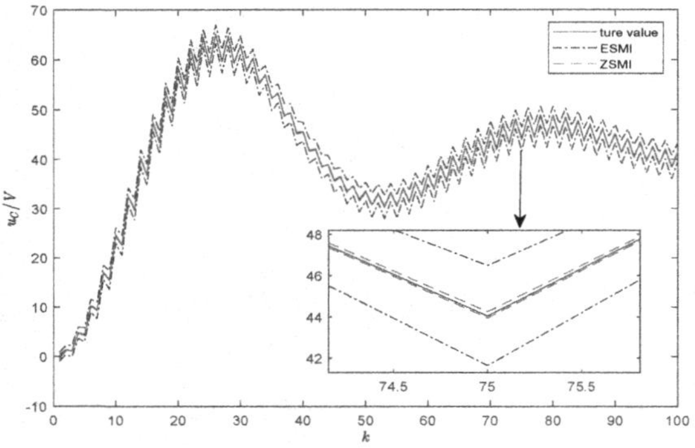

Figure 3.9: Comparison between ESMI and ZSMI methods on capacitive voltage estimation in boost mode.

In practical engineering field, many nonlinear factors, such as switching loss of MOSFET or temperature dependence of electronic devices, are imposed to system inevitably. The future works include the nonlinear design of lithium battery formation estimator, and further to reduce the conservativeness of the algorithm.

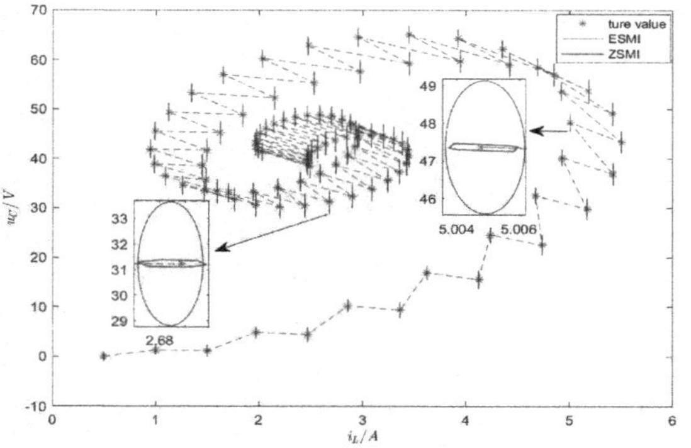

Figure 3.10: Comparison on comprehensive trajectories of inductive current and capacitive voltage in boost mode.

3.2 A novel set-valued observer based state estimation algorithm for nonlinear systems

3.2.1 System description

First, certain preliminary notations are introduced. An interval $[a,b]$ is the set $\{x : a \leqslant x \leqslant b\}$. $\mathrm{B} = [-1,1]$ represents the unitary interval, and a box of order l is denoted as B^l that is composed by l unitary intervals [4]. The notation $f^{(i)}$ is the derivative of order i, H^{T} represents the transpose of matrix H and $H^{i,\mathrm{T}}$ represents the transpose of matrix H^i. $\|f(x)\|_1$ denotes the 1 norm of function $f(x)$, If function $f(x)$ is a vector $f(x) = [x_1, x_2, \ldots, x_n]$, 1 norm of $f(x)$ is $|x_1| + |x_2| + \cdots + |x_n|$. If function $f(x)$ is a matrix, 1 norm of $f(x)$ is $\max\limits_{1 \leq j \leq n} \sum\limits_{i=1}^{n} |x_{ij}|$, where n is the number of columns and x_{ij} represents the element in the ith row and jth column.

Definition 3.3 The Minkowski sum of two zonotopes is defined as $\Psi_s = \{\boldsymbol{x} : \boldsymbol{x} = \boldsymbol{x}_1 + \boldsymbol{x}_2, \boldsymbol{x}_1 \in \mathcal{Z}_1, \boldsymbol{x}_2 \in \mathcal{Z}_2\}$ and it can also be expressed as $\Psi_s = \mathcal{Z}_1 \oplus \mathcal{Z}_2$.

Definition 3.4 $\mathcal{Z} = p \oplus H\mathrm{B}^l = \{p + Hz : z \in \mathrm{B}^l\}$, simplified as $\mathcal{Z}(p,H)$, is defined as a zonotope of order l, where $p \in \mathbb{R}^n$ is the center of the zonotope and matrix $H \in \mathbb{R}^{n \times l}$.

Consider an uncertain nonlinear discrete-time system,

$$\begin{aligned} \boldsymbol{x}_{k+1} &= \boldsymbol{f}(\boldsymbol{x}_k) + \boldsymbol{w}_k, \\ \boldsymbol{y}_k &= \boldsymbol{q}(\boldsymbol{x}_k) + \boldsymbol{v}_k, \end{aligned} \tag{3.25}$$

where $\boldsymbol{x}_{k+1} \in \mathbb{R}^{n_w}$ is the state of the system and $\boldsymbol{y}_k \in \mathbb{R}$ is the measured output vector. The vector $\boldsymbol{w}_k \in \mathbb{R}^{n_w}$ represents the process noise and the vector $\boldsymbol{v}_k \in \mathbb{R}^{n_v}$ is the measurement noise. $\boldsymbol{f}(\cdot)$ and $\boldsymbol{q}(\cdot)$, assumed as nonlinear functions. Assuming that $\boldsymbol{x}_0 \in \mathcal{Z}_0$, and process noise and measurement noise terms satisfy $\boldsymbol{w}_k \in \mathcal{Z}(0, \boldsymbol{r}_w)$ and $\boldsymbol{v}_k \in \mathcal{Z}(0, \boldsymbol{r}_v)$, respectively.

3.2.2 Central difference zonotopic set-valued observer

In the process of the iterative update, errors will accumulate as the number of iterations increases, and in the sequence update, if the zonotope is only considered in the last step, the errors will be amplified. Therefore, to avoid the decrease in estimation accuracy caused by the accumulation of errors, the intersection of the strip and zonotope should be tightened first, and all zonotopes in the previous step need to be considered when updating in the present step. Next we consider the support planes of the zonotope and propose a method for strip tightening in the observer.

3.2.2.1 Nonlinear model linearization

Due to the computational complexity of Jacobian matrix and Hessian matrix of Taylor series, polynomial interpolation can be used to approximate nonlinear functions in an interval. Furthermore, most interpolation equations do not require differentiation, thus it is considerably easier to obtain approximate values. Moreover, the accuracy of the interpolation equation can be set higher than the Taylor series of the same order by setting an appropriate step size. After transformation [65], the Stirling interpolation equation for the function $\boldsymbol{f}(\boldsymbol{x})$ centered on the point $\boldsymbol{x} = \bar{\boldsymbol{x}}$ can be transformed into

$$\begin{aligned}\boldsymbol{f}(\boldsymbol{x}) \approx &\boldsymbol{f}(\bar{\boldsymbol{x}}) + \boldsymbol{f}'(\bar{\boldsymbol{x}})(\boldsymbol{x}-\bar{\boldsymbol{x}}) + \frac{\boldsymbol{f}''(\bar{\boldsymbol{x}})}{2!}(\boldsymbol{x}-\bar{\boldsymbol{x}})^2 + \\ &\left(\frac{\boldsymbol{f}^{(3)}(\bar{\boldsymbol{x}})}{3!}h^2 + \frac{\boldsymbol{f}^{(5)}(\bar{\boldsymbol{x}})}{5!}h^4 + \cdots\right)(\boldsymbol{x}-\bar{\boldsymbol{x}}) + \\ &\left(\frac{\boldsymbol{f}^{(4)}(\bar{\boldsymbol{x}})}{4!}h^2 + \frac{\boldsymbol{f}^{(6)}(\bar{\boldsymbol{x}})}{6!}h^4 + \cdots\right)(\boldsymbol{x}-\bar{\boldsymbol{x}})^2,\end{aligned} \tag{3.26}$$

where h denotes a selected interval length.

Assuming that only the second-order polynomial Stirling interpolation equation is considered and extended to the high-dimensional case, the Stirling interpolation equation at $\boldsymbol{x} = \bar{\boldsymbol{x}}$ of $\boldsymbol{f}(\boldsymbol{x})$ can be expressed as:

$$\boldsymbol{f}(\boldsymbol{x}) = \boldsymbol{f}(\bar{\boldsymbol{x}}) + \tilde{D}_{\Delta x}\boldsymbol{f} + \frac{1}{2!}\tilde{D}^2_{\Delta x}\boldsymbol{f} + H.O.T., \tag{3.27}$$

where $H.O.T.$ is the higher-order co-item of the Stirling interpolation equation of $\boldsymbol{f}(\cdot)$; the difference operators can be expressed as follows:

$$\tilde{D}_{\Delta x}\boldsymbol{f} = \frac{1}{h}\left(\sum_{i=1}^{n} \Delta x_i \mu_i \delta_i\right)\boldsymbol{f}(\bar{\boldsymbol{x}}), \tag{3.28}$$

$$\begin{aligned}\tilde{D}^2_{\Delta x}\boldsymbol{f} = &\frac{1}{h^2}\left(\sum_{i=1}^{n} \Delta x_i^2 \delta_i^2 + \right. \\ &\left.\sum_{i=1}^{n}\sum_{q=1,q\neq i}^{n} \Delta x_i \Delta x_q (\mu_i\delta_i)(\mu_q\delta_q)\right)\boldsymbol{f}(\bar{\boldsymbol{x}}),\end{aligned} \tag{3.29}$$

where μ_i is the ith average operator and δ_i is the ith difference operator. And the parameters $\delta_i\boldsymbol{f}(\bar{\boldsymbol{x}}), \mu_i\boldsymbol{f}(\bar{\boldsymbol{x}})$ and the first-order polynomial Stirling interpolation equation can be found in [65]. For the convenience of calculation, the second-order Stirling interpolation equation is simplified as

$$\boldsymbol{f}(\boldsymbol{x}) \approx \boldsymbol{f}(\bar{\boldsymbol{x}}) + \boldsymbol{f}'_{DD}(\bar{x})(\boldsymbol{x}-\bar{\boldsymbol{x}}) + \frac{\boldsymbol{f}''_{DD}(\bar{\boldsymbol{x}})}{2!}(\boldsymbol{x}-\bar{\boldsymbol{x}})^2, \tag{3.30}$$

where

$$f'_{DD}(\bar{x}) = \frac{f(\bar{x}+he_i)-f(\bar{x}-he_i)}{2h},$$

$$f''_{DD}(\bar{x}) = \frac{f(\bar{x}+he_i)+f(\bar{x}-he_i)-2f(\bar{x})}{h^2}.$$

Expanding $f(x_k)$ in Eq. (3.25) into the form of Eq. (3.30) at state $\hat{x}_k$:

$$f(x_k) \approx f(\hat{x}_k) + F'_k(x_k-\hat{x}_k) + F''_k(x_k-\hat{x}_k)^2, \tag{3.31}$$

where

$$F'_k = \frac{1}{2h}\begin{bmatrix} (f(\hat{x}_k^{1+})-f(\hat{x}_k^{1-}))^{\mathrm{T}} \\ (f(\hat{x}_k^{2+})-f(\hat{x}_k^{2-}))^{\mathrm{T}} \\ \vdots \\ (f(\hat{x}_k^{n+})-f(\hat{x}_k^{n-}))^{\mathrm{T}} \end{bmatrix}^{\mathrm{T}},$$

$$F''_k = \frac{1}{2h^2}\begin{bmatrix} (f(\hat{x}_k^{1+})+f(\hat{x}_k^{1-})-2f(\hat{x}_k^{1}))^{\mathrm{T}} \\ (f(\hat{x}_k^{2+})+f(\hat{x}_k^{2-})-2f(\hat{x}_k^{2}))^{\mathrm{T}} \\ \vdots \\ (f(\hat{x}_k^{n+})+f(\hat{x}_k^{n-})-2f(\hat{x}_k^{n}))^{\mathrm{T}} \end{bmatrix}^{\mathrm{T}},$$

and $\hat{x}_k^{i+} = \hat{x}_k + he_i, \hat{x}_k^{i-} = \hat{x}_k - he_i, \hat{x}_k^{i} = \hat{x}_k$.

3.2.2.2 Bounded linearization error

Assuming that $f(x)$ is the difference of convex function on a convex set S, then there are two convex functions $g_1(x)$ and $g_2(x)$ such that $f(x) = g_1(x) - g_2(x)$. The convex hypothesis of $f(x)$ is easy to satisfy, because $f(x)$ is a second-order continuous differentiable function, and each continuous function can be approximated by the function with arbitrary precision [5].

Assuming that $\frac{\partial^2 f(x)}{\partial x^2} \geqslant -2\alpha I, \alpha \geqslant 0$ and selecting $g_1(x) = f(x) + \alpha x^{\mathrm{T}} x$, $g_2(x) = \alpha x^{\mathrm{T}} x$. The Stirling interpolation equation for the functions g_1 and g_2 at the current state point x_k can be expressed as

$$g_1(x_k) \geqslant g_1(\hat{x}_k) + G'_{1,k}(x_k-\hat{x}_k) + G''_{1,k}(x_k-\hat{x}_k)^2, \tag{3.32}$$

$$g_2(x_k) \geqslant g_2(\hat{x}_k) + G'_{2,k}(x_k-\hat{x}_k) + G''_{2,k}(x_k-\hat{x}_k)^2. \tag{3.33}$$

Define

$$f_L(x) := f(\bar{x}) + f'_{DD}(\bar{x})(x-\bar{x}) + \frac{f''_{DD}(\bar{x})}{2!}(x-\bar{x})^2, \tag{3.34}$$

$$\bar{g}_1(x_k) := g_1(\hat{x}_k) + G'_{1,k}(x_k-\hat{x}_k) + G''_{1,k}(x_k-\hat{x}_k)^2, \tag{3.35}$$

$$\bar{g}_2(x_k) := g_2(\hat{x}_k) + G'_{2,k}(x_k-\hat{x}_k) + G''_{2,k}(x_k-\hat{x}_k)^2. \tag{3.36}$$

The linearization error $e(k) = \boldsymbol{f}(\boldsymbol{x}_k) - \boldsymbol{f}_L(\boldsymbol{x}_k) = g_1(\boldsymbol{x}_k) - g_2(\boldsymbol{x}_k) - \boldsymbol{f}_L(\boldsymbol{x}_k)$, and $\bar{g}_1 - g_2 - \boldsymbol{f}_L$ is a concave function while $g_1 - \bar{g}_2 - \boldsymbol{f}_L$ is a convex function. Thus, the range of $e(k)$ is

$$\begin{bmatrix} \min\limits_{\boldsymbol{x}_k \in V_S} \{\bar{g}_1(\boldsymbol{x}_k) - g_2(\boldsymbol{x}_k) - \boldsymbol{f}_L(\boldsymbol{x}_k)\}, \\ \max\limits_{\boldsymbol{x}_k \in V_S} \{g_1(\boldsymbol{x}_k) - \bar{g}_2(\boldsymbol{x}_k) - \boldsymbol{f}_L(\boldsymbol{x}_k)\} \end{bmatrix}, \tag{3.37}$$

where V_S involves the vertices of the feasible parameter set (FPS), and then the linearization error is wrapped with the minimum volume zonotope $\mathcal{Z}(\boldsymbol{a}_e, \boldsymbol{r}_e)$, where

$$\begin{aligned} \boldsymbol{a}_e &= (\boldsymbol{e}_{k,\max} + \boldsymbol{e}_{k,\min})/2, \\ \boldsymbol{r}_e &= (\boldsymbol{e}_{k,\max} - \boldsymbol{e}_{k,\min})/2. \end{aligned}$$

Then the total process noise is the Minkowski sum of the linearization noise and process noise:

$$\mathcal{Z}(\boldsymbol{a}_e, \boldsymbol{r}_e) \oplus \mathcal{Z}(\boldsymbol{0}, \boldsymbol{r}_w) = \mathcal{Z}(\boldsymbol{a}_e, [\boldsymbol{r}_e + \boldsymbol{r}_w]). \tag{3.38}$$

3.2.2.3 *Time update*

Assuming that the feasible parameter set at step k is $\mathcal{Z}_k = p_k \oplus H_k \mathrm{B}^l$. According to the zonotope equation, the vertices of the FPS can be obtained, and the predicted zonotope$\tilde{\mathcal{Z}}_k = f(p_k) \oplus \tilde{H}_k \mathrm{B}^l$ can be obtained by predicting the vertices and center point of the zonotope, respectively. Then the time updated zonotope is the Minkowski sum of the total process noise and predicted zonotope:

$$\begin{aligned} \mathcal{Z}_{k+1,k} &= p_{k+1,k} \oplus H_{k+1,k} \mathrm{B}^{l+r_{ew}} \\ &= f(p_k) \oplus [\tilde{H}_k, \boldsymbol{r}_e + \boldsymbol{r}_w] \mathrm{B}^{l+r_{ew}}, \end{aligned} \tag{3.39}$$

where r_{ew} is the number of columns in the matrix $\boldsymbol{r}_e + \boldsymbol{r}_w$.

3.2.2.4 *Observation update*

The function $\boldsymbol{q}(\cdot)$ of Eq. (3.25) can be approximated by the first-order polynomial Stirling interpolation equation at a predicted state $\hat{\boldsymbol{x}}_{k+1,k}$ as

$$\boldsymbol{q}(\boldsymbol{x}_{k+1,k}) = \boldsymbol{q}(\hat{\boldsymbol{x}}_{k+1,k}) + \boldsymbol{Q}'_{k+1}(\boldsymbol{x}_{k+1,k} - \hat{\boldsymbol{x}}_{k+1,k}) + \boldsymbol{o}_k, \tag{3.40}$$

where

$$\boldsymbol{Q}'_{k+1} = \frac{1}{2h} \begin{bmatrix} (\boldsymbol{q}(\hat{\boldsymbol{x}}^{1+}_{k+1,k}) - \boldsymbol{q}(\hat{\boldsymbol{x}}^{1-}_{k+1,k}))^{\mathrm{T}} \\ (\boldsymbol{q}(\hat{\boldsymbol{x}}^{2+}_{k+1,k}) - \boldsymbol{q}(\hat{\boldsymbol{x}}^{2-}_{k+1,k}))^{\mathrm{T}} \\ \vdots \\ (\boldsymbol{q}(\hat{\boldsymbol{x}}^{n+}_{k+1,k}) - \boldsymbol{q}(\hat{\boldsymbol{x}}^{n-}_{k+1,k}))^{\mathrm{T}} \end{bmatrix}^{\mathrm{T}}, \tag{3.41}$$

and $\hat{\boldsymbol{x}}^{i+}_{k+1,k} = \hat{\boldsymbol{x}}_{k+1,k} + h\boldsymbol{e}_i, \hat{\boldsymbol{x}}^{i-}_{k+1,k} = \hat{\boldsymbol{x}}_{k+1,k} - h\boldsymbol{e}_i, \hat{\boldsymbol{x}}^{i}_{k+1,k} = \hat{\boldsymbol{x}}_{k+1,k}$ and $\boldsymbol{o}_k$ is the linearization error.

Similarly, according to the processing described in the first few sections, the total measurement noise can be obtained as

$$\mathcal{Z}(\boldsymbol{a}_o,\boldsymbol{r}_o)\oplus\mathcal{Z}(\boldsymbol{0},\boldsymbol{r}_v)=\mathcal{Z}(\boldsymbol{a}_o,[\boldsymbol{r}_o+\boldsymbol{r}_v]). \tag{3.42}$$

where $\mathcal{Z}(\boldsymbol{a}_o,\boldsymbol{r}_o)$ is the linearization error of the measurement equation, and $\mathcal{Z}(\boldsymbol{0},\boldsymbol{r}_v)$ is the input error of the measurement equation.

The observation set can be expressed as

$$S_{k+1}=\left\{\boldsymbol{x}:|\boldsymbol{y}_{k+1}-\boldsymbol{q}(\hat{\boldsymbol{x}}_{k+1,k})+\boldsymbol{Q}'_{k+1}(\hat{\boldsymbol{x}}_{k+1,k}-\boldsymbol{x})|\leqslant\boldsymbol{r}_o+\boldsymbol{r}_v\right\}. \tag{3.43}$$

Eq. (3.43) can be viewed as the intersection of m independent strips:

$$\bigcap_{i=1}^{m}S_{k+1,i}=\bigcap_{i=1}^{m}\{\boldsymbol{x}:|y^a_{k+1,i}-\boldsymbol{Q}'_{k+1,i}\boldsymbol{x}|\leqslant r^a_i\}, \tag{3.44}$$

where $y^a_{k+1,i}$ and r^a_i are the ith components of $\boldsymbol{y}_{k+1}-\boldsymbol{q}(\hat{\boldsymbol{x}}_{k+1,k})+\boldsymbol{Q}'_{k+1}\hat{\boldsymbol{x}}_{k+1,k}$ and $\boldsymbol{r}_o+\boldsymbol{r}_v$, respectively.

Thus, the feasible set of parameters is derived as

$$\mathcal{Z}_{k+1}=\mathcal{Z}_{k+1,k}\bigcap\left(\bigcap_{i=1}^{m}S_{k+1,i}\right). \tag{3.45}$$

As $\bigcap_{i=1}^{m}S_{k+1,i}$ is a polyhedron, it is usually difficult to solve the intersection of a polyhedron and zonotope. Eq. (3.45) is decomposed into the intersection of a zonotope with m strips. Initialize iteration zonotope as in Eq. (3.39), where $p^0=p_{k+1,k}$, $H^0=H_{k+1,k}$.

Assuming that the ith strip is $\{\boldsymbol{x}:|c^{\mathrm{T}}\boldsymbol{x}-d|\leqslant\sigma\}$. Then the iterative equations are:

$$v(j)=\begin{cases}p^{i-1}+\left(\frac{d-c^{\mathrm{T}}p}{c^{\mathrm{T}}H_j^{i-1}}\right)H_j^{i-1}, & \text{if } 1\leqslant j\leqslant r \text{ and } c^{\mathrm{T}}H_j^{i-1}\neq 0\\ p^{i-1}, & \text{otherwise}\end{cases} \tag{3.46}$$

$$T(j)=\begin{cases}\left[T_1^jT_2^j\ldots T_r^j\right], & \text{if } 1\leqslant j\leqslant r \text{ and } c^{\mathrm{T}}H_j^{i-1}\neq 0\\ H^{i-1}, & \text{otherwise}\end{cases} \tag{3.47}$$

$$T_l^j=\begin{cases}H_l^{i-1}-\left(\frac{c^{\mathrm{T}}H_l^{i-1}}{c^{\mathrm{T}}H_j^{i-1}}\right)H_j^{i-1}, & \text{if } l\neq j\\ \left(\frac{\sigma}{c^{\mathrm{T}}H_j^{i-1}}\right)H_j^{i-1}, & \text{if } l=j\end{cases} \tag{3.48}$$

where r is the number of columns in the matrix H^{i-1} and

$$\begin{aligned}j^*&=\arg\min_{0\leqslant j\leqslant r}2^{n_w}\sqrt{\det(T(j)T(j)^{\mathrm{T}})}\\&=\arg\min_{0\leqslant j\leqslant r}\det\left(T(j)T(j)^{\mathrm{T}}\right).\end{aligned} \tag{3.49}$$

Then $p^i = v(j^*), H^i = T(j^*)$ and $\mathcal{Z}_{k+1} = p^m \oplus H^m \mathbf{B}^{l+r_{ew}}$.

The specific process of central difference zonotopic set-valued observer based state estimation algorithm (CDZSVO) is given in the following Algorithm.

3.2.3 Numerical examples

Example 1: The following Van der Pol nonlinear discrete-time system is studied [89].

$$\begin{aligned} \begin{bmatrix} x_{1,k+1} \\ x_{2,k+1} \end{bmatrix} &= \begin{bmatrix} x_{1,k} + hx_{2,k} \\ x_{2,k} + h\delta_{2,k} \end{bmatrix} + \boldsymbol{w}_k, \\ \boldsymbol{y}_k &= \begin{bmatrix} 0 & 1 \\ 1 & 0 \end{bmatrix} \begin{bmatrix} x_{1,k} \\ x_{2,k} \end{bmatrix} + \boldsymbol{v}_k. \end{aligned} \tag{3.50}$$

where $\delta_{2,k} = -9x_{1,k} + \mu\left(1 - x_{1,k}^2\right)x_{2,k}$. The initial conditions are $h = 0.02$, $\mu = 2$, $\boldsymbol{x}_0 = (1,2)^{\mathrm{T}}$, $p_0 = (1,2)^{\mathrm{T}}$ and $H_0 = \mathrm{diag}(0.1, 0.1)$. The process and measurement disturbances are uniformly distributed and $|w_{k,i}| \leqslant 0.01$, $|v_{k,i}| \leqslant 0.001$.

In comparison to the central difference set-membership filtering (CDSMF) algorithm that is described in [86], the simulation results are shown in Figs. 3.12–3.14.

Fig. 3.12 shows the state trajectories and changes in the FPS of the two algorithms. As can be seen from Fig. 3.12, both algorithms track the true trajectory well. It should be noted that the state trajectory is only a mathematical probability, not a real motion trajectory. In the upper right corner, lower right corner and middle position of Fig. 3.12, an enlarged graph of the FPS at time $k = 200$, time $k = 1$ and random time is given. The FPS of CDZSVO is always smaller than the one of CDSMF, which illustrates that CDZSVO is less conservative than CDSMF. In Figs. 3.13 and 3.14, the state boundary of both algorithms can contain true values. However, compared with CDSMF, the CDZSVO proposed in this section can obtain tighter boundaries that can be also verified from Fig. 3.12. Again, the conservative improvement of this algorithm is demonstrated, showing the superiority of this algorithm.

Example 2: A spring-mass-damper nonlinear system estimation example in [81, 86, 127] is also given to illustrate the effectiveness of the proposed algorithm and its physical map and diagram are shown as Figs. 3.15 and 3.16. The discrete-time system of the Duffing equation can be expressed as

$$\begin{aligned} \begin{bmatrix} x_{1,k+1} \\ x_{2,k+1} \end{bmatrix} &= \begin{bmatrix} x_{1,k-1} + \Delta T x_{2,k-1} \\ x_{2,k-1} + \Delta T \delta_{2,k-1} \end{bmatrix} + \boldsymbol{w}_k, \\ \boldsymbol{y}_k &= \begin{bmatrix} 1 & 0 \end{bmatrix} \begin{bmatrix} x_{1,k} \\ x_{2,k} \end{bmatrix} + \boldsymbol{v}_k, \end{aligned} \tag{3.51}$$

where $\delta_{2,k-1} = -k_0 x_{1,k-1}(1 + k_d x_{1,k-1}^2) - cx_{2,k-1}$. The process and measurement disturbances are uniformly distributed. For the simulations to follow the system parameters, the initial setting parameters are shown in Table. 3.1 [81]. In comparison to ESMF that is described in [81] and CDSMF that is described in [86], the results are shown in Figs. 3.17–3.22.

	Algorithm : Framework of the CDZSVO algorithm

Input: The initial zonotope $\mathscr{Z}_0 = p_0 \oplus H_0 \mathbf{B}^l$ and system output y_k.

Output: The parameter estimate x_{k+1} and final zonotope $\mathscr{Z}_k = p_k \oplus H_k \mathbf{B}^l$

1    $k \leftarrow 0$, $L \leftarrow$ Constant;

2   Initialization: Selected initial length h, $\mathscr{Z}_0 = p_0 \oplus H_0 \mathbf{B}^l$, initial state $x_0 = p_0$;

3   **for** $k = 1 : L$ **do**

4   Linearize the function $f(\cdot)$ at the state point x_k;

5   Set the DC functions $g_1(x)$, $g_2(x)$, and linearize to obtain $\bar{g}_1(x)$, $\bar{g}_2(x)$;

6   According to the Eq. (3.37), the bounded linearization error is obtained;

7   According to the Eq. (3.38), calculate the total process noise;

8   Update time and obtain $\mathscr{Z}_{k+1,k} = p_{k+1,k} \oplus H_{k+1,k} \mathbf{B}^{l+r_{ew}}$;

9   Obtain the linearization error of the function $q(x)$ according to the same method;

10   Observation update: Initialize iteration zonotope $p^0 = p_{k+1,k}$, $H^0 = H_{k+1,k}$;

11   **for** $j = 1 : m$ **do**

12   Obtain the ith strip from Eq. (3.44) and calculate the support planes of the zonotope: $q_{u,0} = c^{\mathrm{T}} p^0 + \|H^{0,\mathrm{T}} c\|_1, q_{l,0} = c^{\mathrm{T}} p^0 - \|H^{0,\mathrm{T}} c\|_1$;

13   **if** $y + \sigma > q_{u,0} > y - \sigma > q_{l,0}$ **then**

14          $y = \frac{q_{u,0} + y - \sigma}{2}, \sigma = \frac{q_{u,0} - y + \sigma}{2}$;

15   **else if** $y + \sigma > q_{u,0} > q_{l,0} > y - \sigma$ **then**

16          $y = \frac{q_{u,0} + q_{l,0}}{2}, \sigma = \frac{q_{u,0} - q_{l,0}}{2}$;

17   **else if** $q_{u,0} > y + \sigma > q_{l,0} > y - \sigma$ **then**

18            $y = \frac{y + \sigma + q_{l,0}}{2}, \sigma = \frac{y + \sigma - q_{l,0}}{2}$;

19   **end**

20   **end**

21   **end**

22   **if** $j \geqslant 1$ **then**

23   **for** $i = 1 : j$ **do**

24   Calculating the support planes of the zonotope: $q_{u,i} = c^{\mathrm{T}} p^i + \|H^{i,\mathrm{T}} c\|_1, q_{l,i} = c^{\mathrm{T}} p^i - \|H^{i,\mathrm{T}} c\|_1$;

25   **if** $y + \sigma > q_{u,i} > y - \sigma > q_{l,i}$ **then**

26            $y = \frac{q_{u,i} + y - \sigma}{2}, \sigma = \frac{q_{u,i} - y + \sigma}{2}$;

27   **else if** $y + \sigma > q_{u,i} > q_{l,i} > y - \sigma$ **then**

28            $y = \frac{q_{u,i} + q_{l,i}}{2}, \sigma = \frac{q_{u,i} - q_{l,i}}{2}$;

29   **else if**

           $q_{u,i} > y + \sigma > q_{l,i} > y - \sigma$ **then**

30            $y = \frac{y + \sigma + q_{l,i}}{2}, \sigma = \frac{y + \sigma - q_{l,i}}{2}$;

31   **end**

32   **end**

33   **end**

34   **end**

35   **end**

36   According to Eq. (3.46), Eq. (3.47), Eq. (3.48) and Eq. (3.49), obtain the ith iteration zonotope $p^i = v(j^*), H^i = T(j^*)$;

37   **end**

38 **end**

39 Using SVD to reduce the dimension of the zonotope;

40 **return** kth zonotope $\mathscr{Z}_k = p_k \oplus H_k \mathbf{B}^l$ and the state x_{k+1};

Figure 3.11: The CDZSVO Algorithm.

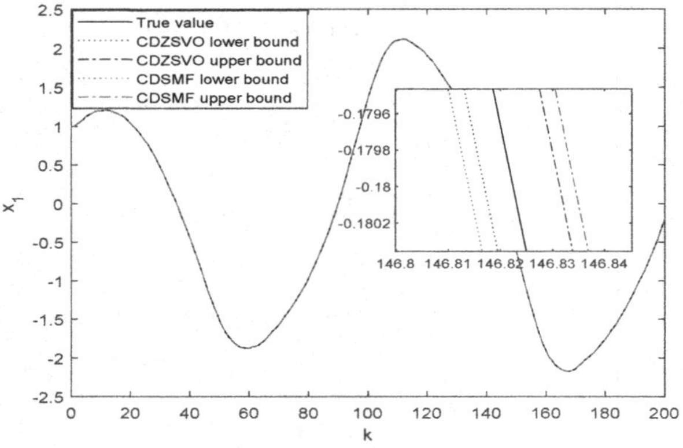

Figure 3.12: Comparison of state trajectories and feasible parameter sets between CDZSVO and CDSMF algorithms.

Table 3.1: Parameters of the spring-mass-damper nonlinear system

Parameter	Value
ΔT	0.1
k_d	3
k_0	1.5
c	1.24
$\boldsymbol{x}_0$	$(1,2)^{\mathrm{T}}$
w_k	0.002
v_k	10.001
p_0	$(1,2)^{\mathrm{T}}$
H_0	diag(0.06, 0.06)

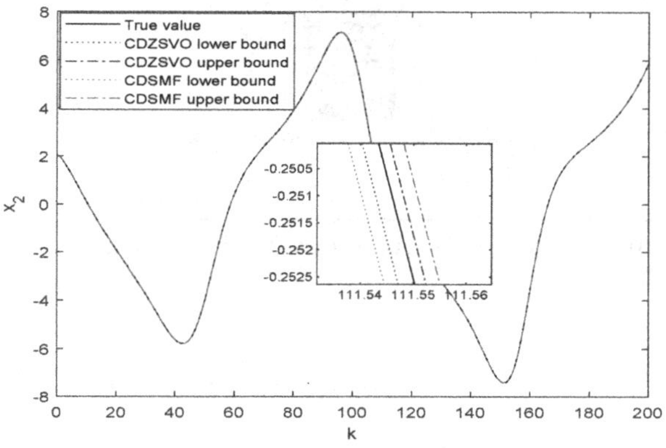

Figure 3.13: Comparison of guaranteed bounds and centers of state x_1 between CDZSVO and CDSMF algorithms.

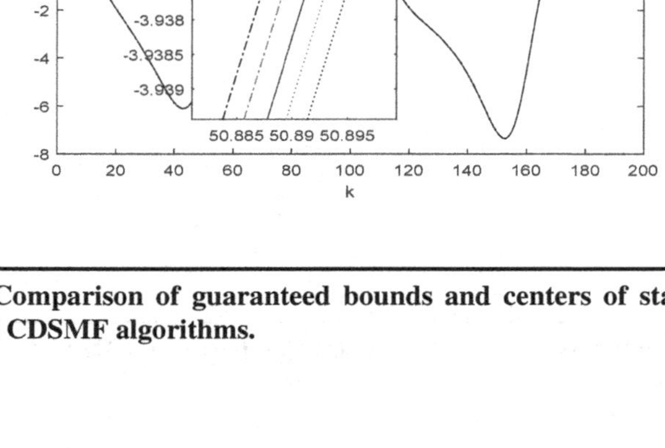

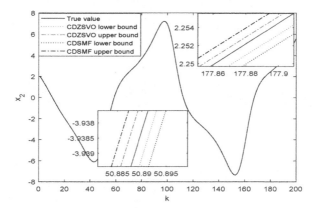

Figure 3.14: Comparison of guaranteed bounds and centers of state x_2 between CDZSVO and CDSMF algorithms.

Figure 3.15: The physical map of spring-mass-damper system.

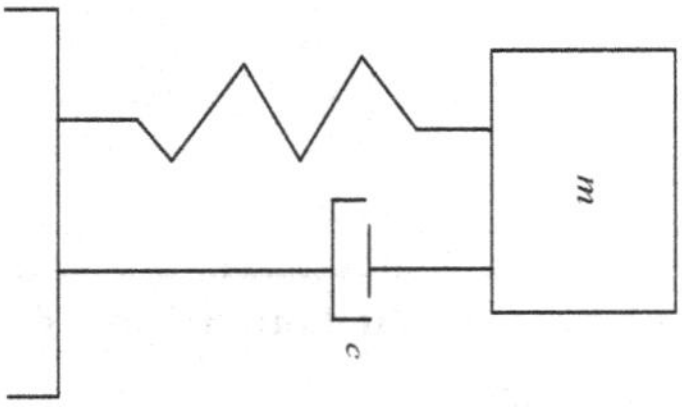

Figure 3.16: The diagram of spring-mass-damper system.

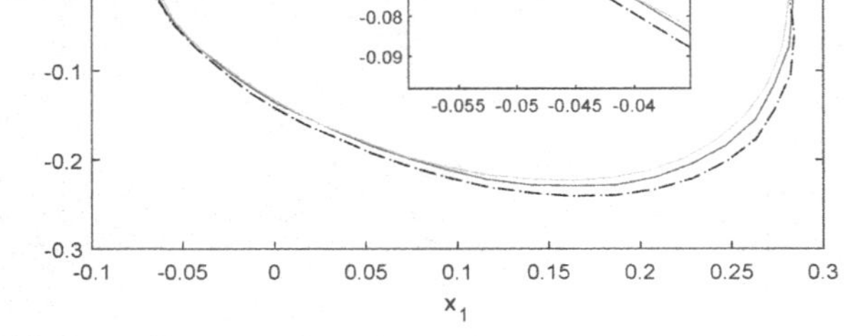

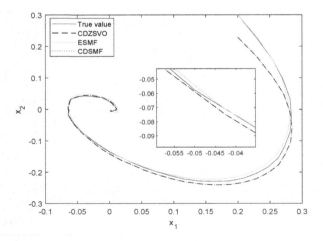

Figure 3.17: Comparison of state trajectories between CDZSVO, CDSMF and ESMF algorithms.

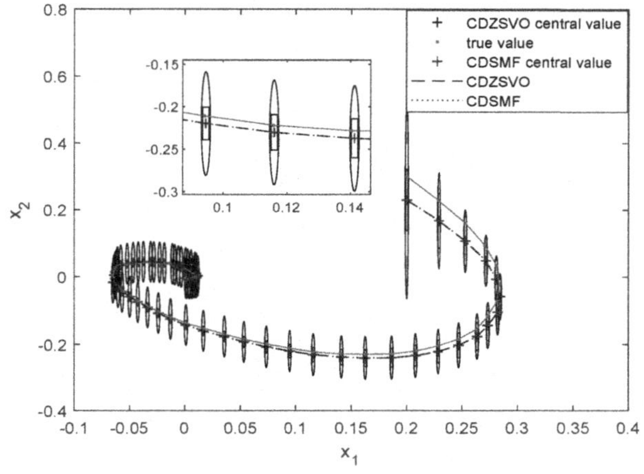

Figure 3.18: Comparison of feasible sets between CDZSVO and CDSMF algorithms.

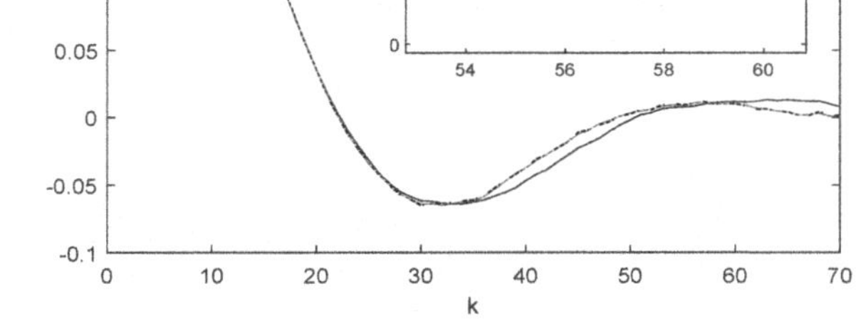

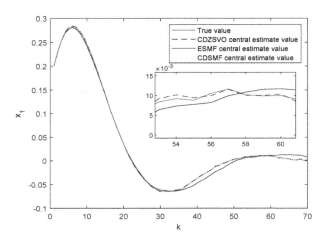

Figure 3.19: Comparison of state estimation of x_1 between CDZSVO, CDSMF and ESMF algorithms.

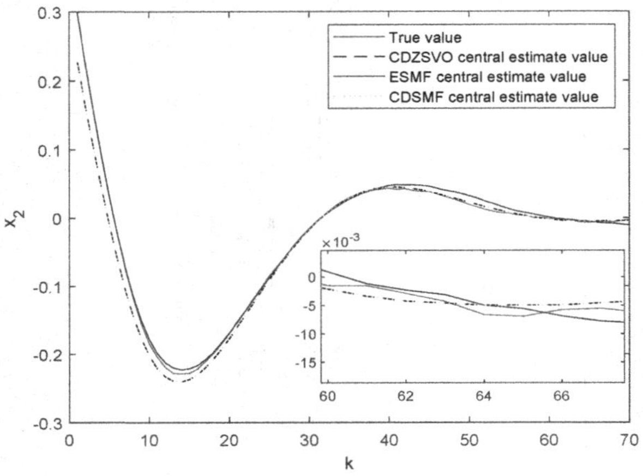

Figure 3.20: Comparison of state estimation of x_2 between CDZSVO, CDSMF and ESMF algorithms.

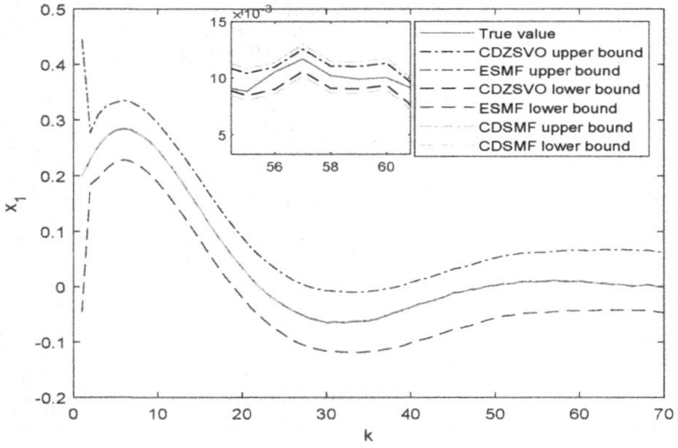

Figure 3.21: Comparison of guaranteed bounds of state x_1 between CDZSVO, CDSMF and ESMF algorithms.

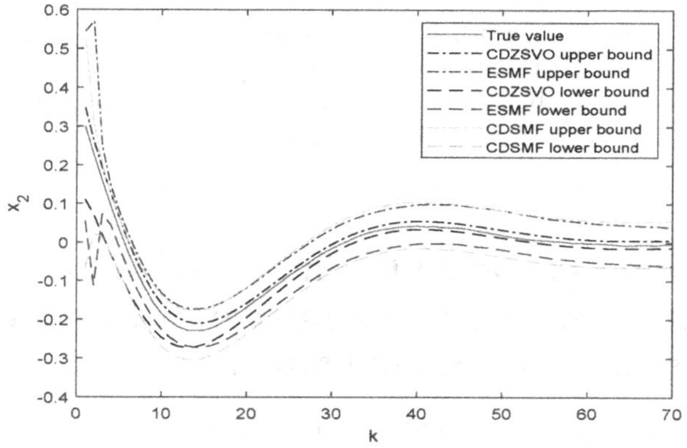

Figure 3.22: Comparison of guaranteed bounds of state x_2 between CDZSVO, CDSMF and ESMF algorithms.

- As can be seen from Fig. 3.17, all three algorithms can track the true value trajectory very well. Although CDZSVO and CDSMF did not work well at the beginning of the algorithm, they still converged to the true value at the end. Fig. 3.18 shows the changes of the feasible parameter set of CDZSVO and CDSMF. The feasible parameter sets of the two algorithms always wrap the true value, and the feasible parameter set of CDZSVO is smaller than CDSMF.

- Figs. 3.19 and 3.20 show the variation curves of the center estimates and true values of the three algorithms. Although the CDZSVO and CDSMF center estimates are the same, the performance of the three algorithms is similar irrespective of state x_1 or x_2.

- The guaranteed bounds for states x_1 and x_2 are shown in Figs. 3.21 and 3.22. It can be seen that the state bounds of the three algorithms can contain true value; however, CDZSVO is more compact than CDSMF and ESMF, indicating that the algorithm proposed in this section has made a large improvement on conservativeness.

3.2.4 Conclusions

A new set-member filtering method is proposed to solve the state estimation problem of unknown but bounded noise nonlinear systems. First, the nonlinear system is linearized by the second-order polynomial Stirling interpolation to further reduce linearization error. Simultaneously, the uncertainty caused by the linearization error is considered, and its boundary is determined by the difference of convex function. Next, the observation set is decomposed into the intersection of multiple bands and a serialized update method is used to determine the FPS. To avoid errors caused by iteration, a method of tight strips and zonotope is proposed. To avoid greater computational complexity, using SVD technology to reduce the dimensionality of zonotopes. Finally, the performance advantages of the proposed algorithm in point estimation and boundary estimation are verified by simulations.

3.3 Zonotopic particle filtering based state estimation algorithm and its application on temperature recognition for lithium battery

3.3.1 Problem equation and preliminaries

3.3.1.1 Problem formulation

Consider the following linear system

$$\begin{cases} x_{k+1} = Ax_k + Bu_k + F\omega_k, \\ y_k = c^{\mathrm{T}}x_k + v_k, \end{cases} \tag{3.52}$$

where x_k, y_k, u_k are the system state, output and input vectors, respectively. A, B, c, F are the system matrices. $\omega_k \in \mathbb{R}^{n_x}$ is the state disturbance. $v_k \in \mathbb{R}^{n_y}$ is the system measurement noise. Assume that both disturbance and noise are bounded, i.e., $|\omega_k| \leqslant \sigma$, $|v_k| \leqslant \gamma$.

3.3.1.2 Preliminaries

Some definitions and properties are adopted in this section.

Definition 3.5 Given two sets X, Y, their vector sum is recorded as

$$X \oplus Y = \{x + y : x \in X, y \in Y\}. \tag{3.53}$$

This vector sum is called Minkowski sum.

Definition 3.6 Define unit interval as $\mathrm{B} = [-1, 1]$, m unit intervals can form a unit box B^m. Similarly, m-dimensional zonotope is defined as

$$p \oplus H\mathrm{B}^m = \{p + H\Delta : \Delta \in \mathrm{B}^m\}, \tag{3.54}$$

where $p \in \mathbb{R}^n$ is the center point of the zonotope, and $H \in \mathbb{R}^{n\times m}$ is the shape matrix of the zonotope.

Property 2 *The Minkowski sum can decompose a zonotope into several parts:* $p \oplus H\mathrm{B}^m = p \oplus H_1\mathrm{B} \oplus H_2\mathrm{B} \oplus \cdots \oplus H_m\mathrm{B}$, *where* H_i *is the ith column of* H, $H_i\mathrm{B} = \{H_i b : b \in \mathrm{B}\} = \{H_i b : b \in [-1, 1]\}$ *is a part of the sum [4].*

Property 3 *The Minkowski sum of the two zonotopes $Z_1 = p_1 \oplus H_1 \mathrm{B}^{m_1} \in \mathbb{R}^n$ and $Z_2 = p_2 \oplus H_2 \mathrm{B}^{m_2} \in \mathbb{R}^n$ is [4]*

$$Z = Z_1 \oplus Z_2 = (p_1 + p_2) \oplus [H_1 \quad H_2] \mathrm{B}^{m_1+m_2}. \tag{3.55}$$

Property 4 *For the zonotope $Z = p \oplus H\mathrm{B}^m$ and the matrix $A \in \mathbb{R}^{n\times n}$, through the linear transformation of matrix A, the zonotope Z can be expressed as [4]:*

$$A \cdot Z = (A \cdot p) \oplus (A \cdot H)\mathrm{B}^m. \tag{3.56}$$

3.3.2 Zonotopic particle filtering based state estimation algorithm

Particle filter is a nonparametric method based on Bayesian filter. For the linear Gaussian system, the posterior probability distribution of the system can be obtained by Bayesian estimation. However, in non-Gaussian system, it is difficult to get the analytic solution of posterior probability by Bayesian estimation. Particle filter is based on Monte Carlo method. In the posterior probability distribution, a large number of samples $x_k^i \sim q(x_k|x_{0:k-1}, y_{1:k})$ are extracted according to the importance probability density distribution $q(x_{0:k}|y_{1:k})$, and a group of weighted particle samples $\{x_k^{i,-}\}_{i=1}^N$ are obtained. In this section, we introduce the zonotopic set-membership algorithm to improve the quality of particles and make it closer to the real value.

The zonotopic algorithm uses a zonotope to contain the original state variables for iterative calculation, and obtains the possible distribution of the state variables at the next moment by finding the minimum volume zonotope which can contain the intersection of the predicted zonotope and the parameter feasible set. The algorithm is divided into prediction step and update step.

3.3.2.1 Prediction step

Suppose the system state variable of the ith particle at time k satisfies

$$x_k \in Z_k^i = \hat{p}_k^i \oplus \hat{H}_k^i \mathrm{B}^m, \tag{3.57}$$

where Z_k^i is the zonotope where the ith particle is located, $\hat{p}_k^i$ is the center of the zonotope and $\hat{H}_k^i$ is the shape matrix of the zonotope.

According to Property 3, Property 4 and system state equation, the feasible set of state variables at time $k+1$ is

$$x_{k+1} \in Z_{k+1|k}^i = \hat{p}_{k+1|k}^i \oplus \hat{H}_{k+1|k}^i \mathrm{B}^{m+n_F}, \tag{3.58}$$

where $\hat{p}_{k+1|k}^i = A\hat{p}_k^i + Bu$, $\hat{H}_{k+1|k}^i = [A\hat{H}_k^i \quad F]$.

According to the system observation value at the current time, the strip feasible set of the system state variable distribution is as follows:

$$S_{k+1} = \{x_{k+1} \in \mathbb{R}^n : |c^{\mathrm{T}} x_{k+1} - y_{k+1}| \leqslant \gamma\}. \tag{3.59}$$

3.3.2.2 Update step

In the update step, the feasible set of state variable parameters can be obtained by finding the intersection of the strip space and the zonotope obtained by Eq. (3.58).

From [21], for the zonotope and strip space: $Z = p \oplus H\mathrm{B}^r = p \oplus [H_1 H_2 \dots H_r] \subset \mathbb{R}^n$, $S = \{x \in \mathbb{R}^n : |c^{\mathrm{T}}x - y| \leqslant \gamma\}$, then we can obtain a minimum volume zonotope $q \oplus L\mathrm{B}^r$ which contains the intersection of the strip space $F = \{z \in \mathbb{R}^r : |c^{\mathrm{T}}Hz - (y - c^{\mathrm{T}}p)| \leqslant \gamma\}$ and the box B^r, q is a vector satisfying $q \in \mathbb{R}^r$, L is a diagonal matrix satisfying $L \in \mathbb{R}^{r \times r}$.

Thus, we can obtain the smallest zonotope that contains the zonotope Z and the strip space S:

$$Z \cap S \subseteq \bar{X}(j) = \bar{\tau}(j) \oplus \bar{T}(j)\mathrm{B}^r, j = 0, 1, \dots, r, \tag{3.60}$$

where

$$\bar{\tau}(j) = \begin{cases} \bar{p} + \left(\frac{y - c^{\mathrm{T}}\bar{p}}{c^{\mathrm{T}}H_j}\right), & \text{if } 1 \leqslant j \leqslant r \text{ and } c^{\mathrm{T}}H_j \neq 0, \\ \bar{p}, & \text{otherwise,} \end{cases} \tag{3.61}$$

$$\bar{p} = p + Hq, \tag{3.62}$$

$$\bar{T}(j) = \begin{cases} [\bar{T}_1^j \ \bar{T}_2^j \ \dots \ \bar{T}_r^j], & \text{if } 1 \leqslant j \leqslant r \text{ and } c^{\mathrm{T}}H_j \neq 0, \\ HL, & \text{otherwise,} \end{cases} \tag{3.63}$$

$$\bar{T}_i^j = \begin{cases} L_i\left(H_i - \left(\frac{c^{\mathrm{T}}H_i}{c^{\mathrm{T}}H_j}\right)H_j\right), & \text{if } i \neq j, \\ \left(\frac{\gamma}{c^{\mathrm{T}}H_j}\right)H_j, & \text{if } m = j, \end{cases} \tag{3.64}$$

The vector q and the diagonal matrix L are obtained by the following equation:

$$q_i = \frac{\alpha_i^+ - \alpha_i^-}{2}\mathrm{sign}(c^{\mathrm{T}}H_i), \tag{3.65}$$

$$L_i = \frac{\alpha_i^+ + \alpha_i^-}{2}, \tag{3.66}$$

where

$$\alpha_i^+ = \begin{cases} \min\left(1, \frac{\gamma + y - c^{\mathrm{T}}p + \sum_{l=1}^{r}|c^{\mathrm{T}}H_l|}{|c^{\mathrm{T}}H_i|} - 1\right), & \text{if } c^{\mathrm{T}}H_i \neq 0, \\ 1, & \text{otherwise,} \end{cases}$$

$$\alpha_i^- = \begin{cases} \min\left(1, \frac{\gamma - y + c^{\mathrm{T}}p + \sum_{l=1}^{r}|c^{\mathrm{T}}H_l|}{|c^{\mathrm{T}}H_i|} - 1\right), & \text{if } c^{\mathrm{T}}H_i \neq 0, \\ 1, & \text{otherwise,} \end{cases} \quad \text{for } i = 1, \dots, r.$$

Then, we can obtain the smallest zonotope of the ith particle which contains the zonotope $\bar{Z}_{k+1|k}^i$ and the strip space S_{k+1}. It can be further determined whether the particle falls in the region, and a point randomly generated from the zonotope is used to replace the particles outside the region. A new particle set is obtained $\{x_{k+1}^{i,-}\}_{i=1}^{N}$. Thus, the weight of each particle can be calculated as follows:

$$w_{k+1}^{i} \propto p(y_{k+1}|x_{k+1}^{i}). \tag{3.67}$$

After calculating the weight of all particles, they can be normalized as follows:

$$\tilde{w}_{k+1}^{i} = \frac{w_{k+1}^{i}}{\sum_{i=1}^{N} w_{k+1}^{i}}. \tag{3.68}$$

Due to the particle degradation in the iterative process, resampling technique is usually used to suppress the particle degradation. After obtaining the particle set with weight, the particles with smaller weight will be removed and replaced by the particles with larger weight to obtain the posterior particle set$\{x_{k+1}^{i}\}_{i=1}^{N}$. The general steps of resampling method are as follows:

1. Generate a number u_i randomly in $(0,1]$, $i = 1,2,\ldots,N$.
2. Find the particles that meet the conditions in the particle set by the following equation:

$$\sum_{m=1}^{j-1} \tilde{w}_k^m < u_i < \sum_{m=1}^{j} \tilde{w}_k^m. \tag{3.69}$$

3. Put the particles that meet the conditions into the new particle set $\{x_{k+1}^{1}, x_{k+1}^{2}, \ldots, x_{k+1}^{N}\}$, and their weights are reset to $1/N$.

Finally, the final estimated value at time $k+1$ can be obtained by weighting the particle set $\{x_{k+1}^{1}, x_{k+1}^{2}, \ldots, x_{k+1}^{N}\}$.

To sum up, the steps using the zonotopic particle filtering based state estimation algorithm can be summarized as follows:

Step 1 Collect the input and output sampling data of the system, and extract N particles from the prior distribution to form a particle set $\{x_0^i\}_{i=1}^{N}$, and wrap each particle by a zonotope $\{\hat{Z}_0^i\}_{i=1}^{N}$. Set the center point set $\{\hat{p}_0^i\}_{i=1}^{N}$ and initial shape matrix $\{\hat{H}_0^i\}_{i=1}^{N}$ of the initial zonotope. Set L as the maximum number of iterations.

Step 2 Obtain the prediction particles $\{x_k^{i,+}\}_{i=1}^{N}$ at time k based on the importance density function, and the prediction step zonotope set $\{\hat{Z}_{k|k-1}^{i}\}_{i=1}^{N}$ at that time is calculated based on the zonotope set at the previous time.

Step 3 Update the particle prediction step zonotope set $\{\hat{Z}_k^i\}_{i=1}^{N}$ by feasible intensive strip S_k based on the time k parameters.

Step 4 Judge whether each particle falls in the corresponding zonotope region, and replace the particles outside the region by the point generated randomly from the zonotope to get the particle set $\{x_k^{i,-}\}_{i=1}^{N}$.

Step 5 Calculate the weights of all particles and normalize them based from Eq. (3.67).

Step 6 Resample the particles in the particle set $\{x_k^{i,-}\}_{i=1}^N$ and obtain the posterior particle set $\{x_k^i\}_{i=1}^N$, then reset the particle weight to $1/N$.

Step 7 Calculate the estimated value $\bar{x}_k$ of the state variable corresponding to the time by calculating the mean value of the particle set $\{x_k^i\}_{i=1}^N$, return to step 2; when the number of iterations is equal to L, output the state estimation result.

The algorithm flow chart of ZPFSE is shown in Fig. 3.23.

3.3.3 Simulation

The electrothermal coupling model of lithium battery is shown in Eq. (3.70):

$$\begin{cases} x_{k+1} = A_d x_k + B_d u_{q,k} + F\omega_k, \\ y_k = c^{\mathrm{T}} x_k + v_k, \end{cases} \tag{3.70}$$

where $x = [T_{c,k} \quad T_{s,k}]^{\mathrm{T}}$ is the system state variable at time k, $T_{c,k}$ is the battery core temperature, $T_{s,k}$ is the battery surface temperature, $u_{q,k} = [Q_{gen,k}, T_{e,k}]^{\mathrm{T}}$ is the system input, $Q_{gen,k}$ is the battery core heating power, $T_{e,k}$ is the environment temperature, y_k is the sum of $T_{c,k}$ and $T_{s,k}$. A_d, B_d, c, F are the given matrices of the system, which are $A_d = \begin{bmatrix} 1-\frac{\Delta t}{R_cC_c} & \frac{\Delta t}{R_cC_c} \\ \frac{\Delta t}{R_cC_s} & 1-\frac{\Delta t}{R_cC_s}-\frac{\Delta t}{R_uC_s} \end{bmatrix}$, $B_d = \begin{bmatrix} \frac{\Delta t}{C_c} & 0 \\ 0 & \frac{\Delta t}{R_uC_s} \end{bmatrix}$, $c = [1 \quad 1]^{\mathrm{T}}$, $F = [1 \quad 1]^{\mathrm{T}}$, C_c is the heat capacity coefficient of the internal material of the battery, C_s is the heat capacity coefficient of the surface of the battery, R_c is the thermal resistance between the core and the surface of the battery and R_u is the convective resistance between the surface of the battery and the cooling air. Define $T_{w,k}$ as the width between upper and lower bounds of two methods. Considering it is difficult to describe the width of ZPFSE, the width of the zonotope at each step is used instead of the width of particles.

The core heating power of lithium battery $Q_{gen,k}$ can be obtained by the second-order Thevenin equivalent circuit. Fig. 3.24 shows the second order Thevenin equivalent circuit model of lithium battery, where I, R and C represent the current, resistance and capacitance respectively, and U represents the voltage at both ends of the capacitor. From [99], we have

$$Q_{gen,k} = I(rI + U_1 + U_2), \tag{3.71}$$

where

$$\begin{cases} \dot{U}_1 = -\frac{1}{R_1C_1}U_1 + \frac{1}{C_1}I, \\ \dot{U}_2 = -\frac{1}{R_2C_2}U_2 + \frac{1}{C_2}I. \end{cases} \tag{3.72}$$

The main parameters of the electrothermal coupling model are shown in Table 3.2. In this simulation, the initial number of particles is $N = 100$, the initial value

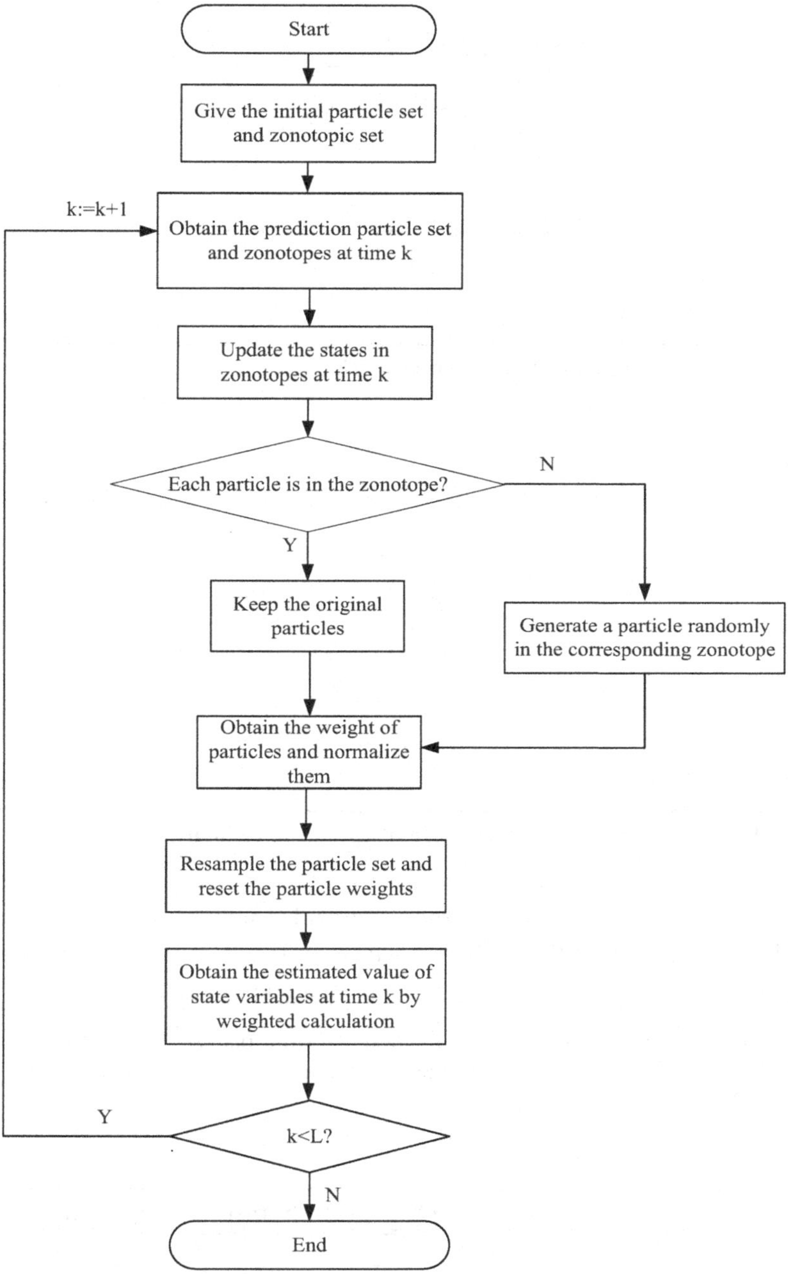

Figure 3.23: The algorithm flow chart of ZPFSE.

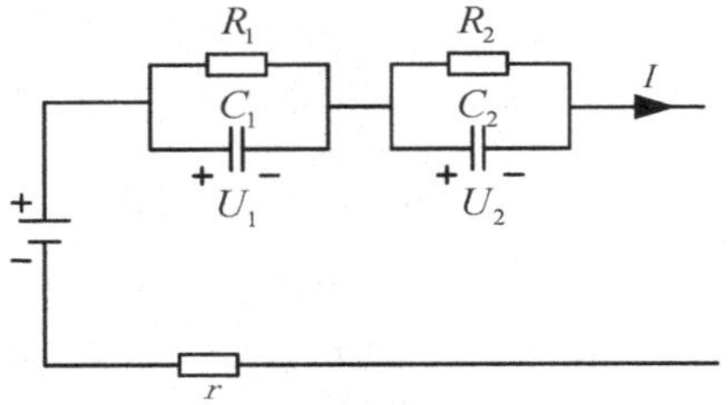

Figure 3.24: Thevenin equivalent circuit model.

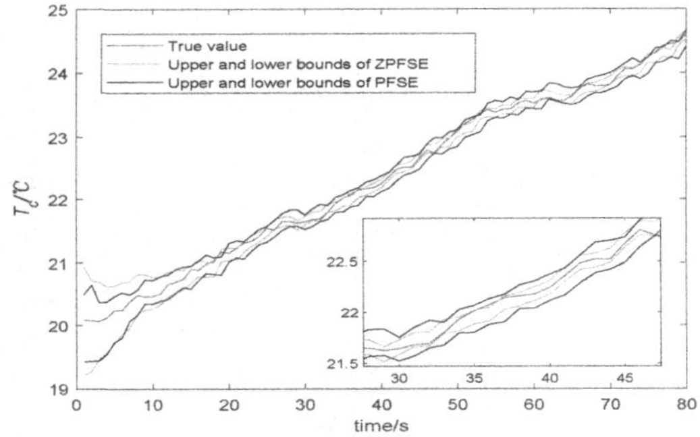

Figure 3.25: Comparison of battery core temperature estimates.

is set as $x_0 = [20,20]^{\mathrm{T}}$, the initial particle distribution satisfies $N \sim (x_0, \sigma_p^2)$, where $\sigma_p = 0.2$. The corresponding initial zonotopic center is the particle itself, the shape matrix is $\hat{H}_0^i = \begin{bmatrix} 1 & 0 \\ 0 & 1 \end{bmatrix}$, $i = 1,2,\ldots,N$, and the noise boundary is $\sigma_n = 0.1, \gamma = 0.1$. By comparing the proposed ZPFSE method with the particle filter based on state estimation (PFSE) method in [100], the simulation results are shown in Figs. 3.25, 3.26 and 3.27.

Table 3.2: Main parameters of lithium battery electrothermal coupling model

R_c/Ω	$C_c/J\ K^{-1}$	$C_s/J\ K^{-1}$	R_u/Ω	R_1/Ω	C_1/F
1.98	63.5	4.5	1.718	0.0298	1787.7
R_2/Ω	C_2/F	r/Ω	I/A	$\Delta t/s$	
0.03819	5.26	0.0501	2	1	

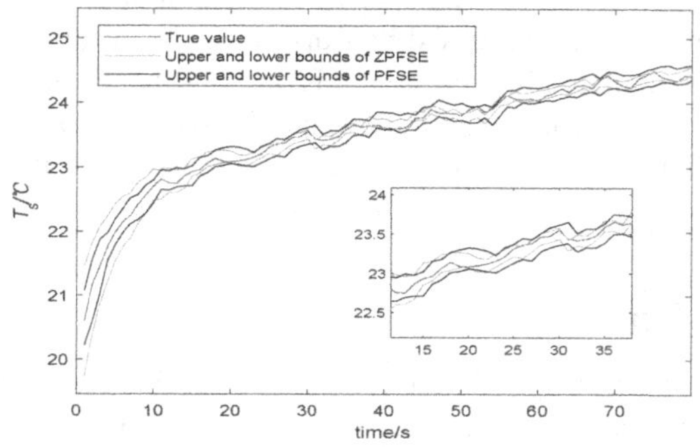

Figure 3.26: Comparison of battery surface temperature estimates.

Figs. 3.25 and 3.26 illustrate the comparison between the ZPFSE and PFSE algorithms on temperature recognition. The horizontal axes are the sampling times, and the vertical axes are the battery surface and core temperature. The purple line represents the change of the real temperature value, the black lines represent the upper and lower bounds of the particle filter algorithm and the blue-green lines represent the upper and lower bounds of the ZPFSE algorithm proposed in this section. It can be seen from the figure that the estimated values calculated by the two algorithms have good following property for the true value, and the upper and lower bounds of the algorithm can wrap the true value of the system. Compared with PFSE algorithm, ZPFSE algorithm limits the diffusion range of particles, has narrower upper and lower bounds, and is less conservative.

Fig. 3.27 is a schematic diagram of the state estimation region width of the two methods with respect to the battery core temperature. The width of upper and lower bounds of PFSE is the difference between the largest particle and the smallest particle

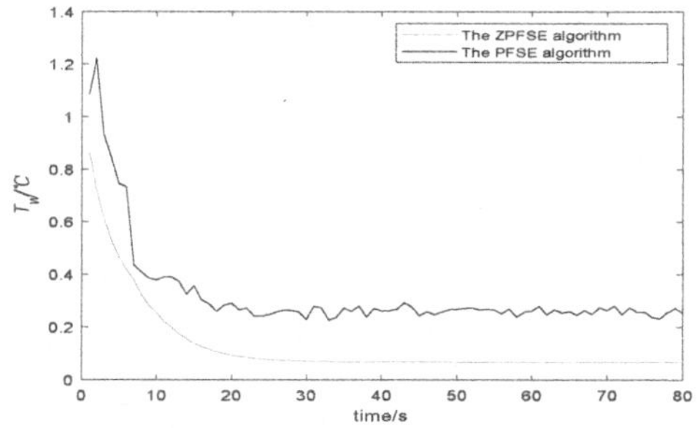

Figure 3.27: Comparison on the width of estimating cell core temperature.

before the particle is defined in each step, and the width of upper and lower bounds of the ZPFSE algorithm is the average width of upper and lower bounds of zonotope in the zonotopic group calculated in each step. It is obvious that the state estimation area of the proposed ZPFSE algorithm is smaller than that of the particle filter algorithm, which limits the diffusion of particles and verifies the superiority of the proposed method.

3.3.4 Conclusion

Aiming at the situation that the system noise is non Gaussian distribution or unknown distribution, this section proposes a system state estimation method based on zonotopic particle filter. The particles are extracted from the given initial distribution, and a new group of particles which are closer to the real value is obtained by constructing a multi cell group according to the zonotopic particle group at the previous time and the system observation value at that time to restrict the diffusion distribution of particles, and to replace the particles outside the restricted area, so as to improve the particle diversity in the resampling process of particle filter. The zonotopic particle filter algorithm proposed in this section can be combined with the idea of filtering in fault state, and can be extended to solve the problem of fault diagnosis in practical system.

Chapter 4

State estimation based on convex spatial structure

4.1 Hyperparallel space set-membership filtering based state estimation algorithm for nonlinear system

4.1.1 Problem description

The mathematical model of the given discrete nonlinear system is as follows,

$$\begin{cases} x_{k+1} = f(x_k) + w_k, \\ y_k = h(x_k) + v_k, \end{cases} \tag{4.1}$$

where $x_k \in X \subseteq \mathbb{R}^n$, x_k is the system states at time k, X represents feasible parameter set (FPS), $y_k \subseteq \mathbb{R}^m$ represents output measurement at time k. w_k and v_k represent process noise and measurement noise at time k respectively, $f(x_k)$ and $h(x_k)$ are discrete nonlinear function of x_k.

Definition 4.1 Suppose $\omega \in \mathbb{R}^{l\times 1}$, $\varepsilon \in \mathbb{R}^{l\times 1}$, define the following symbol [127],

$$\|\omega^i\|_\infty^{\varepsilon^i} = \max\left(|\frac{\omega^i}{\varepsilon^i}|\right), \; i = 1,\ldots,l, \tag{4.2}$$

where ω^i and ε^i represent the ith component of ω and ε.

DOI: 10.1201/b23146-4

Assumption 2 *Aiming at system (4.1), provided that w_k and v_k are both unknown but bound, and satisfy the following equations,*

$$\|w_k^i\|_\infty^{\varepsilon_w^i} \leqslant 1, i = 1,2,\dots,n, \tag{4.3}$$

$$\|v_k^j\|_\infty^{\varepsilon_v^j} \leqslant 1, j = 1,2,\dots,m, \tag{4.4}$$

where ε_w^i and ε_v^j represent the ith boundary value of w and v at time k respectively.

In this section, a state estimation method for nonlinear systems based on hyperparallel space set-membership filtering is proposed. First, $f(x_k)$ and $h(x_k)$ are linearized by Stirling expansion, The linearization error is bounded by difference of convex programming [5, 86], then the parallelotope is used to integrate the set operation of higher-order error term, and the state estimation problem of nonlinear system (4.1) is solved. Compared with Central Difference Set-Membership Filter (CDSMF) [86] and Minimum Segments Zonotopic Method (MSZM) [4], this algorithm has less conservatism.

4.1.2 Preknowledge

4.1.2.1 Parallelotope and orthotope

Definition 4.2 Define the ith observation strip as [94]

$$S_i = \{\theta \in \mathbb{R}^n \mid \|p_i\theta - c_i\|_\infty \leqslant 1\},\ i = 1,2,\dots,m, \tag{4.5}$$

where $p_i \in \mathbb{R}^{1\times n}$, θ are state parameters, $c_i \in \mathbb{R}^{1\times 1}$ is central point of S_ith strip.

Definition 4.3 The parallelotope $\mathcal{P}$ containing parametric feasible set is defined as [18]

$$\mathcal{P}(\theta_c, T) = \{\theta : \theta = \theta_c + T\alpha, \|\alpha\|_\infty \leqslant 1\} \triangleq \theta_c \oplus T\mathrm{B}^n, \tag{4.6}$$

where $\theta_c \in \mathbb{R}^{n\times 1}$ represent the central point of parallelotope, $T = [t_1,\ \dots,\ t_n] \in \mathbb{R}^{n\times n}$represent shape matrix, $\alpha \in \mathbb{R}^{n\times 1}$, B^n is the united box of n dimensions.

For example, the spatial structure of parallelotope $\mathcal{P}(\theta_c,\ T)$ in three dimensions is shown in Fig. 4.1.

Definition 4.4 The orthotope $\mathcal{O}$ containing feasible parameter set is defined as [107]

$$\mathcal{O}(\theta_c, d) = \{\theta : \theta = \theta_c + \mathrm{diag}(d)w, \|w\|_\infty \leqslant 1\}, \tag{4.7}$$

where θ_c, d, $w \in \mathbb{R}^n$.

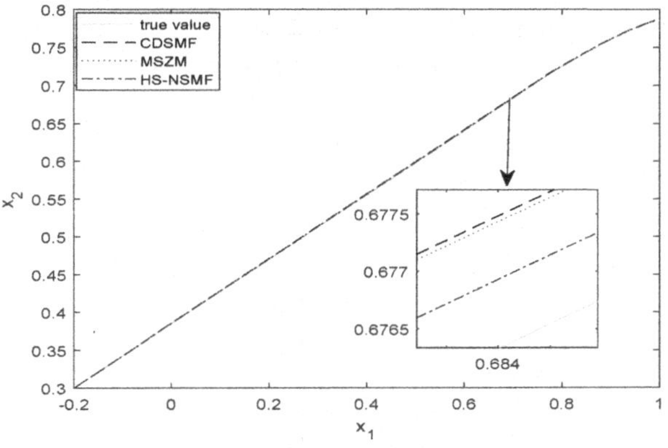

Figure 4.1: Spatial structure of parallelotope ($n = 3$).

When matrix T mentioned in Definition 4.3 satisfies $T = \text{diag}(d), d \in \mathbb{R}^n$, then the Definition 4.3 is equivalent to Definition 4.4. Thus the orthotope is a special form of the parallelotope.

4.1.2.2 Property

Property 5 *The parallelotope satisfies the following operational process,*

$$\mathcal{L}\mathcal{P}(\hat{\theta}, T) + k = \mathcal{P}(\mathcal{L}\hat{\theta} + k, \mathcal{L}T), \tag{4.8}$$

$$\mathcal{P}(\hat{\theta}_1, T_1) \oplus \mathcal{P}(\hat{\theta}_2, T_2) = \mathcal{P}(\hat{\theta}_1 + \hat{\theta}_2, [T_1\ T_2]), \tag{4.9}$$

Eq. (4.8) is the linear transformation property, and Eq. (4.9) is its Minkowski sum property. It can be seen from Eq. (4.9) that the Minkowski sum of two parallelotopes represents a zonotope. With the increase of the number of columns of the shape matrix T_1 and T_2, its dimension becomes larger and larger. In space, there are more and more faces representing the zonotope, and its shape becomes more and more complex. The outer polyhedron of the zonotope can be obtained by reducing the dimension.

4.1.3 Nonlinear set-membership filtering based on parallelotope

4.1.3.1 Outer bound of linearization error

Because the operation of Jacobian matrix or Hessian matrix involved in Taylor expansion requires that each point of the function is continuous and differentiable and the differentiation leads to large amount of calculation and time-consuming program operation. Stirling [86] method is used in this section. First, the nonlinear functions

$f(x_k)$ and $h(x_{k+1})$ in model (4.1) are linearized, and the functions are expanded at the estimated values $\hat{x}_{k|k}$ and $\hat{x}_{k|k}$ respectively. The high-order error term (H.O.T) is wrapped in the form of set and participates in the operation,

$$f(x_k) = f(\hat{x}_{k|k}) + F_k(x_k - \hat{x}_{k|k}) + H.O.T_1, \tag{4.10}$$

$$h(x_{k+1}) = h(\hat{x}_{k+1|k}) + H_{k+1}(x_{k+1} - \hat{x}_{k+1|k}) + H.O.T_2, \tag{4.11}$$

where

$$F_k = \frac{1}{2h}\begin{bmatrix} (f(\hat{x}^{1+}_{k|k}) - f(\hat{x}^{1-}_{k|k}))^{\mathrm{T}} \\ (f(\hat{x}^{2+}_{k|k}) - f(\hat{x}^{2-}_{k|k}))^{\mathrm{T}} \\ \vdots \\ (f(\hat{x}^{n+}_{k|k}) - f(\hat{x}^{n-}_{k|k}))^{\mathrm{T}} \end{bmatrix}^{\mathrm{T}}, \tag{4.12}$$

$$H_{k+1} = \frac{1}{2h}\begin{bmatrix} (h(\hat{x}^{1+}_{k+1|k}) - h(\hat{x}^{1-}_{k+1|k}))^{\mathrm{T}} \\ (h(\hat{x}^{2+}_{k+1|k}) - h(\hat{x}^{2-}_{k+1|k}))^{\mathrm{T}} \\ \vdots \\ (h(\hat{x}^{n+}_{k+1|k}) - h(\hat{x}^{n-}_{k+1|k}))^{\mathrm{T}} \end{bmatrix}^{\mathrm{T}}, \tag{4.13}$$

and

$$\begin{aligned} &\hat{x}^{i+}_{k|k} = \hat{x}_{k|k} + he_i, \hat{x}^{i-}_{k|k} = \hat{x}_{k|k} - he_i, \\ &\hat{x}^{i+}_{k+1|k} = \hat{x}_{k+1|k} + he_i, \hat{x}^{i-}_{k+1|k} = \hat{x}_{k+1|k} - he_i, \\ &e_1 = (1,0,\ldots,0)^{\mathrm{T}}_{n\times 1}, e_2 = (0,1,\ldots,0)^{\mathrm{T}}_{n\times 1}, \ldots, \\ &e_n = (0,0,\ldots,1)^{\mathrm{T}}_{n\times 1}, \end{aligned}$$

define h as step size, note that $f_L^{(i)}(x_k) = f^{(i)}(\hat{x}_{k|k}) + F_k^{(i)}(x_k - \hat{x}_{k|k})$ represents the ith linearization part of the function, the expression of linearization error can be obtained from Eq. (4.10):

$$H.O.T_1 = \begin{bmatrix} f^{(1)}(x_k) - f_L^{(1)}(x_k) \\ f^{(2)}(x_k) - f_L^{(2)}(x_k) \\ \vdots \\ f^{(n)}(x_k) - f_L^{(n)}(x_k) \end{bmatrix}. \tag{4.14}$$

Using DC programming, the function $f(x_k)$ can be represented by the subtraction between two convex functions. Let $g_1(x_k)$ and $g_2(x_k)$ are convex functions, we have $f(x_k) = g_1(x_k) - g_2(x_k)$. Suppose that $g_1(x_k) = f(x_k) + \alpha x_k^{\mathrm{T}} x_k$, $g_2(x_k) = \alpha x_k^{\mathrm{T}} x_k$, $\alpha > 0$. Then the linearization error expression is $H.O.T_1 = g_1(x_k) - g_2(x_k) - f_L(x_k)$. Similarity, supposed that $\bar{g_1}(x_k) = g_1(\hat{x}_k) + G_k(x_k - \hat{x}_k)$, $\bar{g_2}(x_k) = g_2(\hat{x}_k) + H_k(x_k - \hat{x}_k)$

represent linearization of $g_1(x_k)$ and $g_2(x_k)$ respectively. Using the differential properties of convex functions $\bar{g_1}(x_k) < g_1(x_k)$, $\bar{g_2}(x_k) < g_2(x_k)$, expand and contract $H.O.T_1$, we have

$$H.O.T_1^{(i)} \leqslant \max_{x_k \in X_k} \{g_1^{(i)}(x_k) - \bar{g_2}^{(i)}(x_k) - f_L^{(i)}(x_k))\}, \tag{4.15}$$

$$H.O.T_1^{(i)} \geqslant \min_{x_k \in X_k} \{\bar{g_1}^{(i)}(x_k) - g_2^{(i)}(x_k) - f_L^{(i)}(x_k))\}, \tag{4.16}$$

where X_k is feasible parameter set of x_k, using parallelotope $\mathcal{P}_E$ to wrap it up. Using vertex value relaxation of the parallelotope to solve Semi-Definite Problem (SDP) [127], that is, there exists $x_k \in X_k \subseteq \mathcal{P}_E$, the maximum or minimum value of $H.O.T_1^i$ is obtained at the vertexes of $\mathcal{P}_E$.

After solving the boundary $[e_{1,\min}^{(i)}, e_{1,\max}^{(i)}]$ of the linearization error $H.O.T_1^{(i)}$ of function $f^{(i)}(x_k)$, using parallelotope $\mathcal{P}_E = p_e \oplus T_e B^n$ to wrap it up. Where p_e is the central point of parallelotope, T_e is shape matrix :

$$p_e = \left\{ \frac{(e_{1,\max}^{(1)} + e_{1,\min}^{(1)})}{2}, \frac{(e_{1,\max}^{(2)} + e_{1,\min}^{(2)})}{2}, \ldots, \frac{(e_{1,\max}^{(n)} + e_{1,\min}^{(n)})}{2} \right\}^{\mathrm{T}}, \tag{4.17}$$

$$T_e = \mathrm{diag} \left\{ \frac{(\mathrm{e}_{1,\max}^{(1)} - \mathrm{e}_{1,\min}^{(1)})}{2}, \frac{(\mathrm{e}_{1,\max}^{(2)} - \mathrm{e}_{1,\min}^{(2)})}{2}, \ldots, \frac{(\mathrm{e}_{1,\max}^{(\mathrm{n})} - \mathrm{e}_{1,\min}^{(\mathrm{n})})}{2} \right\}. \tag{4.18}$$

4.1.3.2 Predictive step

Assume that there is a feasible parameter set $X_{k|k}$, at this time, the system state $X(K|K)$is wrapped in the feasible set, which means $x(k|k) \in X_{k|k}$. In the equation of state of model (4.1), the nonlinear function $f(x_k)$is linearized by means of linearization and the feasible set can be obtained by using the operation properties of linearized set, noted as $L(X_{k|k})$. Assume that the surrounding set of $w(k)$ is W_k, there exists $X_{k+1|k} = L(X_{k|k}) \oplus W_k$.

In combination with Eq. (4.10), the equation of state of model (4.1) can be written as

$$x_{k+1} = f_L(x_k) + H.O.T_1 + w_k, \tag{4.19}$$

where $f_L(x_k)$ is linearization part of $f(x_k)$, assume that $x_k \in X_k$, $f_L(x_k)$ can be regarded as linear affine transformation of feasible parameter set X_k, it is surrounded by parallelotope $\mathcal{P}_L$; $H.O.T_1$ is surrounded by parallelotope $\mathcal{P}_E$; w_k is surrounded by parallelotope $\mathcal{P}_W$, then the predictive state value $x_{k+1|k}$ at time $k+1$ satisfies $x_{k+1|k} \in X_{k+1|k} = \mathcal{P}_L \oplus \mathcal{P}_E \oplus \mathcal{P}_W$, from here we can see that the set $X_{k+1|k}$ is a zonotope.

Theorem 4.1

Assume that the parallelotope contains x_k at time k is $\mathcal{P}(\hat{x}_{k|k}, T_{k|k})$, the parallelotope contains $H.O.T$ is $\mathcal{P}_E = p_e \oplus T_e \mathrm{B}^n$, the parallelotope contains w_k is $\mathcal{P}_W = p_w \oplus T_w \mathrm{B}^n$.

Then the zonotope contains $x_{k+1|k}$ at time $k+1$ is represented as $\mathcal{Z}(\hat{x}_{k+1|k}, T_{k+1|k})$, where

$$\hat{x}_{k+1|k} = f(\hat{x}_{k|k}) + p_e + p_w, \tag{4.20}$$

$$T_{k+1|k} = [F_k T_{k|k} \quad T_e \quad T_w]. \tag{4.21}$$

Proof 4.1 Apply Property 5, we can get

$$\begin{aligned}
x_{k+1|k} \subseteq & \mathcal{Z}(\hat{x}_{k+1|k}, T_{k+1|k})\mathcal{P}_L \oplus \mathcal{P}_E \oplus \mathcal{P}_W \\
=&(f(\hat{x}_{k|k}) + F_k(\mathcal{P}(\hat{x}_{k|k}, T_{k|k}) - \hat{x}_{k|k})) \oplus (p_e \oplus T_e\mathbf{B}^n) \oplus (p_w \oplus T_w\mathbf{B}^n) \\
=&(f(\hat{x}_{k|k}) + (\mathcal{P}(F_k\hat{x}_{k|k}, F_kT_{k|k}) - F_k\hat{x}_{k|k})) \oplus (p_e \oplus T_e\mathbf{B}^n) \oplus (p_w \oplus T_w\mathbf{B}^n) \\
=&(f(\hat{x}_{k|k}) \oplus F_kT_{k|k}\mathbf{B}^n) \oplus (p_e \oplus T_e\mathbf{B}^n) \oplus (p_w \oplus T_w\mathbf{B}^n) \\
=&(f(\hat{x}_{k|k}) + p_e + p_w) \oplus ([F_kT_{k|k} \quad T_e \quad T_w)]\mathbf{B}^{3n} \\
=&\hat{x}_{k+1|k} \oplus T_{k+1|k}\mathbf{B}^{3n},
\end{aligned}$$

where

$$\hat{x}_{k+1|k} = f(\hat{x}_{k|k}) + p_e + p_w,$$
$$T_{k+1|k} = [F_kT_{k|k} \quad T_e \quad T_w].$$

After obtaining the parameter feasible set $\mathcal{Z}(\hat{x}_{k+1|k}, T_{k+1|k})$ of the prediction state at time $k+1$, the minimum outer parallelotopic feasible set $\mathcal{P}(\hat{x}_{k+1|k}, T_{k+1|k})$ is obtained using the dimension reduction method of literature [25]. The process noise term is a orthotope whose center point is the origin, that is, $w_k \in \mathcal{P}_W = \mathcal{O}_W = \mathbf{0} \oplus \mathrm{diag}\{w_{\mathrm{k},1}, w_{\mathrm{k},2}, \ldots, w_{\mathrm{k,n}}\}$, due to the dimensions of orthotope $\mathcal{P}_E$ is also n, when calculating $\mathcal{Z}(\hat{x}_{K+1|k}, t_{K+1|k})$in Theorem 4.1, the following Theorem can be used to solve the direct sum of two orthotopes $\mathcal{P}_{E+W}$.

Theorem 4.2

For two n dimensional orthotopes $\mathcal{O}_1 = p_1 \oplus \mathrm{diag}\{\mathrm{a}_1, \mathrm{a}_2, \ldots, \mathrm{a}_\mathrm{n}\}$, $\mathcal{O}_2 = p_2 \oplus \mathrm{diag}\{\mathrm{b}_1, \mathrm{b}_2, \ldots, \mathrm{b}_\mathrm{n}\}$, the Minkowski sum of these can be expressed as

$$\mathcal{O}_1 \oplus \mathcal{O}_2 = (p_1 + p_2) \oplus \begin{bmatrix} \mathrm{a}_1 + \mathrm{b}_1 & & & \\ & \mathrm{a}_2 + \mathrm{b}_2 & & \\ & & \ddots & \\ & & & \mathrm{a}_\mathrm{n} + \mathrm{b}_\mathrm{n} \end{bmatrix}.$$

Proof 4.2 Let $\mathrm{B}^n = [\mathrm{B}_1, \mathrm{B}_2, \ldots, \mathrm{B}_{2n}]$, each of these elements has a value of 1 or -1. Define that the shape matrix of two orthotopes as $\mathrm{diag}\{a_1, a_2, \ldots, a_n\}\mathrm{B}_1^n$ and $\mathrm{diag}\{b_1, b_2, \ldots, b_n\}\mathrm{B}_2^n$ respectively. Then the Minkowski sum of the obtained shape matrix is

$$[\mathrm{diag}\{a_1, a_2, \ldots, a_n\} \ \mathrm{diag}\{b_1, b_2, \ldots, b_n\}]\mathrm{B}^{2n},$$

The shape matrix expression of the vertex is $[a_1\mathrm{B}_1 + b_1\mathrm{B}_{n+1}; a_2\mathrm{B}_2 + b_2\mathrm{B}_{n+2}; \cdots ; a_n\mathrm{B}_n + b_n\mathrm{B}_{2n}]$. Choose the maximum and minimum values for each line, the vertex boundary is

$$[\mathrm{B}_1(a_1 + b_1); \mathrm{B}_2(a_2 + b_2); \cdots ; \mathrm{B}_n(a_n + b_n)].$$

After figuring out $\mathcal{P}_{E+W}$, calculate the Minkowski sum of $\mathcal{P}_{E+W}$ and affine transformation parallelotope $\mathcal{P}_L$, the outcome of which is a zonotope. Then calculate its minimum outer parallelotope $\mathcal{P}(\hat{x}_{k+1|k}, T_{k+1|k})$, that is

$$\mathcal{Z}(\hat{x}_{k+1|k}, T_{k+1|k}) \subseteq \mathcal{P}(\hat{x}_{k+1|k}, T_{k+1|k}) = \mathcal{P}_L \oplus \mathcal{P}_{E+W}. \tag{4.22}$$

For example, when $\mathcal{P}_L = \begin{bmatrix} 0.35 \\ 0.1 \end{bmatrix} \oplus \begin{bmatrix} 0.15 & 0.08 \\ 0.09 & 0.18 \end{bmatrix}$, $\mathcal{P}_{E+W} = \begin{bmatrix} 0.05 \\ 0.05 \end{bmatrix} \oplus \begin{bmatrix} 0.025 & 0 \\ 0 & 0.025 \end{bmatrix}$, we get $\mathcal{Z} = \begin{bmatrix} 0.4 \\ 0.15 \end{bmatrix} \oplus \begin{bmatrix} 0.15 & 0.08 & 0.025 & 0 \\ 0.09 & 0.18 & 0 & 0.025 \end{bmatrix}$, $\mathcal{P}(\hat{x}_{k+1|k}, T_{k+1|k}) = \begin{bmatrix} 0.4 \\ 0.15 \end{bmatrix} \oplus \begin{bmatrix} 0.1992 & 0.1042 \\ 0.1195 & 0.2345 \end{bmatrix}$. At this time, the comparison of feasible parameter sets in the prediction process are shown in Fig. 4.2.

4.1.3.3 Update step

For the linearization of $h(x_k)$ in the output equation, the linearized result is defined as $y_{k+1} = L(h(x_{k+1})) + v_{k+1}$, the observation set is defined as $S_{k+1} = \left\{ x \in \mathbb{R}^n \mid \|y_{k+1} - L(h(x_{k+1}))\|_\infty^{\varepsilon_v^i} \leqslant 1 \right\}$, then the updated state estimation $X_{k+1|k+1} = S_{k+1} \cap X_{k+1|k}$.

Since $X_{k+1|k} \subseteq \mathcal{P}(\hat{x}_{k+1|k}, T_{k+1|k})$ at time $k+1$, the purpose of the update step is to solve the minimum outer parallelotope $\mathcal{P}(\hat{x}_{k+1|k+1}, T_{k+1|k+1})$ of the intersection of

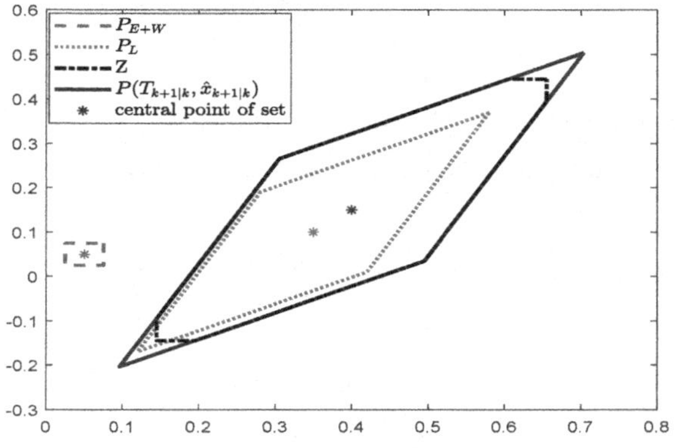

Figure 4.2: Parcel area of the set-membership space in the predictive process.

observation set S_{k+1} and $\mathcal{P}(\hat{x}_{k+1|k}, T_{k+1|k})$ obtained in the previous prediction step, that is

$$S_{k+1} \cap \mathcal{P}(\hat{x}_{k+1|k}, T_{k+1|k}) \subseteq \mathcal{P}(\hat{x}_{k+1|k+1}, T_{k+1|k+1}).$$

Theorem 4.3

Suppose there exits observation equation $y_{k+1} = h(x_{k+1}) + v_{k+1}$, where $y_{k+1} \in \mathbb{R}^{m\times 1}$, $h(x_{k+1})$ is m dimensional nonlinear function of x_{k+1}, $y_{k+1}^{(i)}$ and $h^{(i)}(x_{k+1})$ represent the ith element of y_{k+1} and $h(x_{k+1})$ respectively, $1 \leqslant i \leqslant m$. Let $\varepsilon_v^{(i)}$ represents boundary value of the ith element of v_{k+1}, $H.O.T_2^{(i)}$ represents the ith element of $H.O.T_2$, the upper and lower boundary of which is $H.O.T_2^{(i)} \in [e_{2,\min}^{(i)}, e_{2,\max}^{(i)}]$. Then the i strip $S_{k+1,i}$ of linearized observation set S_{k+1} can be expressed as

$$S_{k+1,i} = \{x \in \mathbb{R}^n \mid \|p_i x_{k+1|k+1} - c_i\|_\infty \leqslant 1\}, \tag{4.23}$$

where

$$p_i = \frac{H_{k+1}}{\varepsilon_v^{(i)} + \frac{e_{2,\max}^{(i)} - e_{2,\min}^{(i)}}{2}}, \tag{4.24}$$

$$c_i = \frac{y_{k+1}^{(i)} - h^{(i)}(\hat{x}_{k+1|k}) + H_{k+1}\hat{x}_{k+1|k} - \frac{e_{2,\max}^{(i)} + e_{2,\min}^{(i)}}{2}}{\varepsilon_v^{(i)} + \frac{e_{2,\max}^{(i)} - e_{2,\min}^{(i)}}{2}}. \tag{4.25}$$

Proof 4.3 The first-order Stirling expansion of the observation equation of model (4.1) at time $k+1$ is

$$\begin{aligned} y_{k+1} = {} & h(\hat{x}_{k+1|k}) + H_{k+1}(x_{k+1|k+1} - \hat{x}_{k+1|k}) \\ & + H.O.T_2 + v_{k+1}. \end{aligned} \tag{4.26}$$

Consider the any component $y_{k+1}^{(i)}$ of $y_{k+1} = [y_{k+1}^{(1)}, y_{k+1}^{(2)}, \dots, y_{k+1}^{(m)}]^{\mathrm{T}}$, $S_{k+1,i}$ can be written as follows based on the boundedness of $v_{k+1}^{(i)}$ and $H.O.T_2^{(i)}$.

$$\begin{aligned} S_{k+1,i} = \Bigg\{ & x \in \mathbb{R}^n \mid \|y_{k+1}^{(i)} - h^{(i)}(\hat{x}_{k+1|k}) + H_{k+1}\hat{x}_{k+1|k} \\ & - H_{k+1}x_{k+1|k+1}\|_\infty^{\varepsilon_v^{(i)} + \frac{e_{2,\max}^{(i)} - e_{2,\min}^{(i)}}{2}} \leqslant 1 \Bigg\}, \end{aligned} \tag{4.27}$$

make the following identical deformation,

$$\|y_{k+1}^{(i)} - h^{(i)}(\hat{x}_{k+1|k}) + H_{k+1}\hat{x}_{k+1|k} - H_{k+1}x_{k+1|k+1}\|_{\infty}^{\varepsilon_v^{(i)} + \frac{e_{2,\max}^{(i)} - e_{2,\min}^{(i)}}{2}}$$

$$= \|\frac{H_{k+1}}{\varepsilon_v^{(i)} + \frac{e_{2,\max}^{(i)} - e_{2,\min}^{(i)}}{2}} x_{k+1|k+1} - \frac{y_{k+1}^{(i)} - h^{(i)}(\hat{x}_{k+1|k})}{\varepsilon_v^{(i)} + \frac{e_{2,\max}^{(i)} - e_{2,\min}^{(i)}}{2}}$$

$$- \frac{H_{k+1}\hat{x}_{k+1|k} - \frac{e_{2,\max}^{(i)} + e_{2,\min}^{(i)}}{2}}{\varepsilon_v^{(i)} + \frac{e_{2,\max}^{(i)} - e_{2,\min}^{(i)}}{2}} \|_{\infty}.$$

Referring to Definition 4.2, the structural parameter of ith strip $S_{k+1,i}$ is rewrited as

$$p_i = \frac{H_{k+1}}{\varepsilon_v^{(i)} + \frac{e_{2,\max}^{(i)} - e_{2,\min}^{(i)}}{2}}, \tag{4.28}$$

$$c_i = \frac{y_{k+1}^{(i)} - h^{(i)}(\hat{x}_{k+1|k}) + H_{k+1}\hat{x}_{k+1|k} - \frac{e_{2,\max}^{(i)} + e_{2,\min}^{(i)}}{2}}{\varepsilon_v^{(i)} + \frac{e_{2,\max}^{(i)} - e_{2,\min}^{(i)}}{2}}. \tag{4.29}$$

Considering that $S_{k+1} = \bigcap_{i=1}^{m} S_{k+1,i}$ is a polytope, $\mathcal{P}(\hat{x}_{k+1|k}, T_{k+1|k}) \cap S_{k+1}$ is also an irregular polyhedron with more complex shape, which is not conducive to iterative calculation. To solve this problem, parallelotope $\mathcal{P}(\hat{x}_{k+1|k}, T_{k+1|k})$ intersects only one strip at a time in this section, that is, $\mathcal{P}(\hat{x}_{k+1|k}, T_{k+1|k}) \cap S_{k+1} = \{\cdots\{\mathcal{P}(\hat{x}_{k+1|k}, T_{k+1|k}) \cap S_{k+1,1}\}, \ldots, \cap S_{k+1,m}\}$. Calculate $\mathcal{P}(\hat{x}_{k+1|k}, T_{k+1|k}) \cap S_{k+1,1}$ at time $k+1$, then calculate the minimum outer parallelotope of this polyhedron. Next, intersect the outer parallelotope with $S_{k+1,2}$. Repeat the process until meeting $s_{k+1,m}$.

Take the calculation of $\mathcal{P}(\hat{x}_{k+1|k}, T_{k+1|k}) \cap S_{k+1,1}$ for example, we set $S_{k+1,1} = S_{k+1,1}(p_{n+1}, c_{n+1})$. First, set the initial parameters of the iteration like

$$x_c = \hat{x}_{k+1|k},\ T = [t_1, t_2, \ldots, t_n] = T_{k+1|k}. \tag{4.30}$$

According to Eqs. 4.28 and 4.29 to calculate p_{n+1} and c_{n+1}, set two parameters

$$\varepsilon_0^{+} = (p_{n+1}x_c - c_{n+1}) + \sum_{i=1}^{n} |p_{n+1}t_i|, \tag{4.31}$$

$$\varepsilon_0^{-} = (p_{n+1}x_c - c_{n+1}) - \sum_{i=1}^{n} |p_{n+1}t_i|. \tag{4.32}$$

When $\varepsilon_0^+ < -1$ or $\varepsilon_0^- > 1$, it indicates that the intersection is empty, then exits current cycle. If the intersection is not an empty set, start iterative calculation. To ensure that direction of the vector $t_1, t_2, \ldots, t_n$ is positive direction. Define directivity operation $p_{n+1}t_i \geqslant 0$, otherwise reverse t_i.

Each of the following three steps needs to do the directional operation of t_i, $\bar{t}_i$ and t_i^*. The first step is to tighten the strip [18, 94],

$$r_{n+1}^+ = \min(1, \varepsilon_0^+), \tag{4.33}$$

$$r_{n+1}^- = \min(1, -\varepsilon_0^-), \tag{4.34}$$

$$\bar{p}_{n+1} = \frac{2}{r_{n+1}^+ + r_{n+1}^-} p_{n+1}, \tag{4.35}$$

$$\bar{c}_{n+1} = \frac{2}{r_{n+1}^+ + r_{n+1}^-} \left(c_{n+1} + \frac{r_{n+1}^+ - r_{n+1}^-}{2} \right), \tag{4.36}$$

then $\bar{S}_{k+1,1}$ is tight.

Second, tighten parallelotope $\mathcal{P}(x_c, T)$ to obtain $\mathcal{P}(\bar{x}_c, \bar{T})$, as far as $i = 1, 2, \ldots, n$, when $p_{n+1}t_i \neq 0$,

$$\begin{cases} r_i^+ = \min\left(1, \dfrac{1-\varepsilon_0^-}{|p_{n+1}t_i|} - 1\right), \\ r_i^- = \min\left(1, \dfrac{1+\varepsilon_0^+}{|p_{n+1}t_i|} - 1\right), \\ \bar{t}_i = \dfrac{r_i^+ + r_i^-}{2} t_i, \\ \bar{x}_c = x_c + \displaystyle\sum_{i=1}^{n} \dfrac{r_i^+ - r_i^-}{2} t_i, \end{cases} \tag{4.37}$$

when $p_{n+1}t_i = 0$, current parallelotope is not tightened.

Finally, choose minimum outer parallelotope $\mathcal{P}(x_c^*, T^*)$ of $\mathcal{P}(\bar{x}_c, \bar{T}) \cap \bar{S}_{k+1,1}$, that is, select n from $(n+1)$ compact strips to form the minimum volume parallelotope.

$$\bar{t}_{n+1} = \frac{\bar{p}_{n+1}^{\mathrm{T}}}{|\bar{p}_{n+1}|^2}, \tag{4.38}$$

$$i^* = \mathop{\arg\max}_{j=1,2,\ldots,n+1} \{\bar{p}_{n+1}\bar{t}_j\}, \tag{4.39}$$

$$\text{if } i^* = n+1, \mathcal{P}(x_c^*, T^*) = \mathcal{P}(\bar{x}_c, \bar{T}), \tag{4.40}$$

$$\text{if } i^* \neq n+1, \begin{cases} t_i^* = \bar{t}_i - \dfrac{\bar{p}_{n+1}\bar{t}_i}{\bar{p}_{n+1}\bar{t}_{i^*}} \bar{t}_{i^*}, \ i \neq i^*, \\ t_i^* = \dfrac{\bar{t}_i}{\bar{p}_{n+1}\bar{t}_i}, \ i = i^*, \\ x_c^* = \bar{x}_c + \dfrac{\bar{t}_{i^*}(\bar{c}_{n+1} - \bar{p}_{n+1}\bar{x}_c)}{\bar{p}_{n+1}\bar{t}_{i^*}}. \end{cases} \tag{4.41}$$

According to the Lemma 2 in literature [7], the volume of the parallelotope obtained by discarding the pth or qth strip is the same when $i_p^* = i_q^*$ and $p \neq q$. Since both parallelotopes are the smallest, we can choose one of the two. Set $x_c = x_c^*$, $T = T^*$, iterate again until M times to obtain $\hat{x}_{k+1|k+1}$ and $T_{k+1|k+1}$.

Hyperparallel Space based Nonlinear Set-Membership Filtering (HS-NSMF) proposed in this section can be summarized as:

Step 1: Set the parallelotope $\mathcal{P}(T_0, x_0)$ containing the initial state value and the process noise parallelotope $\mathcal{P}_W$, collect the input and output data of the model and set the maximum number of iterations L;

Step 2: Use Stirling and DC programming to linearize $f(x_k)$ and $h(x_{k+1})$ into Eqs. (4.10) and (4.11), bring the vertex coordinates of the parallelotope into Eqs. (4.15) and (4.16) to calculate the boundary of $H.O.T$ and update the parallelotope $\mathcal{P}_E$ with Eqs. (4.17) and (4.18);

Step 3: Calculate $\mathcal{P}_{E+W}$ according to Theorem 4.2;

Step 4: Calculate the zonotopic feasible parameter set Z that contains states $\hat{x}_{k+1|k}$ after updating according to Theorem 4.1, use the dimensionality reduction method quoted in literature [25] to calculate its minimum outer parallelotope $\mathcal{P}(T_{k+1|k}, \hat{x}_{k+1|k})$;

Step 5: Use Eqs. (4.24) and (4.25) to calculate the strip space after linearization, and perform the parallelotopic update calculation according to Eqs. (4.30)–(4.41);

Step 6: Repeat Step 2 until the number of iterations reaches L, exit the loop and output the estimation result of system state.

4.1.4 Simulation

Example 1: The following nonlinear discrete system is given,

$$\begin{cases} x_{k+1} = \begin{bmatrix} 0.5x_{1,k}x_{2,k} \\ 0.2x_{2,k}^2 \end{bmatrix} + u_k + w_k, \\ y_k = H_k x_k + v_k, \end{cases} \tag{4.42}$$

where input $u_k = 0.6$, $H_k = [1 \quad 0]$, process noise $|w_k|_\infty \leqslant 0.001$, measurement noise $|v_k|_\infty \leqslant 0.001$, the initial parallelotope is $\mathcal{P}_0 = \begin{bmatrix} -0.2 \\ 0.3 \end{bmatrix} \oplus \begin{bmatrix} 0.1 & 0 \\ 0 & 0.1 \end{bmatrix}$, the initial zonotopic setting is consistent with the initial parallelotope and the initial ellipsoid shape matrix is $\text{diag}(0.1^2, 0.1^2)$. The algorithm in this section is compared with the CDSMF algorithm proposed in literature [32] and the MSZM algorithm proposed in literature [127]. The simulation results are shown in Figs. 4.3–4.8.

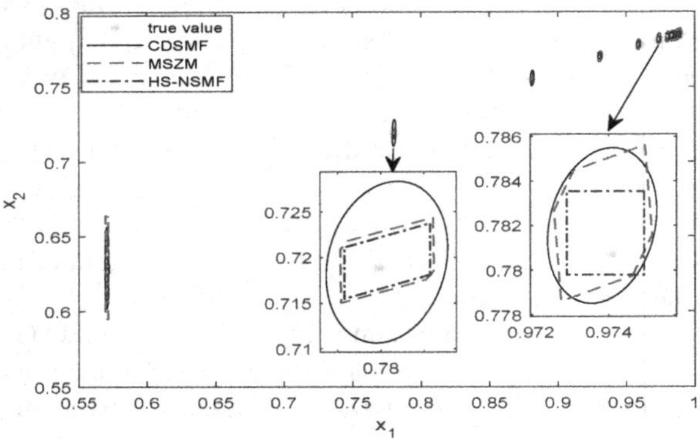

Figure 4.3: Evolution of feasible parameter set.

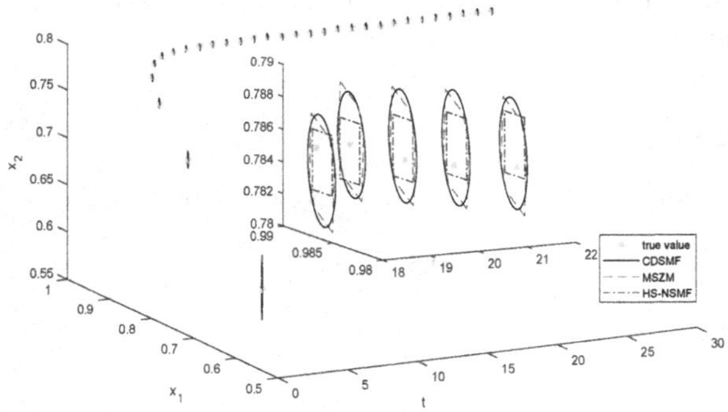

Figure 4.4: Evolution of feasible parameter set with time axis.

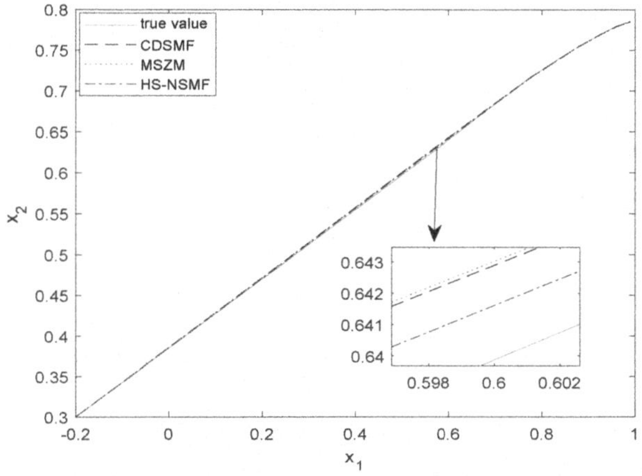

Figure 4.5: Change of states track.

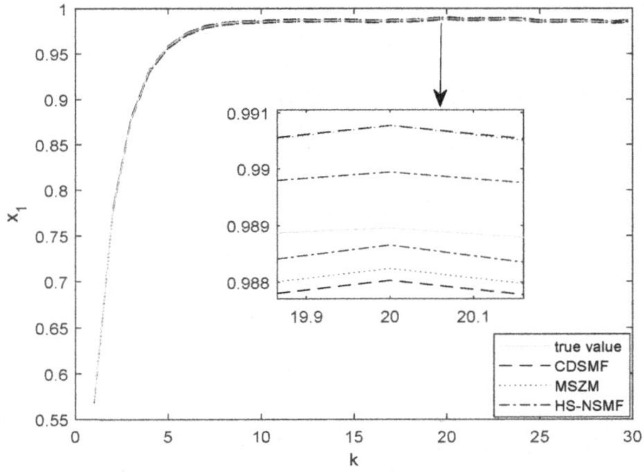

Figure 4.6: Estimated boundary of x_1.

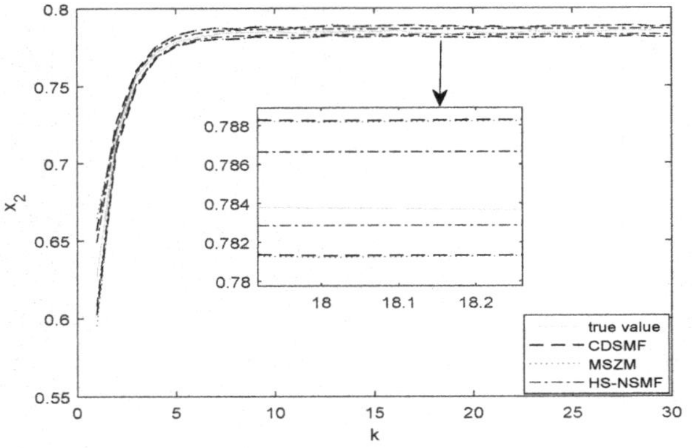

Figure 4.7: Estimated boundary of x_2.

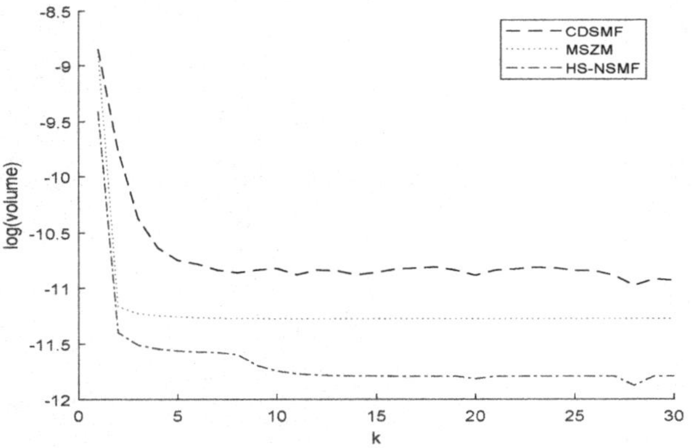

Figure 4.8: Area of feasible parameter set.

Fig. 4.3 shows the evolution of the state feasible set. It can be seen that the three methods wrap the true state value of the system, and the feasible set is shrinking. Also, the area of the feasible set wrapping area of HS-NSMF is smaller than that of CDSMF and MSZM algorithm, indicating that the conservatism of HS-NSMF is much smaller. Fig. 4.4 shows the change of the state feasible set over time. With the increase of the number of iterations, the state feasible sets of the three methods are decreasing, and the wrapping area of the feasible set of HS-NSMF is the smallest.

Fig. 4.5 shows the points trajectory changes of the true and estimate state value in the system. Compared with CDSMF and MSZM algorithm, the HS-NSMF algorithm proposed in this section estimates the system state closer to the true state value, and the filtering effect is more obvious.

Figs. 4.6 and 4.7 are comparison diagram of estimated value boundary of status X_1 and X_2 respectively. It can be seen from the local enlarged diagram that the three algorithms wrap the true state value, but the boundary of HS-NSMF is more compact than the other two algorithms. The boundary compactness of CDSMF and MSZM is approximate, and the latter is slightly smaller. It should be pointed out that the HS-NSMF proposed in this section aims to reduce the wrapping area of state feasible set, Fig. 4.8 shows the geometric area comparison of the state feasible set wrapped by the spatial structure of different filtering algorithms.

It can be seen from Fig. 4.8 that compared with CDSMF and MSZM algorithm, HS-NSMF algorithm has the smallest space area of state feasible set during operation, indicating the least conservatism. The reason is that HS-NSMF algorithm does not need to find out the minimum outer box when calculating the vertices in the process of SDP, instead, it takes $2n$ vertices of parallelotope as selection points into update calculation. For the $H.O.T$ set whose center point and axis radius have been determined, the area of parallelotopic set is smaller than that of the ellipsoidal set.

Example 2: For the spring damping system described in the following mode [86],

$$\begin{cases} x_{k+1} = \begin{bmatrix} x_{1,k} + \Delta T x_{2,k} \\ x_{2,k} + \Delta T(-k_0 x_{1,k}(1 - k_d x_{1,k}^2) - c x_{2,k}) \end{bmatrix} \\ \quad + w_k, \\ y_k = H_k x_k + v_k, \end{cases}$$

where $H_k = [1 \quad 0]$, sampling time $\Delta T = 0.1s$, $k_0 = 1.5$, $k_d = 3$, $c = 1.24$, measurement noise $w_k \in \mathcal{P}_W = \begin{bmatrix} 0 \\ 0 \end{bmatrix} \oplus \begin{bmatrix} 0.002 & 0 \\ 0 & 0.002 \end{bmatrix}$, measurement noise $|v_k|_\infty \leqslant 0.001$. The initial iteration parallelotope is $\mathcal{P}_0 = \begin{bmatrix} 0.2 \\ 0.3 \end{bmatrix} \oplus \begin{bmatrix} 0.2 & 0 \\ 0 & 0.2 \end{bmatrix}$, the size of the initial ellipsoid shape matrix is diag $(0.2^2, 0.2^2)$ and the initial zonotopic setting is consistent with the initial parallelotope. The simulation results are shown in Figs. 4.9–4.14.

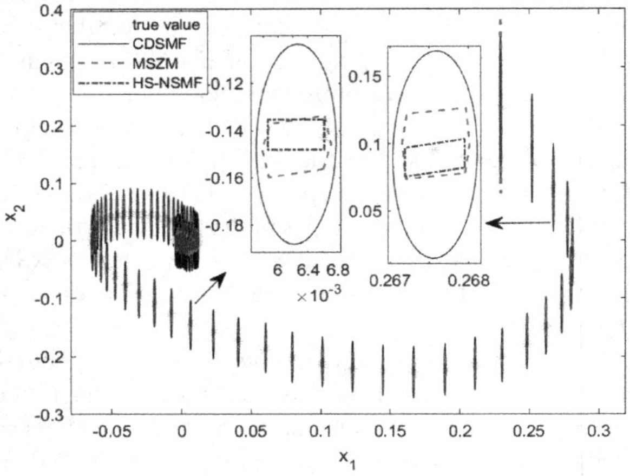

Figure 4.9: Evolution of feasible parameter set.

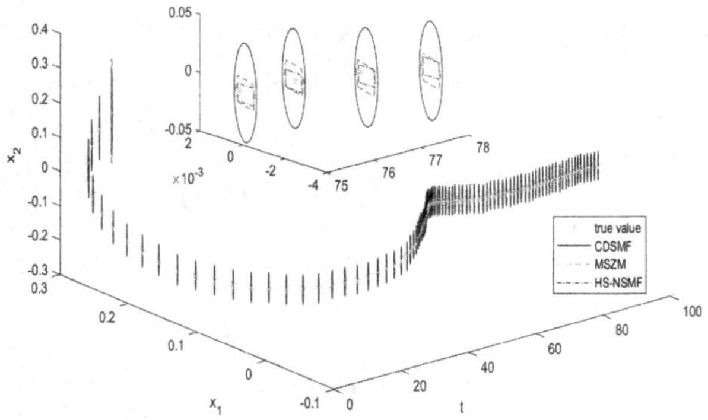

Figure 4.10: Evolution of feasible parameter set with time axis.

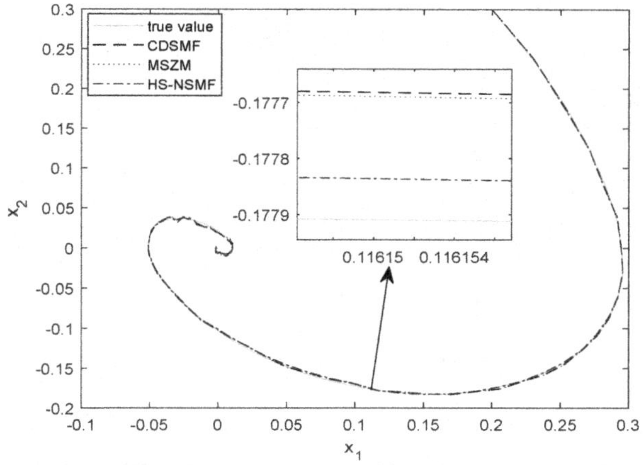

Figure 4.11: Change of states track.

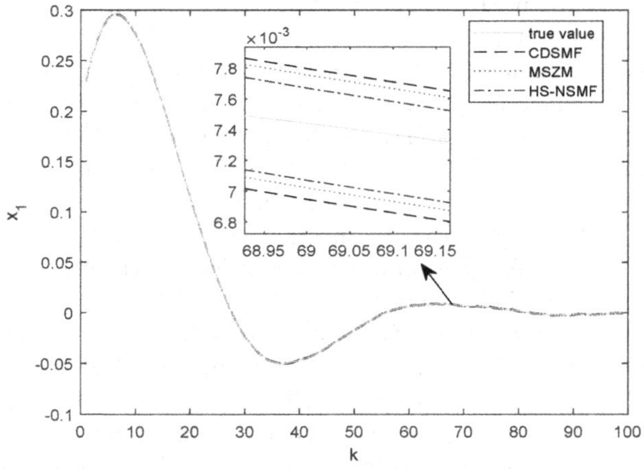

Figure 4.12: Estimated boundary of x_1.

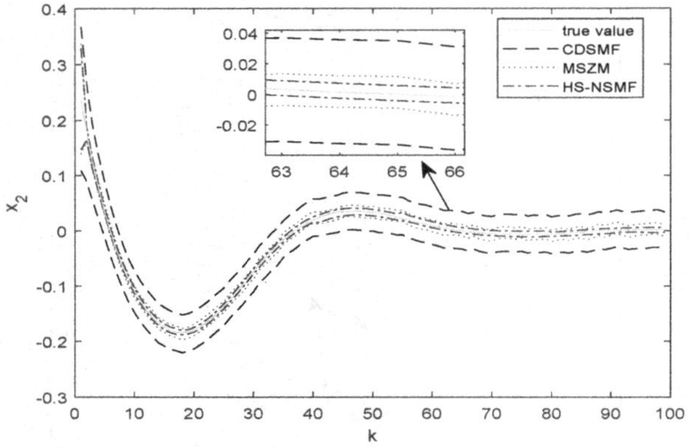

Figure 4.13: Estimated boundary of x_1.

Fig. 4.11 is the points trajectory of true states value and estimate value, state X_1 and X_2 tends to be 0, and the three algorithms can track the change of true value, and the fitness between the estimate value proposed in this algorithm and true value is generally greater than that of CDSMF and MSZM algorithm. Figs. 4.12 and 4.13 are the boundary description of states. The boundary of HS-NSMF algorithm proposed in this section is more compact.

Fig. 4.9 shows the comparison of the state feasible sets of the three algorithms. By amplifying the change of the feasible sets, it can be seen that the HS-NSMF algorithm proposed in this section has less conservative than MSZM and has more obvious advantages than CDSMF algorithm. Fig. 4.10 shows the evolution of the state feasible sets of the three algorithms with the time axis. It can be seen that the feasible set area will decrease with the increase of the steps of iterations, finally, it tends to be stable. The feasible set area of the proposed HS-NSZM algorithm is smaller and more conservative than that of CDSMF and MSZM.

Fig. 4.14 shows the comparison of area of the feasible parameter set which wraps states. Compared with CDSMF and MSZM algorithm, HS-NSMF algorithm has the characteristics of lower conservatism and smaller area. The area of the parallelotope fluctuates greatly in the time period of $k = 7\ 30, 71\ 90$, because the shape of the feasible parameter set contains states obtained at $k = 7$ or $k = 71$ is narrow and long, which leads to a larger upper bound of $H.O.T$ or a smaller lower bound of $H.O.T$ when solving the SDP problem of bound of linearization error at time $k+1$, so that the parallelotope $\mathcal{P}_E$ is large, and the volume of $\mathcal{P}(\hat{x}_{k+1|k+1}, t_{k+1|k+1})$ obtained from the final measurement update is large, resulting in the fluctuation. As can be seen

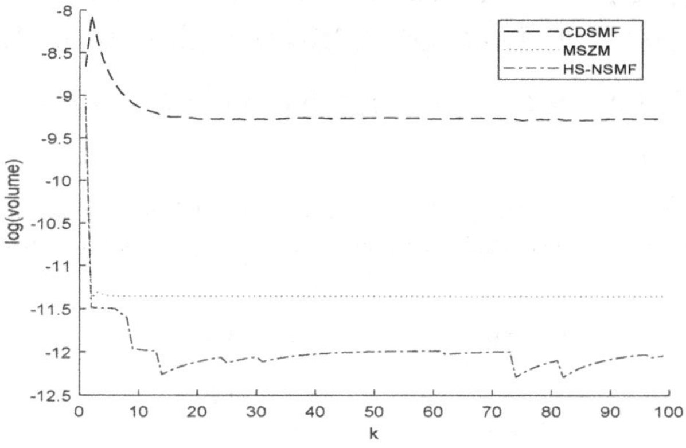

Figure 4.14: Area of feasible parameter set.

from Fig. 4.14, at any time when $k > 3$, the geometric area of feasible set brought by HS-NSMF algorithm is always smaller than that of MSZM and CDSMF.

4.1.5 Conclusion

In this section, a set-membership identification method for nonlinear systems based on parallelotope is proposed. First, the nonlinear function is expanded into first-order linear part and high-order error term by Stirling method, the boundary value of high-order error term is obtained by DC programming and the parallelotope is enveloped to improve its conservatism. Then, the linearized parallelotope is obtained by using the direct sum property of parallelotopes. In the update step, the observation value of M dimension is decomposed into M strips, the linearization error is also calculated, the error is integrated into the strip expression and the strips are successively intersected with the parallelotope to obtain the final feasible set of the system state.

The algorithm proposed in this section is suitable for solving the state estimation problems of systems with nonlinear characteristics, such as Hammerstein system, Wiener system, etc. At the same time, it can be combined with other spatial dimensionality reduction strategies to reduce the amount of calculation, and can be extended to solve the state estimation of other types of nonlinear systems [64], like fault diagnosis and fault-tolerant control [39, 117] and other areas.

4.2 Nonlinear system state estimation based on axisymmetric box space filter under uncertain noise

4.2.1 Related definitions and problem descriptions

4.2.1.1 Related definitions

Define the parameters involved in this section First, such as norm, interval vector, axisymmetric box space, ellipsoid, Minkowski sum of axisymmetric box space:

Definition 4.5 For discrete signals $\boldsymbol{x} \in \mathbb{R}^n$, Define the L_∞:

$$\|\boldsymbol{x}\|_\infty = \sup_{k\geq 0} \|\boldsymbol{x}_k\|, \|\boldsymbol{x}_k\| = \sqrt{\boldsymbol{x}_k^{\mathrm{T}}\boldsymbol{x}_k}. \tag{4.43}$$

Definition 4.6 Define the n-dimensional interval vector as

$$[\boldsymbol{x}_{\min}, \boldsymbol{x}_{\max}] = \{\boldsymbol{x} \in \mathbb{R}^n \mid \boldsymbol{x}_{i,\min} \leqslant \boldsymbol{x}_i \leqslant \boldsymbol{x}_{i,\max}\}, \tag{4.44}$$

where $\boldsymbol{x}_{\min}$ represents the lower bound of the interval vector, and $\boldsymbol{x}_{\max}$ represents the upper bound of the interval vector. Therefore, the interval vector is graphed, and the upper and lower bounds of the interval vector are used as the boundary of the axisymmetric box space, and the axisymmetric box space formed to be

$$B(\boldsymbol{a},\boldsymbol{d}) = \{\boldsymbol{x} \in \mathbb{R}^n \mid \boldsymbol{x} = \boldsymbol{a} + \mathrm{diag}\{\boldsymbol{d}\}\boldsymbol{m}, \| \boldsymbol{m} \|_\infty \leqslant 1\}, \tag{4.45}$$

while, the center of the axisymmetric box space as $\boldsymbol{a} = (\boldsymbol{x}_{\min} + \boldsymbol{x}_{\max})/2$, the radius of the axisymmetric box space as $\boldsymbol{d} = (\boldsymbol{x}_{\max} - \boldsymbol{x}_{\min})/2$, so $\boldsymbol{a}$ and $\boldsymbol{d}$ determine the spatial position and shape of the axisymmetric box .

Definition 4.7 Define the n-dimensional ellipsoid $E(\boldsymbol{a},P)$ as

$$E(\boldsymbol{a},P) = \left\{\boldsymbol{x} \in \mathbb{R}^n \mid (\boldsymbol{x}-\boldsymbol{a})^{\mathrm{T}}P^{-1}(\boldsymbol{x}-\boldsymbol{a}) \leqslant 1\right\}, \tag{4.46}$$

while, $\boldsymbol{a} \in \mathbb{R}^n$ is the center of the ellipsoid, $P \in \mathbb{R}^{n\times n}$ is the envelope matrix that determines the shape of the ellipsoid, which satisfies the symmetric positive definiteness.

Definition 4.8 For the axisymmetric box space $\mathbf{B}_1(\boldsymbol{a}_1,\boldsymbol{d}_1)$ and the axisymmetric box space $\mathbf{B}_2(\boldsymbol{a}_2,\boldsymbol{d}_2)$. Define the Minkowski sum of the two as

$$\mathbf{B}_1 \oplus \mathbf{B}_2 = (\boldsymbol{a}_1 + \boldsymbol{a}_2, [\boldsymbol{d}_1 \quad \boldsymbol{d}_2]). \tag{4.47}$$

4.2.1.2 Problem description

For a nonlinear discrete system as follows:

$$\boldsymbol{x}_{k+1} = f(\boldsymbol{x}_k) + \boldsymbol{w}_k, \tag{4.48}$$

$$\boldsymbol{y}_{k+1} = h(\boldsymbol{x}_{k+1}) + \boldsymbol{v}_k, \tag{4.49}$$

while, $f(\boldsymbol{x}_k)$ and $h(\boldsymbol{x}_{k+1})$ is respectively non-linear state function and measurement function. The $f(\boldsymbol{x}_k)$ and $h(\boldsymbol{x}_{k+1})$ are second-order derivable; $\boldsymbol{x}_k \in \mathbb{R}^n$ is the state variables and $\boldsymbol{y}_k \in \mathbb{R}^m$ is the measurement variables; $\boldsymbol{w}_k \in \mathbb{R}^n$ is bounded unknown process noise and $\boldsymbol{v}_k \in \mathbb{R}^m$ is bounded unknown measurement noise which satisfy $\boldsymbol{w}_k \in \mathbf{B}(0, \boldsymbol{r}_w)$, $\boldsymbol{v}_k \in \mathbf{B}(0, \boldsymbol{r}_v)$. Assumed nonlinear state function $f(\boldsymbol{x})$ is a DC function. That is, every continuous function can be approximated by the difference of two convex functions [5]. So $f(\boldsymbol{x})$ meets the following properties:

Property 6 *If the non-linear state function $f(\boldsymbol{x})$ is a DC function, then each part $f_n(n = 1, 2, 3, \ldots, i)$ of the function $f = [f_1, f_2, f_3, \ldots, f_i]^{\mathrm{T}}$ are all DC functions. The purpose of this section is to propose a nonlinear system state estimation method based on axisymmetric box space filtering under uncertain noise. While accurately estimating the real state $\boldsymbol{x}_k$ of the system, it solves the shortcomings of the traditional set-membership algorithm state estimation set of loose package and large computational complexity.*

4.2.2 State estimation of nonlinear system based on axisymmetric box space filtering

4.2.2.1 Linearization of nonlinear models

Aiming at the linearization problem of the nonlinear model, this section draws on the idea of Kalman filter [58] to obtain the initial linearization model, and uses the axisymmetric box space to wrap the error generated in the linearization process. Then the state function linearization error axisymmetric box space and the process noise axisymmetric box space are subjected to the Minkowski sum operation to obtain the expanded interference error axisymmetric box space.

For nonlinear systems (4.48), expand linearly the state function $f(\boldsymbol{x})$ at the state quantity $\hat{\boldsymbol{x}}_k$ to obtain a linearized model:

$$\begin{aligned} f(\boldsymbol{x}_k) &= f(\hat{\boldsymbol{x}}_k) + J_{1,k}(\boldsymbol{x}_k - \hat{\boldsymbol{x}}_k) + O_{1,k}(\boldsymbol{x}_k - \hat{\boldsymbol{x}}_k) \\ &= f(\hat{\boldsymbol{x}}_k) + J_{1,k}(\boldsymbol{x}_k - \hat{\boldsymbol{x}}_k) + \boldsymbol{e}_{1,k} \\ &= f_L + \boldsymbol{e}_{1,k}, \end{aligned} \tag{4.50}$$

while, $f(\hat{\boldsymbol{x}}_k)$ is the value of the state function when $\boldsymbol{x}_k = \hat{\boldsymbol{x}}_k$, $J_{1,k}$ is the Jacobian matrix of the f function, $O_{1,k}(\boldsymbol{x}_k - \hat{\boldsymbol{x}}_k)$ is the expand higher-order terms for Taylor which is the linearization error of the state function. The interval of the linearization error of the package can be expressed as $\boldsymbol{e}_{1,k} \in E_1 = [\boldsymbol{e}_{1,k,\min}, \boldsymbol{e}_{1,k,\max}]$, while, E_1 is the interval of the linearization error of the state function, $\boldsymbol{e}_{1,k,\min}$ is the lower bound of the linearization error of the state function, $\boldsymbol{e}_{1,k,\max}$ is the upper bound of the linearization error of the state function.

In the same way, linearize the measurement function $h(\boldsymbol{x})$ in the nonlinear system (4.49) to obtain the linearized model:

$$\begin{aligned} h(\boldsymbol{x}_{k+1}) &= h(\hat{\boldsymbol{x}}_{k+1})+J_{2,k}(\boldsymbol{x}_{k+1}-\hat{\boldsymbol{x}}_{k+1})+O_{2,k}(\boldsymbol{x}_{k+1}-\hat{\boldsymbol{x}}_{k+1}) \\ &= h(\hat{\boldsymbol{x}}_{k+1})+J_{2,k}(\boldsymbol{x}_{k+1}-\hat{\boldsymbol{x}}_{k+1})+\boldsymbol{e}_{2,k} \\ &= h_L+\boldsymbol{e}_{2,k}, \end{aligned} \tag{4.51}$$

while, $h(\hat{\boldsymbol{x}}_{k+1})$ is the value of the measurement function when $\boldsymbol{x}_{k+1}=\hat{\boldsymbol{x}}_{k+1}$, $J_{2,k}$ is the Jacobian matrix of the function h, $O_{2,k}(\boldsymbol{x}_{k+1}-\hat{\boldsymbol{x}}_{k+1})$ is the higher-order terms for Taylor which is the linearization error of the measurement function. The interval of the linearization error of the package measurement function can be expressed as $\boldsymbol{e}_{2,k}\in E_2=[\boldsymbol{e}_{2,k,\min},\boldsymbol{e}_{2,k,\max}]$. While E_2 is the interval of the linearization error of the measurement function, $\boldsymbol{e}_{2,k,\min}$ is the lower bound of the linearization error of the measurement function, $\boldsymbol{e}_{2,k,\max}$ is the upper bound of the linearization error of the measurement function.

4.2.2.2 *Interval expression of linearization error*

From the Eq. (4.50), it can be seen that the linearization model obtained is composed of the linearization state function f_L and the linearization error $\boldsymbol{e}_{1,k}$. Compared with the high complexity of directly ignoring the significant deviation caused by the linearization process and the precise calculation error, the error brought by the interval-wrapped linearization $\boldsymbol{e}_{1,k}$, which not only guarantees the calculation accuracy, It also reduces the computational complexity of the algorithm.

Theorem 4.4
Because the linearization error $\boldsymbol{e}_k$ has the interval boundary of $\boldsymbol{e}_k\in[\boldsymbol{e}_{k,\min},\boldsymbol{e}_{k,\max}]$, the axisymmetric box space that wraps the linearization error can be expressed as $\boldsymbol{B}(\boldsymbol{a}_e,\boldsymbol{d}_e)$, while,

$$\boldsymbol{a}_e=(\boldsymbol{e}_{k,\min}+\boldsymbol{e}_{k,\max})/2, \tag{4.52}$$

$$\boldsymbol{d}_e=(\boldsymbol{e}_{k,\max}-\boldsymbol{e}_{k,\min})/2. \tag{4.53}$$

Proof 4.4 Since the non-linear function $f(\boldsymbol{x})$ is a DC function, it can be seen that there are convex functions $g_(i)(\boldsymbol{x})$ and $r_i(\boldsymbol{x})$, satisfy

$$f_i(\boldsymbol{x})=g_i(\boldsymbol{x})-r_i(\boldsymbol{x}),$$

while,

$$g_i(\boldsymbol{x}) = f(\boldsymbol{x})+\alpha\boldsymbol{x}^{\mathrm{T}}\boldsymbol{x}, \tag{4.54}$$

$$r_i(\boldsymbol{x}) = \alpha\boldsymbol{x}^{\mathrm{T}}\boldsymbol{x}. \tag{4.55}$$

According to the properties of convex functions construct functions:

$$\bar{g}_i = g_i(\hat{\boldsymbol{x}}_k) + \boldsymbol{u}_1^{\mathrm{T}}(\boldsymbol{x}_k - \hat{\boldsymbol{x}}_k), \tag{4.56}$$

$$\bar{r}_i = r_i(\hat{\boldsymbol{x}}_k) + \boldsymbol{u}_2^{\mathrm{T}}(\boldsymbol{x}_k - \hat{\boldsymbol{x}}_k), \tag{4.57}$$

in the Eqs. (4.56) and (4.57), $\bar{g}_i \leqslant g_i$, $\bar{r}_i \leqslant r_i$, the $\boldsymbol{u}_1$ and $\boldsymbol{u}_2$ is respectively the subgradient of the point $\hat{\boldsymbol{x}}_k$ in the current state of $g_i(\boldsymbol{x})$ and $r_i(\boldsymbol{x})$, as

$$\boldsymbol{u}_1 = \frac{\partial g_1(\hat{\boldsymbol{x}}_k)}{\partial \boldsymbol{x}}, \quad \boldsymbol{u}_2 = \frac{\partial g_2(\hat{\boldsymbol{x}}_k)}{\partial \boldsymbol{x}}. \tag{4.58}$$

Putting the Eqs. (4.56) and (4.57) into the error relationship, getting:

$$\boldsymbol{e}_k = g_i - r_i - f_{Li} \leqslant g_i - \bar{r}_i - f_{Li}, \tag{4.59}$$

$$\boldsymbol{e}_k = g_i - r_i - f_{Li} \geqslant \bar{g}_i - r_i - f_{Li}. \tag{4.60}$$

At the moment of k, $\boldsymbol{x}_k \in V_B$, V_B is the state axisymmetric box space. From the Eqs. (4.59) $\sim$ (4.60), we can get

$$\boldsymbol{e}_{k,\min} = \min_{\boldsymbol{x}_k \in V_B} \{\bar{g}_i(\boldsymbol{x}_k) - r_i(\boldsymbol{x}_k) - f_{\mathrm{L}}(\boldsymbol{x}_k)\}, \tag{4.61}$$

$$\boldsymbol{e}_{k,\max} = \max_{\boldsymbol{x}_k \in V_B} \{g_i(\boldsymbol{x}_k) - \bar{r}_i(\boldsymbol{x}_k) - f_{\mathrm{L}}(\boldsymbol{x}_k)\}. \tag{4.62}$$

After obtaining the interval range of the linearization error $\boldsymbol{e}_{1,k}$, the error can be further converted into the expression form of the axisymmetric box space to obtain the linearization error axisymmetric box space $\mathbf{B}(\boldsymbol{a}_e, \boldsymbol{d}_e)$, while:

$$\boldsymbol{a}_e = (\boldsymbol{e}_{k,\min} + \boldsymbol{e}_{k,\max})/2, \tag{4.63}$$

$$\boldsymbol{d}_e = (\boldsymbol{e}_{k,\max} - \boldsymbol{e}_{k,\min})/2. \tag{4.64}$$

The certificate is complete.

From Theorem 4.1, the linearization error interval $E_1 = [\boldsymbol{e}_{1,k,\min}, \boldsymbol{e}_{1,k,\max}]$ and $E_2 = [\boldsymbol{e}_{2,k,\min}, \boldsymbol{e}_{2,k,\max}]$ of the nonlinear state function $f(x)$ and the measurement function $h(x)$ can be obtained as

$$\begin{aligned}
\boldsymbol{e}_{1,k,\min} &= \min_{\boldsymbol{x}_{1,k} \in V_B} \{\bar{g}_i(\boldsymbol{x}_{1,k}) - r_i(\boldsymbol{x}_{1,k}) - f_{\mathrm{L}}(\boldsymbol{x}_{1,k})\}, \\
\boldsymbol{e}_{1,k,\max} &= \max_{\boldsymbol{x}_{1,k} \in V_B} \{\bar{g}_i(\boldsymbol{x}_{1,k}) - r_i(\boldsymbol{x}_{1,k}) - f_{\mathrm{L}}(\boldsymbol{x}_{1,k})\}, \\
\boldsymbol{e}_{2,k,\min} &= \min_{\boldsymbol{x}_{2,k} \in V_B} \{\bar{g}_i(\boldsymbol{x}_{2,k}) - r_i(\boldsymbol{x}_{2,k}) - f_{\mathrm{L}}(\boldsymbol{x}_{2,k})\}, \\
\boldsymbol{e}_{2,k,\max} &= \max_{\boldsymbol{x}_{2,k} \in V_B} \{\bar{g}_i(\boldsymbol{x}_{2,k}) - r_i(\boldsymbol{x}_{2,k}) - f_{\mathrm{L}}(\boldsymbol{x}_{2,k})\}.
\end{aligned}$$

From this, the linearized error axisymmetric box space of the nonlinear state function $f(x)$ and the measurement function $h(x)$ is further obtained as $\mathbf{B}(\boldsymbol{a}_{e_f}, \boldsymbol{d}_{e_f})$ and $\mathbf{B}(\boldsymbol{a}_{e_h}, \boldsymbol{d}_{e_h})$.

4.2.2.3 State prediction

Set the axisymmetric box space of the current state quantity $\mathbf{B}_x(\boldsymbol{a}_x, \boldsymbol{d}_x)$, and the state function linearization process axisymmetric box space $\mathbf{B}_e(\boldsymbol{a}_{e_f}, \boldsymbol{d}_{e_f})$ and the axisymmetric box space of prediction noise $\mathbf{B}_w(\boldsymbol{a}_w, \boldsymbol{d}_w)$ performs the Minkowski sum operation to obtain the prediction state set of the current system P_{k+1}:

$$\begin{aligned} P_{k+1} &= \mathbf{B}_x(\boldsymbol{a}_x, \boldsymbol{d}_x) \oplus \mathbf{B}_{e_f}(\boldsymbol{a}_{e_f}, \boldsymbol{d}_{e_f}) \oplus \mathbf{B}_w(\boldsymbol{a}_w, \boldsymbol{d}_w) \\ &= \mathbf{B}(\boldsymbol{a}_x + \boldsymbol{a}_{e_f} + \boldsymbol{a}_w, [\boldsymbol{d}_k \quad \boldsymbol{d}_{e_f} \quad \boldsymbol{d}_w]). \end{aligned} \tag{4.65}$$

Perform affine transformation according to the linear system obtained after linearization, and convert the axisymmetric box space state set $\mathbf{B}_{\boldsymbol{x}_k}(\boldsymbol{a}_{x_k}, \boldsymbol{d}_{x_k})$ is transformed into axisymmetric box space state set $\mathbf{B}_{\boldsymbol{x}_{k+1}}(\boldsymbol{a}_{x_{k+1}}, \boldsymbol{d}_{x_{k+1}})$ at time $k+1$, that is

$$\boldsymbol{a}_{\boldsymbol{x}_{k+1}} = f(\boldsymbol{a}_{\boldsymbol{x}_k}), \tag{4.66}$$

$$\boldsymbol{d}_{\boldsymbol{x}_{k+1}} = J_{1,k}\boldsymbol{d}_{\boldsymbol{x}_k}. \tag{4.67}$$

The linearization error axisymmetric box space set at the next moment is also updated over time. Using the Eqs. (4.63) and (4.64), the linearization error axisymmetric box at each moment can be calculated and then perform the Minkowski sum with the measured noise axisymmetric box space to obtain the error axisymmetric box space. Through the above state prediction process, the interval of the state variable at each moment of the nonlinear system can be obtained and expressed in the form of an axisymmetric box space.

4.2.2.4 Measurement update

In the measurement update step, the state variable axisymmetric box space set is obtained by state prediction as a priori state set, and then the measurement value is further updated posteriorly, so that the original predicted state set is shrunk under certain restricted conditions, and the final result is updated to get a compact, more accurate set package.

Refer to the Kalman filter idea to linearize the non-linear measurement function to obtain the Eq. (4.51), and according to the Eqs. (4.63) and (4.64), the linearity of the axisymmetric box space $\mathbf{B}_{e_h}(\boldsymbol{a}_{e_h}, \boldsymbol{d}_{e_h})$ of measurement function which can be obtained. Then through the Minkowski sum operation, the measured noise axisymmetric box space $\mathbf{B}_v(\boldsymbol{a}_v, \boldsymbol{d}_v)$ and the measurement process noise axisymmetric box space $\mathbf{B}_h(\boldsymbol{a}_h, \boldsymbol{d}_h)$ are obtained.

The axisymmetric box space of the measurement state variable can be expressed according to the measurement set S_{k+1} after the error axisymmetric box space of the measurement process is obtained. The axisymmetric box space of the measurement set at time $\mathbf{B}_{S_{k+1}}(\boldsymbol{a}_{S_{k+1}}, \boldsymbol{d}_{S_{k+1}})$ can be expressed as:

$$\begin{aligned} &\mathbf{B}_{S_{k+1}}(\boldsymbol{a}_{S_{k+1}}, \boldsymbol{d}_{S_{k+1}}) = \{\boldsymbol{x} \in R^n | \boldsymbol{y}_{k+1} - h(\hat{\boldsymbol{x}}_k) + J_{2,k}(\boldsymbol{x}_k - \hat{\boldsymbol{x}}_k) + O_{2,k}(\boldsymbol{x}_k - \hat{\boldsymbol{x}}_k) \\ = \quad &\boldsymbol{a}_{S_{k+1}} + \text{diag}\{\boldsymbol{d}_{S_{k+1}}\}\boldsymbol{m}, \| \boldsymbol{m} \|_\infty \leqslant 1\}, \end{aligned} \tag{4.68}$$

while, $\boldsymbol{a}_{S_{k+1}}$ represents the center of the axisymmetric box space of the measurement set at time $k+1$, $\boldsymbol{d}_{S_{k+1}}$ represents the width of the axisymmetric box space of the measurement set at time $k+1$, $\boldsymbol{m}$ is an arbitrary vector which is in the unit interval with the same dimension as $\boldsymbol{a}_{S_{k+1}}$.

In the traditional set-membership filtering process, the shrinking step of measurement update is obtained by the intersection operation between different set members. Under different spatial structures, there are different intersection calculation methods. For example, due to the diversity of the shape of the ellipsoid, the Minkowski sum operation is often accompanied by large errors. In order to minimize the interference of errors in the process of obtaining intersections, this section proposes an axisymmetric box space intersection strategy. Using the property that the boundaries of the axisymmetric box space $2n$ are orthogonal to each other in the space to perform the Minkowski sum, there is no redundant term in the calculation, so that an accurate axisymmetric box space intersection can be obtained.

Next, split the axisymmetric box space into multiple sets of hyperplanes. It is further transformed into multiple constraints. The predicted state axisymmetric box space is cut through linear programming. While ensuring the correctness of the calculation, the measurement update step is completed.

Theorem 4.5

For the axisymmetric box space, in order to simplify the calculation of the intersection, first split the axisymmetric box space to obtain the expression:

$$\begin{aligned} S'_{k+1} &= \bigcap_{i=n} S'_{k+1,i} \\ &= \bigcap_{i=n} \{\boldsymbol{x} | \boldsymbol{y}_{i,k+1} - \boldsymbol{a}_{i,S_{k+1}} - \boldsymbol{d}_{i,S_{k+1}} \leqslant \boldsymbol{h}_i^{\mathrm{T}} \boldsymbol{x} \leqslant \boldsymbol{y}_{i,k+1} - \boldsymbol{a}_{i,S_{k+1}} + \boldsymbol{d}_{i,S_{k+1}}\}. \end{aligned} \quad (4.69)$$

Then use the hyperplane to split the axisymmetric box space, and find the boundary $[\beta_{i,\min}, \beta_{i,\max}]$ *of the intersection space of the axisymmetric box space. Calculate the intersection of the axisymmetric box space that is the new axisymmetric box space* $\boldsymbol{B}_{X_{k+1}}(\boldsymbol{a}_{X_{k+1}}, \boldsymbol{d}_{X_{k+1}})$ *as follows:*

$$\boldsymbol{a}_{X_{k+1}} = \frac{\beta^i_{i,\max}(k+1) + \beta^i_{i,\min}(k+1)}{2}, \quad (4.70)$$

$$\boldsymbol{d}_{X_{k+1}} = \frac{\beta^i_{i,\max}(k+1) - \beta^i_{i,\min}(k+1)}{2}. \quad (4.71)$$

Proof 4.5 The axisymmetric box space is

$$\mathbf{B}(\boldsymbol{a}, \boldsymbol{d}) = \{\boldsymbol{x} \in R^n \mid \boldsymbol{x} = \boldsymbol{a} + \mathrm{diag}\{\boldsymbol{d}\}\boldsymbol{m}, \| \boldsymbol{m} \|_\infty \leqslant 1\}.$$

Because of the orthogonal nature of the boundary of the axisymmetric box space, each axisymmetric box space can be split when performing the intersection operation of the axisymmetric box space. Split it into n groups of hyperplanes, where n is the spatial dimension of the axisymmetric box space. At this time, the expression of the hyperplane is as follows:

$$\begin{aligned} S'_{k+1} &= \bigcap_{i=n} S'_{k+1,i} \\ &= \bigcap_{i=n} \{\boldsymbol{x} | \boldsymbol{y}_{i,k+1} - \boldsymbol{a}_{i,S_{k+1}} - \boldsymbol{d}_{i,S_{k+1}} \leqslant \boldsymbol{h}_i^{\mathrm{T}} \boldsymbol{x} \leqslant y_{i,k+1} - \boldsymbol{a}_{i,S_{k+1}} + \boldsymbol{d}_{i,S_{k+1}} \}. \end{aligned}$$

Convert a n-dimensional space box S_{k+1} into n sets of hyperplanes, and use each set of hyperplanes to segment another intersecting axisymmetric box space P_{k+1}. The n group of hyperplanes can be understood as $2n$ constraints on the numerical level, and then the hyperplanes are converted into $2n$ constraints for linear programming.

$$\begin{aligned} \beta_{i,\max}(k+1) &= \max h_i^{\mathrm{T}} \boldsymbol{x}, \\ s.t. \quad & \boldsymbol{x} \in P_{k+1} \cap S_{k+1}. \end{aligned} \tag{4.72}$$

$$\begin{aligned} \beta_{i,\min}(k+1) &= \min h_i^{\mathrm{T}} \boldsymbol{x}, \\ s.t. \quad & \boldsymbol{x} \in P_{k+1} \cap S_{k+1}. \end{aligned} \tag{4.73}$$

Among them, the constraint object $h(i)(T)(x)$ is the vertex of the axisymmetric box space.

After segmenting all hyperplanes through linear programming, the new interval boundary of the intersection of the axisymmetric box space can be obtained $[\beta_{i,\min}, \beta_{i,\max}]$. Recalculate the feasible set axisymmetric box space $\mathbf{B}_{X_{k+1}}(\boldsymbol{a}_{X_{k+1}}, \boldsymbol{d}_{X_{k+1}})$ which is the new axisymmetric box space after the intersection of two axisymmetric box spaces at this moment, where

$$\begin{aligned} \boldsymbol{a}_{X_{k+1}} &= \frac{\beta^i_{i,\max}(k+1) + \beta^i_{i,\min}(k+1)}{2}, \\ \boldsymbol{d}_{X_{k+1}} &= \frac{\beta^i_{i,\max}(k+1) - \beta^i_{i,\min}(k+1)}{2}. \end{aligned}$$

The certificate is complete.

By Theorem 4.2, take the intersection of the predicted axisymmetric box space P_{k+1} and the measured axisymmetric box space S_{k+1}, and finally get the feasible set $X(k+1) = P_{k+1} \cap S_{k+1}$ of the state at time $k+1$.

In summary, the steps of the nonlinear system state estimation algorithm based on Axisymmetric Box Space Filtering (ABSF) proposed in this section are summarized as follows:

1. Set the estimated total duration of the state T and the initial moment $k = 1$. Give the initial axisymmetric box space;

2. Use the Eqs. (4.50) and (4.51) to linearize the state function and measurement function of the nonlinear system, and use the axisymmetric box space to wrap the noise generated by the linearization process;

3. Calculate the interval boundary $[e_{k,\min}, e_{k,\max}]$ of the wrapping error in the prediction process by Eqs. (4.61) and (4.62). Calculate the linearized error axisymmetric box space $\mathbf{B}(a_e, d_e)$ by Eqs. (4.63) and (4.64);

4. For the prediction noise axisymmetric box space $\mathbf{B}_w(a_w, d_w)$, the state function linearization error axisymmetric box space $\mathbf{B}_{e_f}(a_{e_f}, d_{e_f})$, prediction state axisymmetric box space $\mathbf{B}_x(a_x, d_x)$ of the prediction process perform the Minkowski sum operation to obtain the prediction state set P_{k+1};

5. In the same way, get the measurement noise axisymmetric box space $\mathbf{B}_v(a_v, d_v)$, the measurement function linearization error axisymmetric box space $\mathbf{B}_{e_h}(a_{e_h}, d_{e_h})$ and the measurement state axisymmetric box space $\mathbf{B}_y(a_y, d_y)$. Then through the Minkowski sum operation to get the measurement state set P_{k+1};

6. Use Theorem 4.2 to split the measurement state set S_{k+1} and convert it into an equivalent n group of hyperplanes, that is, $2n$ restriction conditions are obtained. Use the restriction conditions to measure the state set Carry out linear programming to realize the contraction of the measurement state box, and obtain the feasible set of the state $X_{k+1} = P_{k+1} \cap S_{k+1}$;

7. Set $k = k+1$, return to step 2; when $k = T$, the algorithm ends, and the state estimation result is output.

4.2.3 Simulation

In order to verify the effectiveness of the proposed algorithm, the nonlinear spring-mass-damper system shown in Fig. 4.15 is selected as the simulation object [81], and its motion equation satisfies the following equation:

$$\ddot{x} + k_0 x\left(1 + k_d x^2\right) + c\dot{x} = 0. \tag{4.74}$$

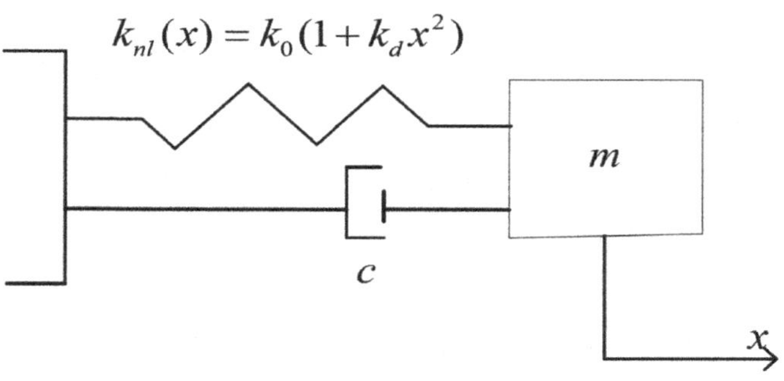

Figure 4.15: Nonlinear spring mass damper system.

After discretizing the motion Eq. (4.74), we can get:

$$x_{k+1} = f(x_k) + w_k, \tag{4.75}$$

$$y_{k+1} = h(x_{k+1}) + v_k, \tag{4.76}$$

while,

$$x_k = \begin{bmatrix} x_{1,k} \\ x_{2,k} \end{bmatrix},$$

$$f(x_k) = \begin{bmatrix} x_{1,k} + \Delta T x_{2,k} \\ x_{2,k} + \Delta T\left(-k_0 x_{1,k}\left(1 + k_d x_{1,k}^2\right) - c x_{2,k}\right) \end{bmatrix},$$

$$h(x_{k+1}) = \begin{bmatrix} x_{1,k+1} \\ x_{2,k+1} \end{bmatrix}.$$

The relevant parameters of the system are as follows: $x_{1,k}$ and $x_{2,k}$ are the spring position and spring speed at k, $\Delta T = 0.1$, $k_0 = 1.5$, $k_d = 3$, $c = 1.24$. The initial state is $\boldsymbol{x}_0 = [0.2, 0.3]^{\mathrm{T}}$, and the initial state feasible set $\mathbf{B} = (\hat{\boldsymbol{x}}_0, \boldsymbol{d}_0)$, where $\hat{\boldsymbol{x}}_0 = [0.2, 0.3]$, $d_0 = \mathrm{diag}\{0.01, 0.01\}$; Predicted process noise w_k and measured noise v_k are respectively satisfying conditions $w_k \in [-0.002, 0.002]$ and $v_k \in [-0.001, 0.001]$ uniform noise. Comparing the ABSF algorithm proposed in this section with the Extended Kalman Filter (EKF) algorithm, the Central Difference Set-Membership Filter (CDSMF) proposed and the Extended Ellipsoidal Set-Membership (EESME) algorithm, the simulation results are shown in Figs. 4.16–4.19.

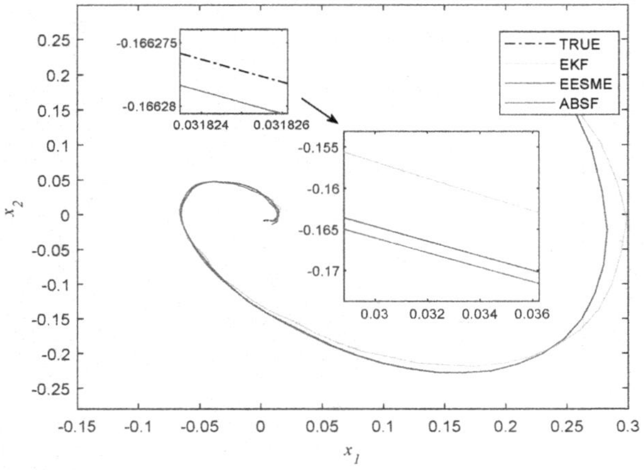

Figure 4.16: State trajectories contrast.

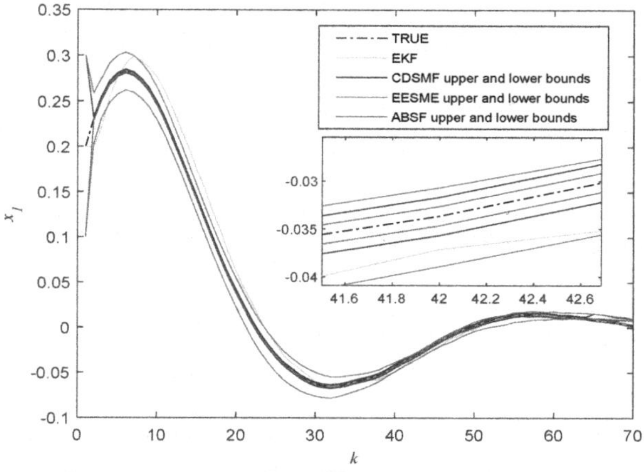

Figure 4.17: Comparison of state estimations x_1.

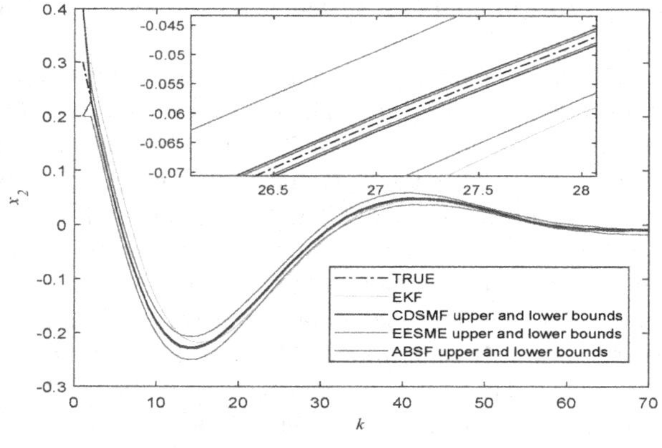

Figure 4.18: Comparison of state estimations x_2.

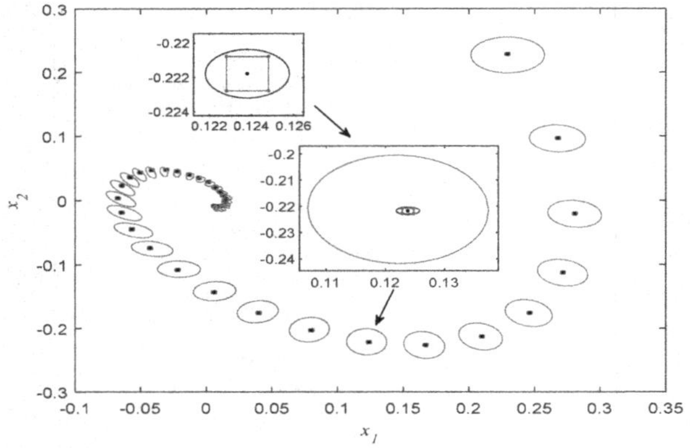

Figure 4.19: Space evolution of feasible sets.

Fig. 4.16 shows the comparison of state trajectory change results when using the ABSF algorithm, EKF algorithm and EEMSE algorithm for nonlinear system state estimation. It can be seen from the figure that compared with EKF algorithm and EESME algorithm, the ABSF algorithm proposed in this section can better realize the state trajectory tracking of nonlinear system and is closer to the real state of the system.

Figs. 4.17 and 4.18 describe the curve changes and interval boundaries of the estimated system state values x_1 and x_2. It can be seen from the results that the ABSF algorithm can realize the estimation of the state quantity x_1 and state quantity x_2 of the system, and the boundary of the axisymmetric box space wraps the true value under the interference at all times. The classic EKF calculation has a larger error when estimating the system state under unknown noise conditions. But the ABSF algorithm proposed in this section is more accurate and stable. At the same time, it can be seen from Figs. 4.17 and 4.18 that although the state estimation set boundary of the ABSF algorithm, CDSMF algorithm and EESME algorithm can wrap the true value, the use of the ellipsoid to wrap is more conservative, that is, the interval range is large. And the state estimation of the EESME algorithm is relatively not stable enough at the beginning, and the boundary of the state set can be continuously contracted after $k = 50$, and the ABSF algorithm proposed in this section can achieve good results in the initial stage, and can always maintain the true value of the compact package state.

Fig. 4.19 shows the comparison of the space evolution of the feasible set of the ABSF algorithm, the CDSMF algorithm and the EESME algorithm. The box area represents the feasible set of states under the operation of the ABSF algorithm. First of all, it can be clearly seen from Fig. 4.19 that the state set estimated by the CDSMF algorithm, the EESME algorithm and the ABSF algorithm proposed in this section can completely wrap the truth value. And as the true value changes, it keeps shrinking during the update process to achieve an accurate estimation of the state. In contrast, the state set shrinkage of the EESME algorithm is significantly slower than the ABSF algorithm. It can be seen that the conservativeness of the EESME algorithm is too large. Although the CDSMF algorithm is closer to the ABSEMF algorithm in general effect, it is still inferior to the ABSF algorithm in accuracy, mainly because it is compared with the ABSF algorithm. In terms of the axisymmetric box space, a certain amount of outer wrapping error will occur when the iterative intersection of the ellipsoid. After an iterative update, the volume of the axisymmetric box space can always be smaller than the ellipsoidal state set.

4.2.4 Conclusion

This section presented a set-membership filtering algorithm for axisymmetric box space wrapping errors to solve the state estimation problem of nonlinear systems under the interference of unknown and bounded uncertain noise. Through the interval estimation of the linearization error, combined with the non-probabilistic distribution characteristics of the bounded unknown noise, the interval estimation of the noise is obtained. Then design the axisymmetric box space wrapping noise and the state feasible set, and the prediction set is updated by the measurement of the axisymmetric box space through the prediction and update steps, and the state feasible set of the nonlinear system is obtained. In the intersection operation of the axisymmetric box space, this section proposes the idea of splitting the interval box, which not only ensures the accuracy, but also improves the boundary contraction speed of the state estimation and reduces the conservativeness of the state estimation algorithm. The feasibility and superiority of the algorithm are verified by the simulation of the given spring-mass-damping system.

Chapter 5

Fault diagnosis based on interval

5.1 Guaranteed fault-estimation algorithm based on interval set inversion observer filtering

5.1.1 Preliminaries and problem description

Selected notations are provided first. $\mathbb{R}^n$ and $\mathbb{R}^{n\times m}$ denote n and $n\times m$ dimensional space, respectively. $\boldsymbol{I}_n$ represents the n-dimensional identity matrix. For a matrix $A\in\mathbb{R}^{n\times m}$, A^{T} and $tr(A)$ represent the transpose and trace, respectively. $\|A\|_F=\sqrt{tr(AA^{\mathrm{T}})}$ are defined as the F-norm. For a vector $M\in\mathbb{R}^n$, $[M]=[\underline{M},\overline{M}]$ denotes the interval vector with a lower bound of $\underline{M}\in\mathbb{R}^n$ and the upper bound of $\overline{M}\in\mathbb{R}^n$, which could also be referred to as an interval box. $W([M])=\max_{i=1,\ldots,n}(\overline{M}_i-\underline{M}_i)$ represents the width of $[M]$, where $\overline{M}_i$ and $\underline{M}_i$ are the upper and lower bounds of the interval of the ith dimension of $[M]$. $Z=\langle p,H\rangle\in\mathbb{R}^n$ denotes a zonotope by center $p\in\mathbb{R}^n$ and generator matrix $H\in\mathbb{R}^{n\times m}$. ∂ is the sign of the partial derivative.

Below are a few definitions used in the following content [102]:

Definition 5.1 The Minkowski sum of two zonotopes $Z_1=\langle p_1,H_1\rangle\in\mathbb{R}^n$ and $Z_2=\langle p_2,H_2\rangle\in\mathbb{R}^n$ is

$$Z_1\oplus Z_2=\langle p_1+p_2,[H_1,H_2]\rangle. \tag{5.1}$$

Definition 5.2 The linear mapping of matrix $L\in\mathbb{R}^{m\times n}$ to zonotope Z_1 is

$$L\odot Z_1=\langle Lp_1,LH_1\rangle. \tag{5.2}$$

DOI: 10.1201/b23146-5

Definition 5.3 For a zonotope $Z = \langle p, H\rangle \in \mathbb{R}^n$, the upper and lower bounds of the interval box $[z] \in \mathbb{R}^n$ containing it are

$$\begin{cases} \overline{z} = p + \sum_{i=1}^{m} |H_i|, \\ \underline{z} = p - \sum_{i=1}^{m} |H_i|. \end{cases} \tag{5.3}$$

Property 7 *For matrices of appropriate dimensions A and B, the rules governing the operation of the matrix trace are as follows:*

$$tr(A+B) = tr(A) + tr(B), \tag{5.4}$$
$$\partial tr(A^{\mathrm{T}}XB^{\mathrm{T}})/\partial X = \partial tr(BX^{\mathrm{T}}A)/\partial X = AB. \tag{5.5}$$

In this section, we consider the following discrete-time linear system:

$$x_{k+1} = Ax_k + Bu_k + Ff_k + Dw_k, \tag{5.6a}$$
$$y_k = Cx_k + Ev_k, \tag{5.6b}$$

where $x_k \in \mathbb{R}^n$ $u_k \in \mathbb{R}^p$ and $y_k \in \mathbb{R}^q$ are the state, input and output vectors, respectively. $f_k \in \mathbb{R}^m$ is an actuator fault with any form. $w_k \in \mathbb{R}^w$ and $v_k \in \mathbb{R}^v$ denote the system disturbance vector and measurement noise vector, respectively. A, B, C, D, E and F are constant matrices with appropriate dimensions. It is assumed that the pair (A, C) is observable, F has full column rank and $q \leqslant m$.

Assume that the unknown but bounded disturbance vector w_k and the measurement noise vector v_k satisfy

$$\underline{w}_k \leqslant w_k \leqslant \overline{w}_k, \underline{v}_k \leqslant v_k \leqslant \overline{v}_k, \tag{5.7}$$

and are equivalent to

$$w_k \in \langle 0, W\rangle, v_k \in \langle 0, V\rangle, \tag{5.8}$$

where $\underline{w}_k$ and $\overline{w}_k$ are the lower and upper bounds of w_k, $\underline{v}_k$ and $\overline{v}_k$ are the lower and upper bounds of v_k, respectively. $W = \mathrm{diag}(\overline{w}_k)$, $V = \mathrm{diag}(\overline{v}_k)$.

According to (5.6), the actuator fault f_k can be expressed as

$$f_k = O_f(y_{k+1} - C(Ax_k + Bu_k + Dw_k) - Ev_{k+1}), \tag{5.9}$$

where $O_f = ((CF)^{\mathrm{T}}CF)^{-1}(CF)^{\mathrm{T}}$.

Substituting the above equation into (5.6a), the state vector of the system can be reformulated as

$$x_{k+1} = \mathcal{A}x_k + \mathcal{B}u_k + \mathcal{F}y_{k+1} + \mathcal{D}w_k - \mathcal{F}Ev_{k+1}, \tag{5.10}$$

with $\mathcal{A} = A - FO_fCA$, $\mathcal{B} = B - FO_fCB$, $\mathcal{D} = D - FO_fCD$ and $\mathcal{F} = FO_f$.

The main purpose of this study is to estimate the upper and lower bounds of the actuator fault satisfying $\underline{f}_k \leqslant f_k \leqslant \overline{f}_k$. The new state expression (5.10) solves the dilemma that the state interval cannot be directly estimated owing to the unknown fault value. By designing the state interval observer based on (5.10), we can obtain the fault estimation interval by combining the observer state estimation interval with (5.9). Increasing the compactness of the estimation intervals would require a more effective interval contraction algorithm.

5.1.2 Main results

In this section, we propose a guaranteed estimation algorithm for system (5.6). The two main processes of the algorithm are: interval estimation and interval contraction. The specific derivation is as follows:

5.1.2.1 Minimal conservative interval observer

Design the dynamic observer for the system (5.10) as

$$\hat{x}_{k+1} = \mathcal{A}\hat{x}_k + \mathcal{B}u_k + \mathcal{F}y_{k+1} + L_k(y_k - C\hat{x}_k), \tag{5.11}$$

where $\hat{x}_k$ is the estimation of x_k, and L_k is the unknown gain matrix determined by Theorem 5.1.

Define the state error as $e_k = x_k - \hat{x}_k$, following which the dynamic error system is obtained by subtracting (5.11) from (5.10),

$$e_{k+1} = (\mathcal{A} - L_kC)e_k + \mathcal{D}w_k - L_kEv_k - \mathcal{F}Ev_{k+1}. \tag{5.12}$$

Theorem 5.1

For the observer in (5.11), when the gain matrix is equal to

$$L_k = \mathcal{A}H_kH_k^{\mathrm{T}}C^{\mathrm{T}}(CH_kH_k^{\mathrm{T}}C^{\mathrm{T}} + EVV^{\mathrm{T}}E^{\mathrm{T}})^{-1}, \tag{5.13}$$

the estimation interval $[x_k^o]$ of the system state x_k obtained by the observer has minimal conservatism.

Proof 5.1 According to the definitions of zonotope, e_{k+1} can be bounded by $\langle 0, H_{k+1}\rangle$, and the generator matrix H_{k+1} is expressed as

$$H_{k+1} = [(\mathcal{A} - L_kC)H_k \;\; \mathcal{D}W \;\; -L_kEV \;\; -\mathcal{F}EV], \tag{5.14}$$

and we can obtain the upper bound $\overline{e}_k$ and the lower bound $\underline{e}_k$ of the state error.

Combined with Eq. (5.11), the upper and lower bounds of the observer estimate interval $[x_{k+1}^o]$ are

$$\begin{cases} \overline{x}_{k+1}^o = \hat{x}_{k+1} + \overline{e}_{k+1}, \\ \underline{x}_{k+1}^o = \hat{x}_{k+1} + \underline{e}_{k+1}. \end{cases} \tag{5.15}$$

This indicates that the conservatism of $[x^o_{k+1}]$ is determined by the upper and lower bounds of its error. Then, define the conservatism as

$$conx_{k+1} = \|H_{k+1}\|_F^2. \tag{5.16}$$

According to Eq. (5.11), the above equation is simplified as

$$\begin{aligned} conx_{k+1} =& tr(H_{k+1}H_{k+1}^{\mathrm{T}}) \\ =& tr((\mathcal{A}-L_kC)H_kH_k^{\mathrm{T}}(\mathcal{A}-L_kC)^{\mathrm{T}}) \\ &+tr(\mathcal{D}WW^{\mathrm{T}}\mathcal{D}^{\mathrm{T}})+tr(L_kEVV^{\mathrm{T}}E^{\mathrm{T}}L_k^{\mathrm{T}}) \\ &+tr(\mathcal{F}EVV^{\mathrm{T}}E^{\mathrm{T}}\mathcal{F}^{\mathrm{T}}). \end{aligned} \tag{5.17}$$

Clearly, the size of $conx_{k+1}$ is related to the unknown observer gain matrix L_k. To minimize the conservatism of the estimation, the optimal gain is obtained by solving the following optimization problem:

$$L_k = \arg\min conx_{k+1}. \tag{5.18}$$

Let $\partial conx_{k+1}/\partial L_k = 0$, we have

$$L_kEVV^{\mathrm{T}}E^{\mathrm{T}} - \mathcal{A}H_kH_k^{\mathrm{T}}C^{\mathrm{T}} - L_kCH_kH_k^{\mathrm{T}}C^{\mathrm{T}} = 0. \tag{5.19}$$

Therefore, according to the above, the gain matrix of the minimum conservative interval observer is

$$L_k = \mathcal{A}H_kH_k^{\mathrm{T}}C^{\mathrm{T}}(CH_kH_k^{\mathrm{T}}C^{\mathrm{T}} + EVV^{\mathrm{T}}E^{\mathrm{T}})^{-1}. \tag{5.20}$$

Rewrite the fault in Eq. (5.9) as the sum of the fault estimation $\hat{f}_k$ and the fault error e_k^f:

$$f_k = \hat{f}_k + e_k^f, \tag{5.21}$$

where

$$\begin{cases} \hat{f}_k = O_f y_{k+1} - O_f CA\hat{x}_k - O_f CBu_k, \\ e_k^f = -O_f CDw_k - O_f Ev_{k+1}. \end{cases} \tag{5.22}$$

In the same way, e_k^f can be bounded by a zonotope

$$e_k^f \in \langle 0, [-O_f CAH_k \quad -O_f CDW \quad -O_f EV]\rangle. \tag{5.23}$$

Then, the upper and lower bounds of the observer fault interval $[f_k^o]$ can be estimated based on the fault error interval $[e_k^f] = [\underline{e}_k^f, \overline{e}_k^f]$ from Eq. (5.23):

$$\begin{cases} \overline{f}_k^o = \hat{f}_k + \overline{e}_k^f, \\ \underline{f}_k^o = \hat{f}_k + \underline{e}_k^f. \end{cases} \tag{5.24}$$

5.1.2.2 Dimensional vector set inversion interval contraction

In this subsection, a new contraction algorithm based on interval analysis is proposed. Based on the estimation of the observer state interval $[x_k^o]$ and the observer fault interval $[f_k^o]$ by the interval observer , we are committed to more compact estimation results. Because the algorithm based on interval analysis can obtain the guaranteed results that are consistent with the uncertain part, the proposed contraction algorithm is suitable for all systems.

According to Eq. (5.9), the contraction of the fault estimation interval $[f_k^o]$ can be regarded as the following set inversion problem:

$$[f_k^s] = O^{-1}[\Pi_k], \tag{5.25}$$

with the new and tighter fault interval $[f_k^s]$ obtained by set inversion contraction and $O = CF$, $[\Pi_k] = y_{k+1} - CA[x_k^o] - CBu_k - CD[w_k] - E[v_{k+1}]$.

Since assuming the invertibility of O is quite restrictive, one of the traditional ways to solve problem (5.25) is the SIVIA algorithm. It is a recursive algorithm based on dichotomy to find a feasible solution set in accordance with the conditions within the initial given infinite interval box. Although the SIVIA algorithm can obtain a high-precision guaranteed solution set, the long computation time and requirement for a large amount of running memory are the main factors limiting its wide application. To overcome these difficulties, we propose an algorithm to accomplish interval contraction via dimensional vector set inversion.

Given an initial box $[f_k^s]_0 \in \mathbb{R}^m$ containing the solution set, transform it into a row vector $\mathcal{L}_0 \in \mathbb{R}^{1\times 2m}$. Select any dimension of $[f_k^s]_0$ and contract the interval of the selected dimension and leave the intervals of the other dimensions unchanged. For each bisection, the new interval box is denoted as $[f_k^s]_i \in \mathbb{R}^m (i = 1,2,\ldots,t)$, and its corresponding row vector is $\mathcal{L}_i \in \mathbb{R}^{1\times 2m}$. All $\mathcal{L}_i$ form a vector group $\mathcal{L} \in \mathbb{R}^{t\times m}$. Use the following test function to judge the inclusion attribute of the row vectors (interval boxes):

$$[t](\cdot) = \begin{cases} in, & \text{if } O\mathcal{L}_i \subset [\Pi_k], \\ out, & \text{if } O\mathcal{L}_i \cap [\Pi_k] = \varnothing, \\ eps, & \text{if } W(\mathcal{L}_i) < \varepsilon. \end{cases} \tag{5.26}$$

The following three situations arise:

1. When $O\mathcal{L}_i \subset [\Pi_k]$ holds, the Boolean variable $in(i) = 1$, then the interval boxes corresponding to the vector group $\mathcal{L}(in)$ are feasible sets, which are subsets of the solution set $[f_k^s]$. The yellow box $[f_k^s]_5$ shown in Fig. 5.1 is a feasible set.

2. When $O\mathcal{L}_i \cap [\Pi_k] = \varnothing$ holds, the Boolean variable $out(i) = 1$, then the interval boxes corresponding to $\mathcal{L}(out)$ do not belong to the solution set and are termed unfeasible sets, such as the white boxes $[f_k^s]_2$ and $[f_k^s]_6$ in Fig. 5.1.

3. When $O\mathcal{L}_i \cap [\Pi_k] \neq \varnothing$ holds, the interval boxes corresponding to $\mathcal{L}(\neg in \wedge \neg out)$ are indeterminate sets. If $W(\mathcal{L}_i)$ is less than the given precision parameter ε, the Boolean variable $eps(i) = 1$, then the interval boxes corresponding

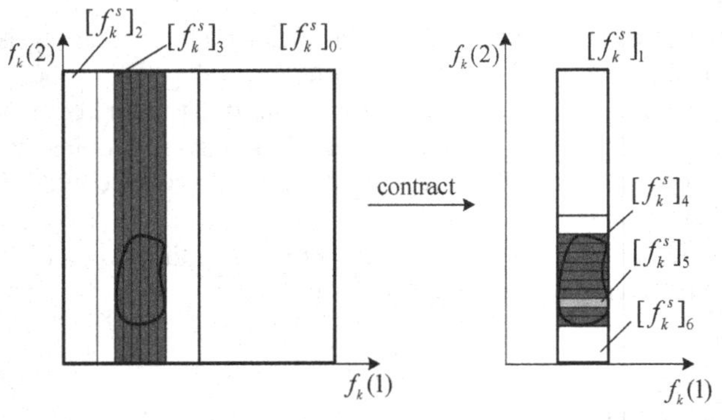

Figure 5.1: Schematic diagram of the interval contraction process.

to $\mathcal{L}(eps)$ are approximate sets of the solution set, such as the blue boxes $[f_k^s]_3$ and $[f_k^s]_4$ shown in Fig. 5.1. Otherwise, the interval boxes are bisected, and a new round of the inclusion attribute judgment of the new boxes begins.

The above process is repeated until the initial interval box is divided into three parts: feasible sets, approximate sets and unfeasible sets. The interval contraction of the selected dimension is accomplished by discarding the infeasible sets, and a compact interval box is obtained, such as the box $[f_k^s]_1$ shown on the right in Fig. 5.1. Next, the interval boxes of the other dimensions are contracted in turn according to the above criteria. Finally, the most compact set inversion contraction fault interval $[f_k^s]$ is obtained. As shown in Fig. 5.1 on the right, the approximate sets (blue boxes) and feasible set (yellow box) form the smallest solution set containing the irregular real solution set.

Remark 5.1 The set inversion problem (5.25) requires the initial interval box $[f_k^s]_0$ to include the solution set. Therefore, the fault estimation interval $[f_k^o]$ obtained by the interval observer in Section 5.1.2.1 is selected as the initial interval box. This further reduces the time and memory requirements. ▮

Similarly, we can solve the following set inversion problem with the initial state $[x_k^o]$ to estimate the state interval $[x_k^s]$ by using the ISIOF algorithm

$$[x_k^s] = C^{-1}(y_k - E[v_k]). \tag{5.27}$$

Remark 5.2 Fig. 5.1 only shows the case in which the solution set is two-dimensional. This contraction algorithm is also applicable when the solution set of the inversion problem is of high dimension. ▮

5.1.2.3 Algorithm analysis

In this section, we analyze the ISIOF algorithm from different perspectives with the aim to further our understanding thereof. The ISIOF algorithm is composed of two parts: a minimal conservative interval observer (MCIO) and another part for dimensional vector set inversion interval contraction (DVSIIC). Because the MCIO has already been studied, the algorithm analysis is mainly focused on DVSIIC here. First, the computational complexity of the algorithm is presented in Theorem 5.2. For simplicity, the judgment rounds of the inclusion attribute are used as the main index to calculate the computational complexity of the DVSIIC algorithm.

Theorem 5.2

(Computational complexity) Assuming that the bisection of the interval box is carried out N times at most in the process of solving the set inversion problem (5.25) by the DVSIIC algorithm, its computational complexity denoted as S satisfies

$$S = \log_2\left(\prod_{i=1}^{m} \mathcal{N}_i\right) + m, \tag{5.28}$$

where $\mathcal{N}_i (i = 1,\ldots,m)$ is an integer multiple of 2 that is greater than and closest to $N_i + 1$, N_i is the number of bisections of the interval box when $[f_k^o](i)$ is contracted, and $[f_k^o](i)$ is the ith dimension of $[f_k^o]$.

Proof 5.2 According to the description in Section 5.1.2.2, the contraction process of $[f_k^o]$ is carried out in the order of the dimensionality. Suppose that the interval box is bisected N_i times during the contraction of $[f_k^o](i)$, we have

$$N = \sum_{i=1}^{m} N_i. \tag{5.29}$$

Because the basic idea of the DVSIIC algorithm is similar to the dichotomy, a binary tree can be used to represent the changes to the interval boxes during contraction. The nodes of the binary tree represent interval boxes, the bifurcations represent the bisections of the interval boxes and the nodes at greater depth in the binary tree represent interval boxes with smaller widths.

When the binary tree is full, the number of bisections of the interval boxes can be expressed as

$$N_i = 2^{s_i - 1} - 1, \tag{5.30}$$

where s_i is equal to the number of levels of the binary tree, representing the total number of inclusion attribute judgments during the contraction of $[f_k^o](i)$ and the computational complexity of the ith dimension. By solving (5.30), we obtain

$$s_i = \log_2(N_i + 1) + 1. \tag{5.31}$$

However, the binary tree is not a full binary tree in most cases, which results in

$$N_i < 2^{s_i - 1} - 1, \tag{5.32}$$

and

$$\log_2(N_i + 1) + 1 < s_i = \log_2 \mathcal{N}_i + 1. \tag{5.33}$$

For the contraction of $[f_k^o]$, the computational complexity of the DVSIIC algorithm is the sum of the complexity of each dimension, namely,

$$S = \sum_{i=1}^{m} s_i = \log_2 \left(\prod_{i=1}^{m} \mathcal{N}_i \right) + m. \tag{5.34}$$

Remark 5.3 If there are N bisections when (5.25) is solved by the SIVIA algorithm, inclusion attribute judgments of each interval box is needed. Then, the computational complexity is equal to the node of the binary tree, which can be expressed as $2N + 1$. ▮

If an interval box is bisected, there is a memory requirement to store the data of two interval boxes. Therefore, the number of bisections is the main determinant that affects the memory requirements of the DVSIIC algorithm. This number is represented by N in Theorem 5.2 for convenience. Then, the specific value of N is given as follows:

Theorem 5.3
(Memory requirements) The DVSIIC algorithm terminates after performing

$$N \leqslant \sum_{i=1,\ldots,m} \left(\frac{2W([f_k^o](i))}{\varepsilon} + 1 \right) \tag{5.35}$$

bisections of the interval boxes at most.

Proof 5.3 For a known interval box width $W([f_k^o](i))$ and precision ε, there is always a positive integer l satisfying the following inequality

$$\frac{W([f_k^o](i))}{2^l} \leqslant \varepsilon. \tag{5.36}$$

Simplify Eq. (5.36) and obtain the inequality about l as follows

$$l \geqslant \log_2 \frac{W([f_k^o](i))}{\varepsilon}. \tag{5.37}$$

Then, we can obtain

$$l \leqslant \log_2 \frac{W([f_k^o](i))}{\varepsilon} + 1. \tag{5.38}$$

Therefore, the number of bisections of $[f_k^o](i)$ in Eq. (5.30) can be expressed as

$$N_i = 2^l - 1 \leqslant \frac{2W([f_k^o](i))}{\varepsilon} + 1. \tag{5.39}$$

Eq. (5.35) can be accumulated according to Eq. (5.29).

Remark 5.4 From the literature [6], it can be seen that the SIVIA algorithm needs less than $\left(\frac{W([f_k^s])}{\varepsilon} + 1\right)^n$ bisections for the same set inversion problem, and its memory requirements increase exponentially with the dimensions of $[f_k^o]$. Compared with (5.35) in Theorem 5.3, it is clear that the DVSIIC algorithm has lower memory requirements and higher practicability. ∎

Theorem 5.4

(Accuracy) Contracting $[f_k^o]$ with the same precision ε by using the SIVIA and DVSIIC algorithms, the accuracy of the solution sets are expressed by the following relationship:

$$\underline{\mathcal{F}}_k^{SIVIA} \subset \mathcal{F}_k \subset [f_k^s] \subset \overline{\mathcal{F}}_k^{SIVIA}, \tag{5.40}$$

where $\mathcal{F}_k$ is the true fault set at time instant k, $\underline{\mathcal{F}}_k^{SIVIA}$ and $\overline{\mathcal{F}}_k^{SIVIA}$, respectively, represent the inner approximate solution set and outer approximate solution set obtained by the SIVIA algorithm, and $[f_k^s]$ is the above-mentioned solution set obtained by the DVSIIC algorithm.

Proof 5.4 All feasible sets obtained by the SIVIA algorithm form the solution set $\underline{\mathcal{F}}_k^{SIVIA}$,

$$\underline{\mathcal{F}}_k^{SIVIA} = \bigcup\{[f_k^s]_i \mid O[f_k^s]_i \subset [\Pi_k]\} \subset \mathcal{F}_k. \tag{5.41}$$

In addition, certain interval boxes only partially belong to the solution set. These boxes are divided among the uncertain sets. Because their width is less than ε, bisection is not needed, and they form the outer approximation of the solution set $\mathcal{E}$ together.

$$\mathcal{E} = \bigcup\{[f_k^s]_i \mid O[f_k^s]_i \cap [\Pi_k] \neq \varnothing, W([f_k^s]_i) < \varepsilon\}. \tag{5.42}$$

Therefore, we have

$$\mathcal{F}_k \subset \overline{\mathcal{F}}_k^{SIVIA} = \underline{\mathcal{F}}_k^{SIVIA} \cup \mathcal{E}. \tag{5.43}$$

Because the contractions in the DVSIIC algorithm are carried out sequentially according to their dimensionality, the following inclusion relation holds:

$$[f_k^s] = \mathcal{U}_m \subset \mathcal{U}_{m-1} \subset \cdots \subset \mathcal{U}_1 \subset [f_k^o], \tag{5.44}$$

where $\mathcal{U}_i(i = 1, \ldots, m)$ is the interval box obtained after the ith dimension has been contracted and $\mathcal{U}_1$ in (5.44) can be seen as $[f_k^s]_1$ in Fig. 5.1.

Because the uncontracted dimensions of the interval box have the initial widths, the boxes $\mathcal{U}_1, \ldots \mathcal{U}_{m-1}$ are composed of feasible sets, unfeasible sets and approximate sets. But for the ith dimension of the box $\mathcal{U}_i(i = 1, \ldots, m-1)$, the interval is the most compact and does not include the approximate part. The final box $\mathcal{U}_m$ consists of feasible sets and approximate sets, similar to the $\overline{\mathcal{F}}_k^{SIVIA}$. As a result, $[f_k^s] \subset \overline{\mathcal{F}}_k^{SIVIA}$ holds.

Remark 5.5 According to Theorem 5.4, part of the true set may be lost while the accuracy is guaranteed if the SIVIA algorithm takes $\underline{\mathcal{F}}_k^{SIVIA}$ as the solution set. The main reason why the width of the solution set $[f_k^s]$ of the DVSIIC algorithm is greater than that of the SIVIA algorithm is that the choice of the uncertainty set with a width smaller than the given precision is different. However, the difference can be ignored because the accuracy given by ε is sufficiently small in actual applications. ■

5.1.3 Simulation analysis

5.1.3.1 Numerical simulation

Consider the following second-order dynamic system:

$$x_{k+1} = \begin{bmatrix} 0.6 & 0.5 \\ -0.2 & -0.3 \end{bmatrix} x_k + \begin{bmatrix} 1 \\ 0.8 \end{bmatrix} u_k + \begin{bmatrix} 0.3 \\ 1 \end{bmatrix} f_k + \begin{bmatrix} -0.85 \\ 0.24 \end{bmatrix} w_k,$$

$$y_k = \begin{bmatrix} 1 & 0 \\ 0 & 1 \end{bmatrix} x_k + \begin{bmatrix} 0.4 \\ 0.4 \end{bmatrix} v_k.$$

The disturbance and noise are assumed to be bounded as $-0.01 \leqslant w_k \leqslant 0.01$, $-0.05 \leqslant v_k \leqslant 0.05$. In the simulation, the initial state $\hat{x}_0 = [2, 1]^{\mathrm{T}}$, $H_0 = 0.1\boldsymbol{I}_2$, and the input signal $u_k = 3\sin(0.02k)$. To verify the effectiveness and superiority of the proposed ISIOF algorithm, we carry out an interval estimation of the state and fault when the system is fault-free and fails, and compare the results with those obtained by the robust fault estimation (RFE) algorithm [36].

When a fault has not occurred, f_k is equal to 0. The state estimation interval of the ISIOF algorithm is shown in Fig. 5.2. Its upper and lower bounds are represented by blue lines, which always surround the true state value represented by the black line. The RFE algorithm designs two sub-observers and realizes the state interval estimation, but it is more conservative than the ISIOF algorithm, as shown by the yellow lines.

Fig. 5.3 shows the state estimation results at time instants $k = 10, 50, 150, 250$. The yellow box is the state interval estimated by the RFE algorithm. Although it contains the real state value represented by the solid point, its accuracy is low. The red zonotope bounds the state interval obtained by the MCIO algorithm proposed in Section 5.1.2.1. The ISIOF algorithm contracts the red zonotope by using the DVSIIC algorithm to obtain the tighter state interval represented by the blue box.

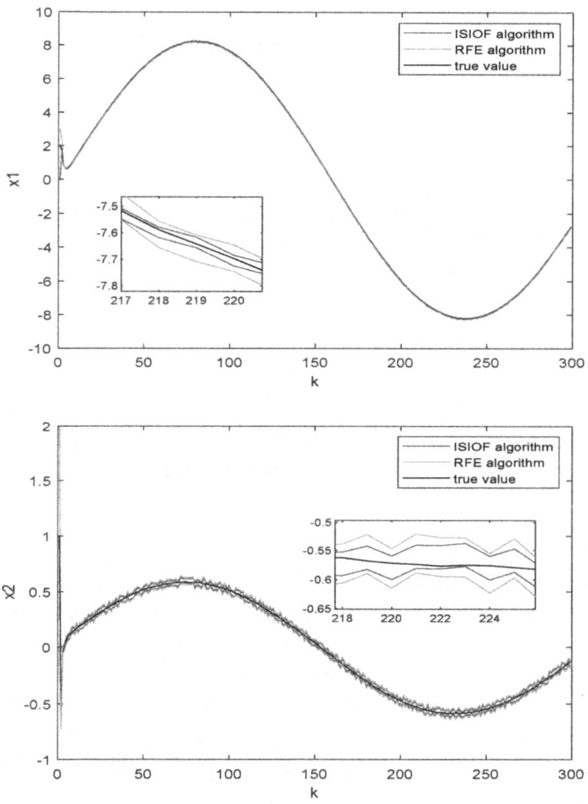

Figure 5.2: Guaranteed state estimation interval in fault-free case.

Fig. 5.4 compares the fault estimation interval obtained by the ISIOF algorithm with that obtained by the RFE algorithm. The blue lines are always closer to the true fault value.

Next, we verify the effectiveness of the ISIOF algorithm when the system fails. Two possible types of additive actuator faults exist and they can occur separately. Fault type 1 is

$$f_{1,k} = \begin{cases} 2, \; k > 100, \\ 0, \; 0 < k \leqslant 100, \end{cases}$$

and fault type 2 is

$$f_{2,k} = \begin{cases} 0.05(k-100), \; 100 < k < 200, \\ 0, \; 0 < k \leqslant 100, k \geqslant 200. \end{cases}$$

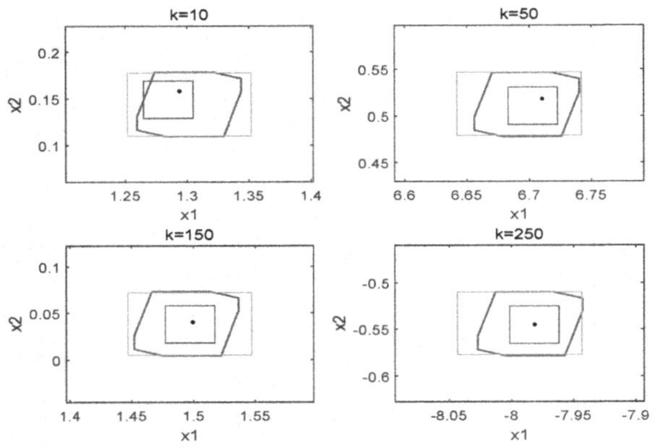

Figure 5.3: Comparison of the partial state estimation results of different methods.

The fault estimation results are shown in Fig. 5.5. For both the instantaneous and time-varying actuator faults, the ISIOF algorithm not only estimates the guaranteed bounds of the faults in the presence of uncertain disturbance and noise but also enables the true fault value to be accurately tracked. In addition, the ISIOF algorithm, in combination with the interval observer and interval contraction function, reduces the redundancy of the estimation interval significantly compared with the RFE algorithm. In summary, the proposed algorithm is superior irrespective of whether the system is fault-free or faulty.

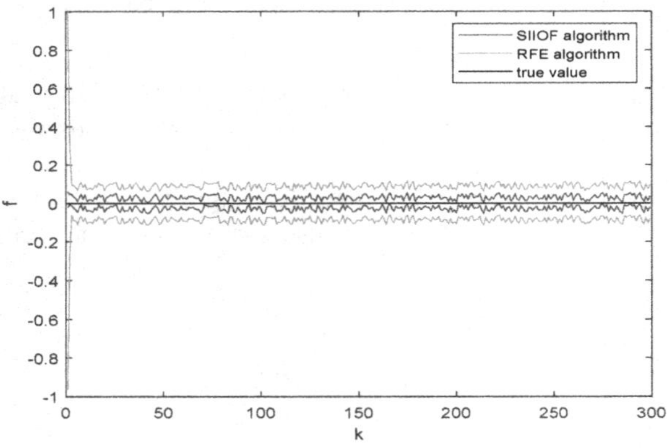

Figure 5.4: Guaranteed fault estimation interval in fault-free case.

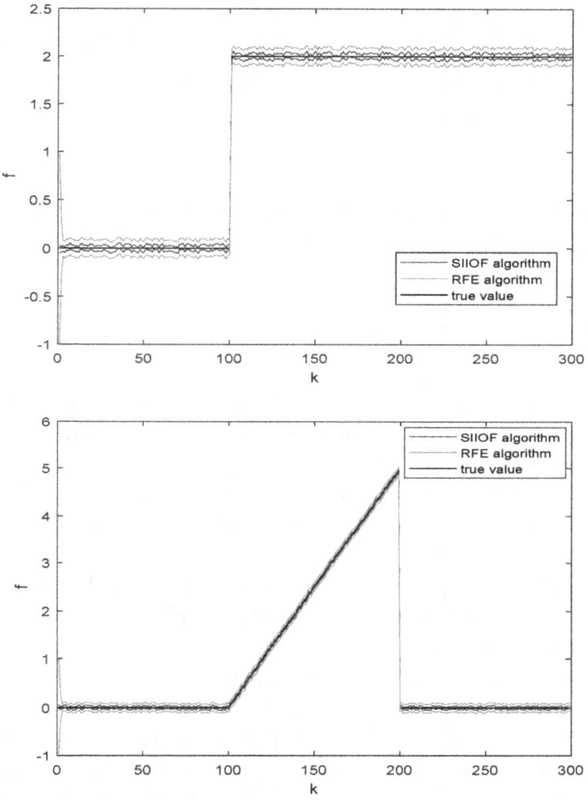

Figure 5.5: Guaranteed fault estimation interval of fault type 1 and fault type 2.

5.1.3.2 DC motor system simulation

The continuous-time DC motor dynamic model is described as follows [16]

$$J\frac{d^2\theta}{dt^2}+b\frac{d\theta}{dt}=Ki(t),$$
$$L\frac{di}{dt}+Ri=u-Kb\frac{d\theta}{dt},$$

where θ, i and u represent the angular position of the motor, armature current and voltage, respectively. The model parameters are the armature moment of inertia $J=0.0985\,\mathrm{kg\,m^2}$, frictional coefficient $b=0.1482\,\mathrm{N\,m\,s}$, motor torque constant, back EMF constant $K=0.4901\,\mathrm{V\,s/rad}$, inductance $L=1.3726H$ and resistance $R=0.0062\Omega$.

Convert the dynamic model into the following form with the state vector $x = [\theta,\ v,\ i]^{\mathrm{T}}$ and the input vector u:

$$\begin{bmatrix} \frac{d\theta}{dt} \\ \frac{dv}{dt} \\ \frac{di}{dt} \end{bmatrix} = \begin{bmatrix} 0 & 1 & 0 \\ 0 & -\frac{b}{J} & \frac{K}{J} \\ 0 & -\frac{K}{L} & -\frac{R}{L} \end{bmatrix} \begin{bmatrix} \theta \\ v \\ i \end{bmatrix} + \begin{bmatrix} 0 \\ 0 \\ \frac{1}{L} \end{bmatrix} u,$$

Take the sampling time $T_s = 0.1s$ and discretize the dynamic DC model into the form (5.6) of with the matrix parameters

$$A = \begin{bmatrix} 1 & 0.1 & 0 \\ 0 & 0.8495 & 0.4977 \\ 0 & -0.0357 & 0.9995 \end{bmatrix}, B = \begin{bmatrix} 0 & 0 & 0.729 \end{bmatrix}^{\mathrm{T}}, C = \boldsymbol{I}_3, F = \begin{bmatrix} 0.2 & 0.2 & 1 \end{bmatrix}^{\mathrm{T}},$$

$D = \begin{bmatrix} 0.5 & 0.5 & 0.5 \end{bmatrix}^{\mathrm{T}}, E = \begin{bmatrix} 0.4 & 0.4 & 0.4 \end{bmatrix}^{\mathrm{T}}$.

In the simulation, we set the input signal $u_k = 2\text{V}$, system disturbance $-0.01 \leqslant w_k \leqslant 0.01$, measurement noise $-0.05 \leqslant v_k \leqslant 0.05$, initial state $\hat{x}_0 = [0.0015, 0.0148, 0]^{\mathrm{T}}$ and $H_0 = \boldsymbol{I}_3$.

When the DC motor system operates normally, the fault value is always equal to zero in the sampling time. The fault estimation interval always contains zero, as shown in the bottom right-hand corner of Fig. 5.6. At the same time, the interval of the angular position, motor speed and armature current of the DC motor can also be estimated in real time, as shown in the subgraph of Fig. 5.6. The robust observers designed by the RFE algorithm can estimate the state and fault interval represented by the yellow lines simultaneously, which always contain the true value represented by the black line. However, the accuracy of the estimation results is low because of uncertain disturbances and noise. The minimum conservative observer designed by the ISIOF algorithm can obtain the guaranteed state and fault estimation interval. After the observer interval is further contracted by the DVSIIC algorithm, the final estimation interval represented by the blue line is more compact.

To verify the effectiveness of the ISIOF algorithm when the DC motor fails, three types of DC actuator faults are considered:

Abrupt fault type 1:

$$f_{1,k} = \begin{cases} 10,\ 100 < k < 200, \\ 0,\ 0 < k \leqslant 100, k \geqslant 200. \end{cases}$$

Sinusoidal time-varying fault type 2:

$$f_{2,k} = \begin{cases} -4.1\sin(0.05k),\ k > 100, \\ 0,\ 0 < k \leqslant 100. \end{cases}$$

Exponential time-varying fault type 3:

$$f_{3,k} = \begin{cases} -10(1 - \exp(-0.05(k - 100))),\ k > 100, \\ 0,\ 0 < k \leqslant 100. \end{cases}$$

In the simulation, three types of faults occur separately, and the estimation results are shown in Figs. 5.7–5.12.

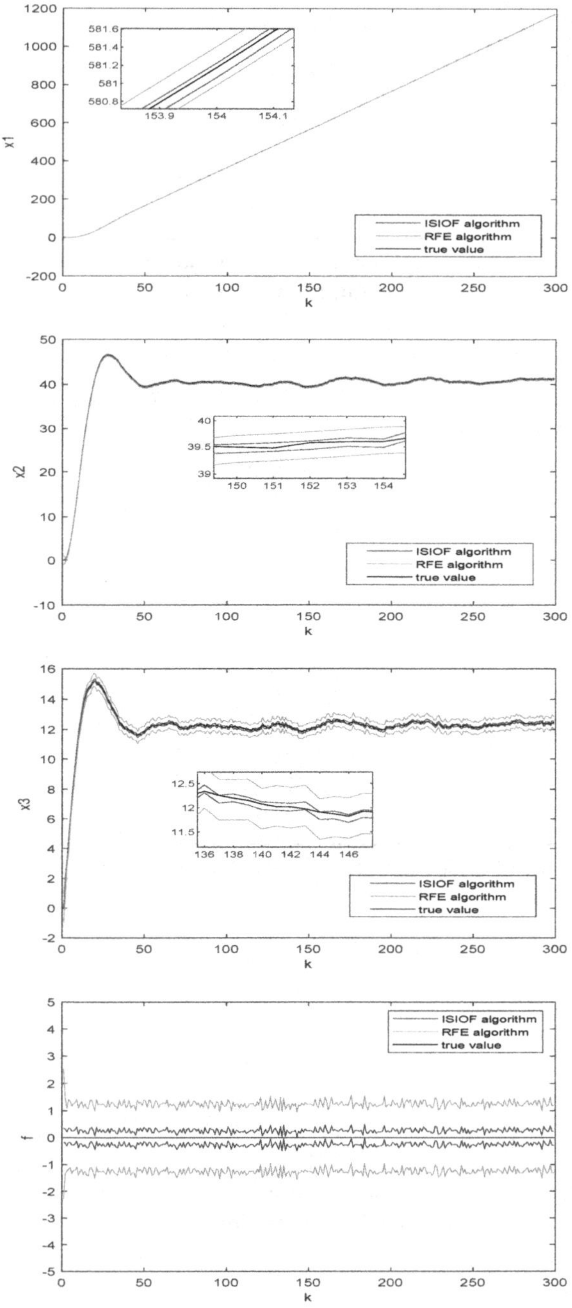

Figure 5.6: Guaranteed state estimation interval and fault estimation interval when the DC motor operates normally.

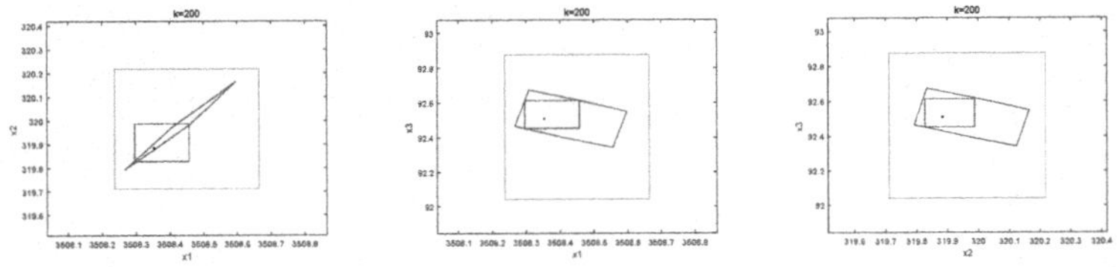

Figure 5.7: Projection of state estimation results at time instant $k = 200$ on the $x_1 - x_2$ axis, $x_1 - x_3$ axis, $x_2 - x_3$ axis when fault type 1 occurs.

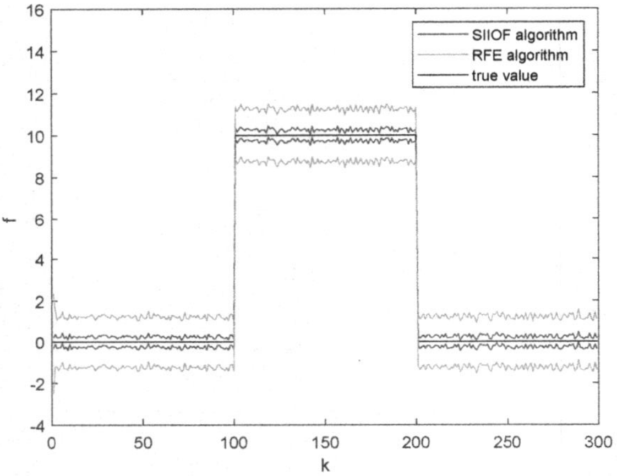

Figure 5.8: Guaranteed fault estimation results for fault type 1.

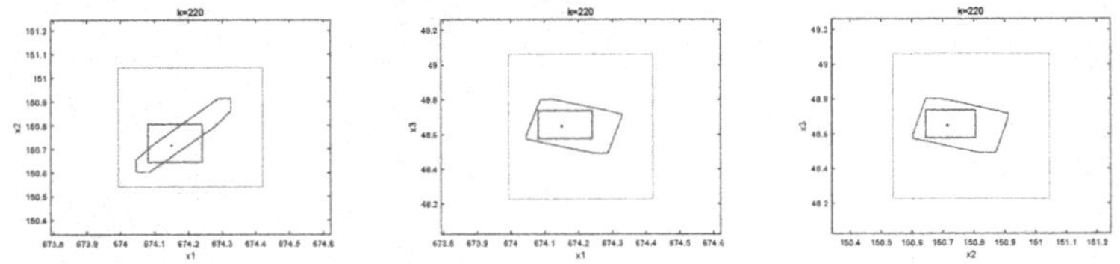

Figure 5.9: Projection of state estimation results at time instant $k = 200$ on the $x_1 - x_2$ axis, $x_1 - x_3$ axis, $x_2 - x_3$ axis when fault type 2 occurs.

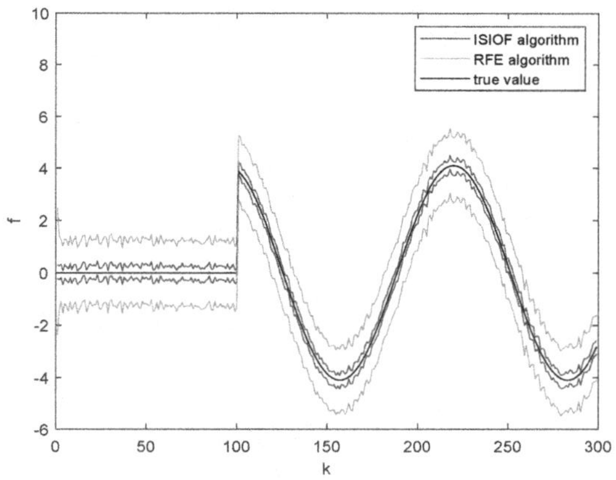

Figure 5.10: Guaranteed fault estimation results for fault type 2.

To compare the results of different methods more intuitively, we project the estimated interval box and zonotope of the state at time instant $k = 200$ (see in Figs. 5.7, 5.9 and 5.11). The projections of the yellow box and the red zonotope represent the state estimation obtained by the interval observer in the RFE algorithm and the MCIO algorithm, respectively. The figures show that both algorithms can track the state of the failed DC motor accurately, but the latter is more effective. Then, a more compact blue box is obtained by the interval contraction step in the ISIOF algorithm to contract the observer interval based on the red zonotope. The estimation results of different types of DC motor actuator faults are given in Figs. 5.8, 5.10 and 5.12. These results show that the ISIOF algorithm is superior to the RFE algorithm for both time-varying and time-invariant faults.

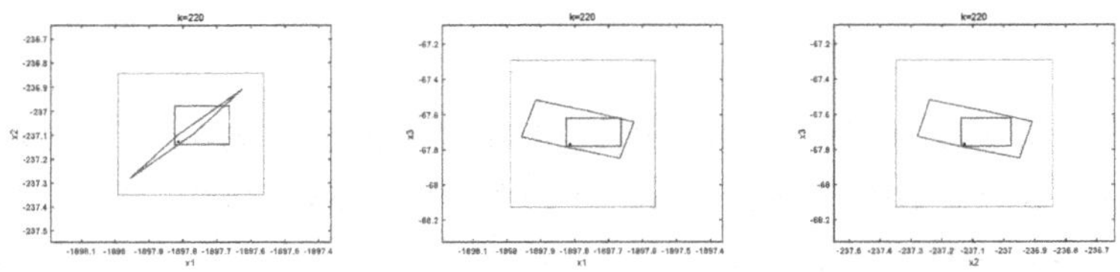

Figure 5.11: Projection of state estimation results at time instant $k = 200$ on the $x_1 - x_2$ axis, $x_1 - x_3$ axis, $x_2 - x_3$ axis when fault type 3 occurs.

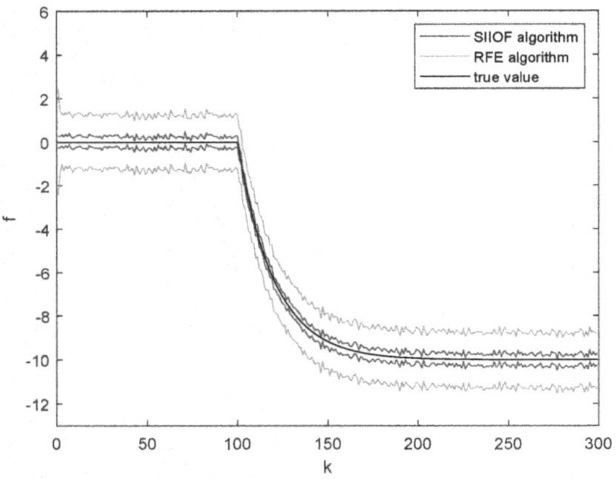

Figure 5.12: Guaranteed fault estimation results for fault type 3.

5.1.4 Conclusions

A fault estimation algorithm for linear discrete-time systems is presented in this section. First, a dynamic interval observer is designed by minimizing the conservatism of the system state to realize the interval estimation of the state and fault. Second, the ISIOF algorithm based on interval analysis is proposed. This algorithm can not only obtain more compact estimation results, but also solve the problem of the exponential increase in the computation time of the traditional interval set inversion algorithm. Then, the superiority of the ISIOF algorithm is proved by using theoretical proofs from different angles. Finally, a numerical system and a DC motor system are simulated to prove the efficiency of the algorithm. The set inversion interval observer filtering-based guaranteed fault estimation algorithm proposed in this section can be extended to investigate fault diagnosis in other industrial processes, such as robot systems [57] and aircraft systems [115].

5.2 Interval observer filtering-based fault diagnosis method for linear discrete-time systems with dual uncertainties

5.2.1 Problem description

Some notations will be explained first. $\mathbb{R}^n$ and $\mathbb{R}^{n\times m}$ denote n and $n \times m$ dimensional space, respectively. $\boldsymbol{I}_n$ represents the n dimensional identity matrix. $\mathbf{0}$ represents zero matrix of the appropriate dimension. The notation $\leqslant$ is understood element-wise.

For a matrix $N \in \mathbb{R}^{n\times m}$, N^{T} and $tr(N)$ stand for its transpose and trace, respectively. The F-norm of N is defined as $\|N\|_F = \sqrt{tr(NN^{\mathrm{T}})}$. For a vector $M \in \mathbb{R}^n$, $|M|$ is its element-wise absolute value, which means that $|M|$ is a vector composed of the absolute value of each element of M. $[M] = [\underline{M}, \overline{M}]$ represents an interval vector with lower bound $\underline{M} \in \mathbb{R}^n$ and upper bound $\overline{M} \in \mathbb{R}^n$. $\langle p, H\rangle$ denotes a zonotope of order m in $\mathbb{R}^n$ by center $p \in \mathbb{R}^n$ and generator matrix $H \in \mathbb{R}^{n\times m}$ and is defined as $\langle p, H\rangle := \mathcal{Z} \in \mathbb{R}^n$. The operator $\oplus$ denotes the Minkowski sum. If a random variable X follows the Gaussian distribution with mean μ and variance σ^2, denote it as $X \sim \mathcal{N}(\mu, \sigma^2)$.

Consider the following linear discrete-time dual uncertain system combining UBB and stochastic uncertainties:

$$x_{k+1} = Ax_k + Bu_k + Ff_k + w_k + d_k, \tag{5.45a}$$

$$y_k = Cx_k + v_k + e_k, \tag{5.45b}$$

where $x_k \in \mathbb{R}^n$, $u_k \in \mathbb{R}^p$, $y_k \in \mathbb{R}^q$ and $f_k \in \mathbb{R}^m$ are the state, input, output and actuator fault vectors, respectively. $w_k \in \mathbb{R}^n$ and $d_k \in \mathbb{R}^n$ denote the system disturbances, and $v_k \in \mathbb{R}^q$ and $e_k \in \mathbb{R}^q$ denote the measurement noises. A, B, C and F are constant matrices with appropriate dimensions. The pair (A, C) is observable.

Assumption 3 *The independent disturbance w_k and noise v_k are unknown but bounded as $|w_k| \leqslant w$ and $|v_k| \leqslant v$, where w and v are known vectors. They are equivalent to*

$$w_k \in \langle 0, W\rangle, v_k \in \langle 0, V\rangle, \tag{5.46}$$

where $W = \mathrm{diag}(w)$ and $V = \mathrm{diag}(v)$.

Assumption 4 *The independent disturbance d_k and noise e_k follow Gaussian distribution*

$$d_k \sim \mathcal{N}(0, R), e_k \sim \mathcal{N}(0, Q), \tag{5.47}$$

with covariance matrices R and Q.

In the fault-free case, consider the following dynamic observer for the system (5.45)

$$\hat{x}_{k+1} = A\hat{x}_k + Bu_k + G_k(y_k - C\hat{x}_k), \tag{5.48}$$

where $\hat{x}_k$ is the estimation of x_k, and G_k is the time-varying gain. Define the state error as $\tilde{x}_k = x_k - \hat{x}_k$. Subtracting Eq. (5.48) from Eq. (5.45), this can be rewritten as

$$\tilde{x}_{k+1} = (A - G_kC)\tilde{x}_k + w_k + d_k - G_k(v_k + e_k). \tag{5.49}$$

Since the system state is affected by UBB and Gaussian disturbances and noises, we define the initial state x_0 as

$$x_0 = \hat{x}_0 + z_0 + g_0, \tag{5.50}$$

where $\hat{x}_0$ is the estimation of x_0 and $z_0 \in \langle 0, H_0 \rangle$, $g_0 \sim \mathcal{N}(0, P_0)$ and H_0 and P_0 are known matrices. According to the properties of the zonotope set and Gaussian set, the interval of the initial state in Eq. (5.50) can be expressed as

$$\begin{cases} \overline{x}_0 = \hat{x}_0 + \overline{z}_0 + \overline{g}_0, \\ \underline{x}_0 = \hat{x}_0 + \underline{z}_0 + \underline{g}_0, \end{cases} \tag{5.51}$$

where $\overline{z}_0$ and $\underline{z}_0$ are the upper and lower bounds of z_0, and $\overline{g}_0$ and $\underline{g}_0$ are the upper and lower bounds of g_0 obtained by 3σ principle, respectively.

Similar to Eq. (5.50), $\tilde{x}_{k+1}$ can be divided into the sum of z_{k+1} and g_{k+1}. Thus, we have

$$z_{k+1} = (A - G_k C) z_k + w_k - G_k v_k, \tag{5.52}$$

$$g_{k+1} = (A - G_k C) g_k + d_k - G_k e_k. \tag{5.53}$$

It can be seen that z_{k+1} in Eq. (5.52) is divided into three parts by operation symbols, which can be delimited by zonotopes $\langle 0, (A - G_k C) H_k \rangle$, $\langle 0, W \rangle$ and $\langle 0, -G_k V \rangle$ respectively according to Eq. (5.50) and Assumption 3. Based on the properties of zonotope [52], the zonotope set bounded z_{k+1} is as follows

$$\begin{aligned} \langle 0, H_{k+1} \rangle &= \langle 0, (A - G_k C) H_k \rangle \oplus \langle 0, W \rangle \oplus \langle 0, -G_k V \rangle, \\ &= \langle 0, [(A - G_k C) H_k \quad W \quad -G_k V] \rangle. \end{aligned} \tag{5.54}$$

Then generator matrix H_{k+1} is expressed as

$$H_{k+1} = [(A - G_k C) H_k \quad W \quad -G_k V]. \tag{5.55}$$

According to Eq. (5.50) and Assumption 4, all three parts divided by operation symbols in Eq. (5.53) obey Gaussian distribution and the variances are equal to $(A - G_k C) P_k (A - G_k C)^{\mathrm{T}}$, R and $G_k Q G_k^{\mathrm{T}}$, respectively. Then, the covariance matrix of $g_{k+1} \sim \mathcal{N}(0, P_{k+1})$ is obtained as

$$P_{k+1} = (A - G_k C) P_k (A - G_k C)^{\mathrm{T}} + R + G_k Q G_k^{\mathrm{T}}. \tag{5.56}$$

Subsequently, the upper and lower bounds of state x_{k+1} can be obtained as follows:

$$\begin{cases} \overline{x}_{k+1} = \hat{x}_{k+1} + \overline{z}_{k+1} + \overline{g}_{k+1}, \\ \underline{x}_{k+1} = \hat{x}_{k+1} + \underline{z}_{k+1} + \underline{g}_{k+1}. \end{cases} \tag{5.57}$$

Remark 5.6 To ensure the accuracy of fault detection and isolation, only one type of fault occurs at the same time. In the fault-free case, Ff_k in system (5.45a) is ignored. ∎

The objective of this section is to estimate the interval of state in the fault-free case and detect and isolate the fault when an actuator fault occurs.

5.2.2 Main results

This section consists of two parts: state estimator and fault diagnosis. Research into designing a state estimator has made great progress, and the interval observer method is one of the most common methods. However, the traditional interval observer is only suitable for systems affected by UBB disturbance and noise. Extending the method proposed by Tang [92] through treating a Gaussian signal as a bounded signal according to the 3σ principle, we can get an extended interval estimation (EIE) method for dual uncertain systems (5.45) to achieve state interval estimation. The classic Kalman filtering method requires the disturbance and noise of the system to satisfy a Gaussian distribution. To make this method feasible for dual uncertain systems, we obtain the mixed Kalman filtering (MKF) method by considering the impact of UBB disturbance and noise on the system in the update step. The last method is an extended zonotopic and Gaussian Kalman filtering (EZGKF) method [28] that is suitable for dual uncertain systems and realizes fault estimation and fault detection. In the simulation analysis in the next section, these methods are used for comparison to show the superiority of the IOF method.

5.2.2.1 State estimator

In the process of estimating the state of a fault-free system, it is meaningful to choose the optimal gain matrix G_k in Eq. (5.48) such that the estimation of Eq. (5.57) is more accurate. When there is only UBB disturbance and noise in the system, the observer gain G_k is obtained by minimizing the Frobenius norm of the generator matrix H_{k+1} in Eq. (5.55), namely $\|H_{k+1}\|_F$. This can be transformed into solving the following optimization problem [4] :

$$\frac{\partial \|H_{k+1}\|_F^2}{\partial G_k} = \frac{\partial tr(H_{k+1}H_{k+1}^{\mathrm{T}})}{\partial G_k} = 0. \tag{5.58}$$

The other is the case where only Gaussian disturbance and noise exist. Similar to the derivation of the standard Kalman gain, G_k is obtained using the derivative rule of the trace of covariance matrix P_{k+1}:

$$\frac{\partial tr(P_{k+1})}{\partial G_k} = 0. \tag{5.59}$$

For a dual uncertain system (5.45) with two types of disturbances and noises at the same time, we let $J_{zk} = tr(H_{k+1}H_{k+1}^{\mathrm{T}})$ and $J_{gk} = tr(P_{k+1})$. We define the optimization criteria

$$J_k = (1-\eta_k)J_{zk} + \eta_k J_{gk}, \tag{5.60}$$

where $\eta_k \in [0,1]$ is the weight coefficient, which represents the influence of the UBB and Gaussian disturbances and noises on the criterion function J_k.

By substituting Eq. (5.55) into J_{zk}, it can be rewritten as

$$\begin{aligned} J_{zk} =& tr(AH_kH_k^{\mathrm{T}}A^{\mathrm{T}} - AH_kH_k^{\mathrm{T}}C^{\mathrm{T}}G_k^{\mathrm{T}} - G_k^{\mathrm{T}}CH_kH_k^{\mathrm{T}}A^{\mathrm{T}} \\ &+ G_kCH_kH_k^{\mathrm{T}}C^{\mathrm{T}}G_k^{\mathrm{T}} + WW^{\mathrm{T}} + G_kVV^{\mathrm{T}}G_k^{\mathrm{T}}). \end{aligned} \tag{5.61}$$

Then, the partial derivative of J_{zk} in the above equation is obtained as

$$\frac{\partial J_{zk}}{\partial G_k} = 2G_kCH_kH_k^{\mathrm{T}}C^{\mathrm{T}} + 2G_kVV^{\mathrm{T}} - 2AH_kH_k^{\mathrm{T}}C^{\mathrm{T}}. \tag{5.62}$$

In the same way, J_{gk} is rewritten by substituting Eq. (5.56)

$$J_{gk} = tr(AP_kA^{\mathrm{T}} - AP_kC^{\mathrm{T}}G_k^{\mathrm{T}} - G_k^{\mathrm{T}}CP_kA^{\mathrm{T}} + G_kCP_kC^{\mathrm{T}}G_k^{\mathrm{T}} + R + G_kQG_k^{\mathrm{T}}), \tag{5.63}$$

and its partial derivative is expressed as

$$\frac{\partial J_{gk}}{\partial G_k} = 2G_kCP_kC^{\mathrm{T}} + 2G_kQ - 2AP_kC^{\mathrm{T}}. \tag{5.64}$$

The optimal gain G_k is obtained when $\partial J_k/\partial G_k = 0$. This means

$$\frac{\partial((1-\eta_k)J_{zk} + \eta_k J_{gk})}{\partial G_k} = (1-\eta_k)\frac{\partial J_{zk}}{\partial G_k} + \eta_k \frac{J_{gk}}{\partial G_k} = 0. \tag{5.65}$$

Combining Eq. (5.62) and Eq. (5.64), the optimization problem $\partial J_k/\partial G_k = 0$ is solved and the gain of the observer Eq. (5.48) is

$$G_k = AP_{\eta k}C^{\mathrm{T}}(CP_{\eta k}C^{\mathrm{T}} + Q_{\eta k})^{-1}, \tag{5.66}$$

where the invertible matrix $P_{\eta k} = (1-\eta_k)H_kH_k^{\mathrm{T}} + \eta_k P_k$ and $Q_{\eta k} = (1-\eta_k)VV^{\mathrm{T}} + \eta_k Q$, and the unknown coefficient η_k is obtained online by the following theorem.

Theorem 5.5

Under Assumptions 3 and 4 and initial state x_0 given in Eq. (5.50), η_k can be determined as follows, according to the average power ratio if the state error $\tilde{x}_k$ is only affected by the UBB disturbance and noise and Gaussian disturbance and noise:

$$\eta_k = \begin{cases} \dfrac{tr(P_{G1})}{tr(Z_1) + tr(P_{G1})}, & k = 1, \\[2ex] \dfrac{tr(P_k)}{tr(H_kH_k^{\mathrm{T}}) + tr(P_k)}, & k > 1, \end{cases} \tag{5.67}$$

where

$$\begin{aligned} P_{G1} &= AP_0A^{\mathrm{T}} + R, \\ Z_1 &= AH_0H_0^{\mathrm{T}}A^{\mathrm{T}} + WW^{\mathrm{T}}. \end{aligned}$$

Proof 5.5 As explained in Assumption 3, w_k and v_k are bounded signals. Thus, the average power equal to their square is bounded in zonotope sets $\langle 0, WW^{\mathrm{T}}\rangle$ and $\langle 0, VV^{\mathrm{T}}\rangle$. For Gaussian signals d_k and e_k under Assumption 4, their power spectral density is uniformly distributed and the average power is equal to the covariance matrix. The average power of other uncertain parts in the system can also be calculated based on the above.

At time instant $k = 1$, the state error $\tilde{x}_1 = A\tilde{x}_0 + w_0 + d_0$ can be divided into two parts: UBB uncertain part

$$z_1 = Az_0 + w_0, \tag{5.68}$$

and Gaussian uncertain part

$$g_1 = Ag_0 + d_0. \tag{5.69}$$

The UBB part z_1 can be viewed as consisting of two boundary signals, where the average power of w_0 has been mentioned above and the average power of the remaining part is bounded in the zonotope set $\langle 0, AH_0H_0^{\mathrm{T}}A^{\mathrm{T}}\rangle$. By combining the two, the average power of z_1 satisfies the following formula:

$$P_{z1} \in \langle 0, Z_1\rangle := \mathcal{Z}_1, \tag{5.70}$$

where $Z_1 = AH_0H_0^{\mathrm{T}}A^{\mathrm{T}} + WW^{\mathrm{T}}$.

Since the power spectral density of the Gaussian signal is uniformly distributed, its average power is equal to its covariance matrix. There are different Gaussian signals included in the Gaussian part g_1. Its average power equal to the sum of the covariance matrices is expressed as

$$P_{G1} = AP_0A^{\mathrm{T}} + R. \tag{5.71}$$

Combining Eq. (5.70) and Eq. (5.71), the weight coefficient is determined as

$$\eta_k = \frac{tr(P_{G1})}{tr(P_{Z1}) + tr(P_{G1})}. \tag{5.72}$$

The state error is given in Eq. (5.49). When $k > 1$, the average power of the Gaussian part g_k can be obtained directly from covariance matrix P_k in Eq. (5.56). For the UBB part z_k containing n bounded signals, its average power satisfies

$$P_{zk} = \sum_{i=1}^{n} z_{ki}z_{ki}^{\mathrm{T}} \in \langle 0, Z_k\rangle := \mathcal{Z}_k, \tag{5.73}$$

where z_{ki} is the ith signal of z_k and Z_k is the generator matrix of zonotope $\mathcal{Z}_k$ that bounds the average power of z_k. Because $\mathcal{Z}_k$ is obtained by the Minkowski sum of zonotopes which bound the square of each signal in z_k, we have $Z_k = H_kH_k^{\mathrm{T}}$. Subsequently, for any $k > 1$, η_k is equivalent to

$$\eta_k = \frac{tr(P_k)}{tr(H_kH_k^{\mathrm{T}}) + tr(P_k)}. \tag{5.74}$$

5.2.2.2 Fault diagnosis

This section proposes a fault detection, isolation and identification method based on the proposed method. According to the dynamic observer (5.48) in the fault-free case, the observer of the faulty system is reconstructed as

$$\hat{x}_{k+1}^f = A\hat{x}_k^f + Bu_k + G_k(y_k - C\hat{x}_k^f), \tag{5.75}$$

where $\hat{x}_k^f$ is the state estimate in the fault case. Subtracting Eq. (5.75) from Eq. (5.45), we can obtain the following state error

$$\tilde{x}_{k+1}^f = (A - G_k C)\tilde{x}_k^f + w_k + d_k - G_k(v_k + e_k) + F f_k. \tag{5.76}$$

Let $\phi_k = \tilde{x}_k^f - \Gamma_k f_k$. By combining it with $\tilde{x}_{k+1}^f$ in Eq. (5.76), ϕ_{k+1} is obtained as

$$\begin{aligned} \phi_{k+1} =& \tilde{x}_{k+1}^f - \Gamma_{k+1} f_{k+1} \\ =& (A - G_k C)(\phi_k + \Gamma_k f_k) + w_k + d_k - G_k(v_k + e_k) + F f_k - \Gamma_{k+1} f_{k+1} \\ =& (A - G_k C)\phi_k + w_k + d_k - G_k(v_k + e_k) + (A - G_k C)\Gamma_k f_k \\ &+ F f_k - \Gamma_{k+1} f_{k+1}. \end{aligned} \tag{5.77}$$

Define the variation of the fault vector as $\triangle f_{k+1} = f_{k+1} - f_k$, the above equation can be simplified to

$$\phi_{k+1} = (A - G_k C)\phi_k + w_k + d_k - G_k(v_k + e_k) - \Gamma_{k+1}\triangle f_{k+1}, \tag{5.78}$$

with $\Gamma_k = (A - G_{k-1}C)\Gamma_{k-1} + F$. Substituting Eq. (5.49) into Eq. (5.78), ϕ_{k+1} can be rewritten as

$$\phi_{k+1} = \tilde{x}_{k+1} + \Omega_{k+1}, \tag{5.79}$$

where

$$\Omega_{k+1} = -\sum_{i=1}^{k}\left(\left(\prod_{j=i}^{k}(A - G_j C)\right)\Gamma_i \triangle f_i\right) - \Gamma_{k+1}\triangle f_{k+1}. \tag{5.80}$$

Due to the cumulative multiplication of $A - G_j C$, the first term of Ω_{k+1} is very small and can be ignored. We approximate the Eq. (5.80) to $\Omega_{k+1} = -\Gamma_{k+1}\triangle f_{k+1}$ and define the measurement residual representing the difference between the real output and the estimated output as

$$\begin{aligned} r_k =& y_k - C\hat{x}_k^f \\ =& C(\phi_k + \Gamma_k f_k) + v_k + e_k \\ =& C\tilde{x}_k + v_k + e_k + C\Gamma_k f_k + C\Omega_k \\ =& C\tilde{x}_k + v_k + e_k + C\Gamma_k f_{k-1}. \end{aligned} \tag{5.81}$$

Subsequently, the above equation can be transformed into the sum of center $c_{r_k}^f$, zonotope part $z_{r_k}^f$ and Gaussian part $g_{r_k}^f$

$$r_k = c_{r_k}^f + z_{r_k}^f + g_{r_k}^f, \tag{5.82}$$

where

$$c_{r_k}^f = C\Gamma_k f_{k-1}, \tag{5.83}$$

$$z_{r_k}^f = C z_k + v_k, \tag{5.84}$$

$$g_{r_k}^f = C g_k + e_k. \tag{5.85}$$

After getting the main factor measurement residual of fault detection and inspired by the generalized likelihood ratio test, we define the fault detection indicator λ_k as

$$\lambda_k = 2\ln \frac{\max\limits_{f_{k-1}} p(r_k|f_{k-1})}{p(r_k|f_{k-1}=0)}, \tag{5.86}$$

where $p(r_k|f_{k-1})$ and $p(r_k|f_{k-1}=0)$ are the probability density functions of r_k when a fault occurs and when there is no fault, respectively.

Similar to the fault-free case, $z^f_{r_k}$ is bounded in a zonotope $\langle 0, H^f_k\rangle$ with generator matrix $H^f_k = CH_k + V$, and $g^f_{r_k}$ obeys Gaussian distribution $\mathcal{N}(0, P^f_k)$ with covariance matrix $P^f_k = CP_kC^T + Q$. If we can estimate the bounds of $z^f_{r_k}$, including $\underline{z}^f_{r_k}$ and $\overline{z}^f_{r_k}$, such that $\underline{z}^f_{r_k} \leqslant z^f_{r_k} \leqslant \overline{z}^f_{r_k}$, then residual r^g_k is regarded as the Gaussian sequence with means $C\Gamma_k f_k + z^f_{r_k}$ and covariance P^f_k that satisfies $\underline{r}^g_k \leqslant r^g_k \leqslant \overline{r}^g_k$. The following two sets of hypotheses are designed:

$$\textbf{Hypothesis 1}: \begin{cases} \mathbb{H}_{u0} : \overline{r}^g_k \sim \mathcal{N}(\overline{z}^f_{r_k}, P^f_k) \\ \mathbb{H}_{u1} : \overline{r}^g_k \sim \mathcal{N}(C\Gamma_k \overline{f}_{k-1} + \overline{z}^f_{r_k}, P^f_k) \end{cases} \tag{5.87}$$

$$\textbf{Hypothesis 2}: \begin{cases} \mathbb{H}_{l0} : \underline{r}^g_k \sim \mathcal{N}(\underline{z}^f_{r_k}, P^f_k) \\ \mathbb{H}_{l1} : \underline{r}^g_k \sim \mathcal{N}(C\Gamma_k \underline{f}_{k-1} + \underline{z}^f_{r_k}, P^f_k) \end{cases} \tag{5.88}$$

among them, $\mathbb{H}_{\cdot 0}$ and $\mathbb{H}_{\cdot 1}$ represent the hypothesis of no-fault condition and fault condition respectively, $\overline{f}_{k-1}$ and $\underline{f}_{k-1}$ are the upper and lower bounds of the fault f_{k-1} respectively.

Theorem 5.6

Based on the hypotheses above, interval indicator $[\underline{\lambda}_k, \overline{\lambda}_k]$ can be obtained as

$$\overline{\lambda}_k = \overline{\Upsilon}^{\mathrm{T}}_k \Lambda^{-1}_k \overline{\Upsilon}_k, \tag{5.89}$$

$$\underline{\lambda}_k = \underline{\Upsilon}^{\mathrm{T}}_k \Lambda^{-1}_k \underline{\Upsilon}_k, \tag{5.90}$$

where Λ_k is a symmetric invertible matrix, specifically expressed as follows

$$\Lambda_k = \Gamma^{\mathrm{T}}_k C^{\mathrm{T}} P^{f^{-1}}_k C\Gamma_k, \tag{5.91}$$

$$\overline{\Upsilon}_k = \Gamma^{\mathrm{T}}_k C^{\mathrm{T}} P^{f^{-1}}_k (\overline{r}^g_k - \overline{z}^f_{r_k}), \tag{5.92}$$

$$\underline{\Upsilon}_k = \Gamma^{\mathrm{T}}_k C^{\mathrm{T}} P^{f^{-1}}_k (\underline{r}^g_k - \underline{z}^f_{r_k}). \tag{5.93}$$

Proof 5.6 From Hypothesis 1 , we have the upper bound of λ_k

$$\overline{\lambda}_k = 2\ln \frac{\max\limits_{f_{k-1}} p(\overline{r}^g_k|f_{k-1} = \overline{f}_{k-1})}{p(\overline{r}^g_k|f_{k-1}=0)}, \tag{5.94}$$

where $p(\overline{r}_k^g|f=\overline{f}_{k-1})$ is the probability density function of $\overline{r}_k^g$ in $\mathbb{H}_{u1}$ and $p(\overline{r}_k^g|f_{k-1}=0)$ is the probability density function obtained from $\mathbb{H}_{u0}$. Then, $\overline{\lambda}_k$ can be further derived into

$$\overline{\lambda}_k = 2\ln \frac{\max\limits_{\overline{f}_{k-1}} \exp \frac{-(\overline{r}_k^g - C\Gamma_k \overline{f}_{k-1} - \overline{z}_{r_k}^f)^{\mathrm{T}}(\overline{r}_k^g - C\Gamma_k \overline{f}_{k-1} - \overline{z}_{r_k}^f)}{2P_k^f}}{\exp \frac{-(\overline{r}_k^g - \overline{z}_{r_k}^f)^{\mathrm{T}}(\overline{r}_k^g - \overline{z}_{r_k}^f)}{2P_k^f}}. \tag{5.95}$$

Let $\Psi_k = (\overline{r}_k^g - C\Gamma_k \overline{f}_{k-1} - \overline{z}_{r_k}^f)^{\mathrm{T}} P_k^{f^{-1}} (\overline{r}_k^g - C\Gamma_k \overline{f}_{k-1} - \overline{z}_{r_k}^f)$ and $\overline{\lambda}_k$ in Eq. (5.95) is at its maximum when $\Psi_k = 0$. We have $\overline{r}_k^g - \overline{z}_{r_k}^f = C\Gamma_k \overline{f}_{k-1}$ and

$$\overline{\lambda}_k = \frac{(\overline{r}_k^g - \overline{z}_{r_k}^f)^{\mathrm{T}}(\overline{r}_k^g - \overline{z}_{r_k}^f)}{P_k^f} = \frac{(C\Gamma_k \overline{f}_{k-1})^{\mathrm{T}}(C\Gamma_k \overline{f}_{k-1})}{P_k^f}. \tag{5.96}$$

To get the unknown $\overline{f}_{k-1}$ in the above equation, the following optimization problem is proposed:

$$\overline{f}_{k-1} = \arg\min \Psi_k. \tag{5.97}$$

For brevity, let $\overline{\Theta}_k = \overline{r}_k^g - \overline{z}_{r_k}^f$ and Ψ_k is rewritten as

$$\begin{aligned}
\Psi_k =& (\overline{\Theta}_k - C\Gamma_k \overline{f}_{k-1})^{\mathrm{T}} P_k^{f^{-1}} (\overline{\Theta}_k - C\Gamma_k \overline{f}_{k-1}) \\
=& \overline{\Theta}_k^{\mathrm{T}} P_k^{f^{-1}} \overline{\Theta}_k - \overline{f}_{k-1}^{\mathrm{T}} \Gamma_k^{\mathrm{T}} C^{\mathrm{T}} P_k^{f^{-1}} \overline{\Theta}_k - \overline{\Theta}_k^{\mathrm{T}} P_k^{f^{-1}} C\Gamma_k \overline{f}_{k-1} \\
& + \overline{f}_{k-1}^{\mathrm{T}} \Gamma_k^{\mathrm{T}} C^{\mathrm{T}} P_k^{f^{-1}} C\Gamma_k \overline{f}_{k-1}.
\end{aligned} \tag{5.98}$$

$\overline{f}_{k-1}$ can be found by solving the following problem:

$$\frac{\partial \Psi_k}{\partial \overline{f}_{k-1}} = -2\Gamma_k^{\mathrm{T}} C^{\mathrm{T}} P_k^{f^{-1}} \overline{\Theta}_k + 2\Gamma_k^{\mathrm{T}} C^{\mathrm{T}} P_k^{f^{-1}} C\Gamma_k \overline{f}_{k-1} = 0. \tag{5.99}$$

Then, the optimal solution of $\overline{f}_{k-1}$ is as follows:

$$\begin{aligned}
\overline{f}_{k-1} =& (\Gamma_k^{\mathrm{T}} C^{\mathrm{T}} P_k^{f^{-1}} C\Gamma_k)^{-1} \Gamma_k^{\mathrm{T}} C^{\mathrm{T}} P_k^{f^{-1}} \overline{\Theta}_k \\
=& \Lambda_k^{-1} \Upsilon_k.
\end{aligned} \tag{5.100}$$

By substituting Eq. (5.100) into Eq. (5.96), we have

$$\overline{\lambda}_k = \Upsilon_k^{\mathrm{T}} \Lambda_k^{-1} \Upsilon_k. \tag{5.101}$$

Similarly, the derivation process of $\underline{\lambda}_k$ can be obtained according to Hypothesis 2 and is omitted here.

Furthermore, because of the influence of the Gaussian disturbance and noise, the dynamics center λ_k^c of interval $[\underline{\lambda}_k, \overline{\lambda}_k]$ is selected as the judgment criterion for fault detection. Given threshold λ_{thr}, if $\lambda_{k-L:k}^c > \lambda_{thr}$ holds, then one can be sure that a fault has occurred where L is the set time interval length. Otherwise, the system is a fault-free case.

In the proof of Theorem 5.6, the fault estimation interval $[\underline{f}_k, \overline{f}_k]$ is obtained as

$$\overline{f}_k = \Lambda_{k+1}^{-1}\overline{\Upsilon}_{k+1}, \tag{5.102}$$

$$\underline{f}_k = \Lambda_{k+1}^{-1}\underline{\Upsilon}_{k+1}. \tag{5.103}$$

To improve the accuracy of the fault interval estimation result, the most common sliding window filtering method is used. The fault estimation interval can be calculated as follows:

$$\overline{f}_k = \frac{1}{l}\sum_{i=k-l+1}^{k+1} \Lambda_i^{-1}\overline{\Upsilon}_i, \tag{5.104}$$

$$\underline{f}_k = \frac{1}{l}\sum_{i=k-l+1}^{k+1} \Lambda_i^{-1}\underline{\Upsilon}_i, \tag{5.105}$$

where l is a known constant. If the fault estimation interval of an actuator includes the zero point, then the actuator is normal at that moment. Otherwise, we can find the actuator fault that causes the system to be abnormal. When the estimation result is stable, the fault type can be judged according to the known fault value wrapped by the fault estimation interval. Fault isolation and identification is now complete.

Remark 5.7 The choice of constant l in Eq. (5.104) and Eq. (5.105) affects the real time and accuracy of the interval estimation $[\underline{f}_k, \overline{f}_k]$. When the value of l is small, the accuracy of the estimation result is low. Increasing the value of l improves the estimation accuracy, but the response time becomes longer, namely, the time when the true fault value starts to be included in the fault estimation interval becomes later after the fault occurs. Therefore, it is important to select the appropriate l. ■

Remark 5.8 The upper and lower bounds of the fault detection indicator λ_k are defined according to Hypothesis 1 and 2, respectively. After simplifying, it can be found that the fault has a direct impact on the interval indicator. Its upper and lower bounds increase as the absolute value of the fault increases when a fault occurs. Therefore, the fault detection process is completed by judging whether center λ^c is always greater than λ_{thr} within the time length of L. In addition, the upper and lower bounds of λ_k is also related to the observer gain G_k and the covariance matrix P_k^f. The detection factor will have a certain amplitude and oscillation and is not exactly zero because the system is affected by Gaussian disturbance and noise. ■

Remark 5.9 In this section, the estimated fault interval is used to complete the isolation and magnitude identification of actuator fault at the same time. The isolability condition for locating an occurred fault by fault estimation interval $[\underline{f}_k, \overline{f}_k]$ is

$$f_{i_k} \in [\underline{f}_k, \overline{f}_k], \tag{5.106}$$

where f_{i_k} refers to the fault value of the ith type of known fault at time instant k. ■

Remark 5.10 The proposed method can realize the state estimation and fault diagnosis without any constraints. In the calculation process, the interval is divided into center, UBB uncertain part and Gaussian uncertain part and operated separately instead of calculating according to the algorithm of the interval box directly, so the computational complexity will not rise sharply with the increase of the system dimension. At the same time, it only needs to store the confidence interval of the fault detection indicator within the length of time L and the fault estimation interval within the length of time l for fault diagnosis, resulting in small memory requirements. ■

Remark 5.11 The proposed method uses prior knowledge such as system model, system disturbances and noises to achieve real-time monitoring of state and fault effectively. The running of data-based or model-free methods do not need the physical model, and achieve fault diagnosis by extracting the key features of the monitoring data. However, a large amount of historical data and real-time data are accompanied by high calculation and high memory requirement, and the selection of history data may affect the diagnosis results. In this section, the unknown disturbances and noises are represented by unknown but bounded signals and Gaussian distribution signals, and the interval observer and fault detection indicator are designed to reduce the influence of model uncertainty. ■

5.2.3 Simulation analysis

The simulation examples of a second-order numerical system and a third-order DC motor system are given in this section. First, they are all observable linear discrete-time systems and used as simulation objects in this field commonly. Second, the simulation analyses of the systems without fault are performed to prove that the selected systems are stable. At the same time, the state interval of the system is estimated. Finally, the simulations when different types of faults occur in the systems is carried out. The proposed method can realize fault detection, isolation and identification, and the obtained results are consistent with the simulation assumptions, which proves its effectiveness.

Algorithm 1 Interval observer filtering-based fault diagnosis algorithm

---

Require: Initial values: $\hat{x}_0$, H_0, P_0, W, V, R, Q, L, l, λ_{thr}, $FaultSignal \leftarrow FALSE$, $FaultType \leftarrow 0$

for $k = 1 : n$ **do**

  Collect input data u_k and output data $y(t)$;

  Update weight coefficient η_k based on Theorem 1;

  Determine observer gain G_k using Eq. (5.66);

  Calculate state estimation $\hat{x}_k$, generator matrix H_k and covariance matrix P_k using Eqs. (5.48), (5.55) and (5.56);

  Obtain $[z_k]$ and $[g_k]$ based on the properties of zonotope and 3σ principle;

  Calculate state estimation interval $[x_k]$ using Eq. (5.57);

  $\Gamma_k \leftarrow (A - G_{k-1}C)\Gamma_{k-1} + F$;

  Calculate Λ_k based on Eq. (5.91);

  $[r_k^g] \leftarrow y_k - C[x_k]$;

  Obtain $[z_{r_k}^f]$ based on Eq. (5.84);

  Calculate $\overline{\Upsilon}_k$ and $\underline{\Upsilon}_k$ based on Eqs. (5.92) and (5.93);

  Update indicator λ_k^c based on interval indicator $[\underline{\lambda}_k, \overline{\lambda}_k]$ obtained from Eqs. (5.89) and (5.90) in Theorem 5.6;

  while $k > L$ **do**

    if $\lambda_{k-L:k}^c > \lambda_{thr}$ **then**

      $FaultSignal \leftarrow TRUE$

    end if

  end while

  if $k < l$ **then**

    Calculate fault interval $[\underline{f}_{k-1}, \overline{f}_{k-1}]$ using Eqs. (5.102) and (5.103);

  else

    Calculate fault interval $[\underline{f}_{k-1}, \overline{f}_{k-1}]$ using Eqs. (5.104) and (5.105);

  end if

  for $type = 1 : total\ number\ of\ fault\ types$ **do**

    if $f_{type} \in [\underline{f}_{k-1}, \overline{f}_{k-1}]$ **then**

      $FaultType \leftarrow type$

    end if

  end for

end for

Ensure: $FaultSignal$, $FaultType$, $[\underline{f}_k, \overline{f}_k]$

5.2.3.1 *Numerical example*

Consider the following second-order dynamic system:

$$\begin{aligned} x_{k+1} &= \begin{bmatrix} 0.6 & 0.5 \\ -0.2 & -0.3 \end{bmatrix} x_k + \begin{bmatrix} 1 \\ 0 \end{bmatrix} u_k + \begin{bmatrix} 1 & 0 \\ 1 & 0.2 \end{bmatrix} f_k + w_k + d_k, \\ y_k &= \begin{bmatrix} 1 & 0 \\ 0 & 1 \end{bmatrix} x_k + v_k + e_k, \end{aligned} \tag{5.107}$$

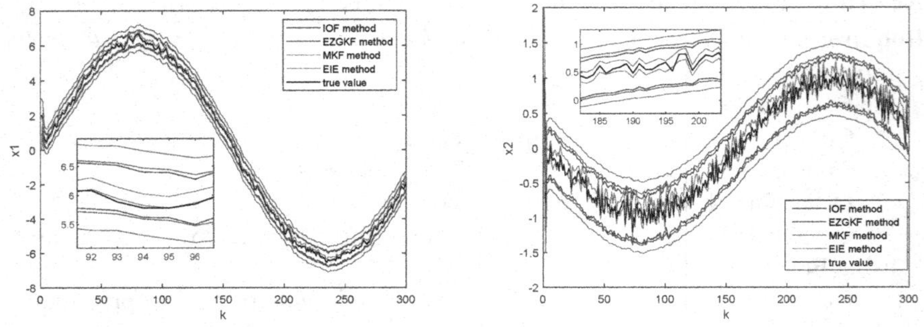

Figure 5.13: State estimation interval of different methods.

where the signal settings are as follows: input signal $u_k = 3\sin(0.02k)$, UBB disturbance signal $|w_k| \leqslant [0.01,\ 0.01]^{\mathrm{T}}$, UBB noise signal $|v_k| \leqslant [0.1,\ 0.1]^{\mathrm{T}}$, Gaussian disturbance signal $d_k \sim \mathcal{N}(0,R)$, $R = 0.1^2\boldsymbol{I}_2$, Gaussian noise signal $e_k \sim \mathcal{N}(0,Q)$, $Q = 0.05^2\boldsymbol{I}_2$, initial state $\hat{x}_0 = [2,1]^{\mathrm{T}}$, $H_0 = \boldsymbol{I}_2$, $P_0 = \boldsymbol{I}_2$ and initial actuator fault $f_k = \mathbf{0}$.

To verify the effectiveness and superiority of the proposed IOF method, we carried out the state estimation when the system is fault-free and the fault detection and isolation when the system fails.

5.2.3.2 *Fault-free case*

When there is no fault, f_k in (5.107) is equal to $[0,\ 0]^{\mathrm{T}}$. The state estimation interval according to the IOF method proposed in this section is shown by the dark red lines in Fig. 5.13. The black line represents the real state value of the system (5.107). It can be seen that the estimation interval always contains the true state value and can effectively suppress disturbances and noises. In addition, the simulation results of the methods for comparison mentioned are shown in Fig. 5.13 with solid lines in other colors. Although the EIE method represented by the brown lines realizes the state interval estimation, converting the Gaussian disturbance and noise into a fixed interval directly makes the estimation result less accurate to ensure that the estimated interval can always contain the true state. The state estimation result obtained according to the EZGKF method is represented by the blue lines. The accuracy of the estimation interval is higher than that of the EIE method and lower than that of the IOF method. Compared with the previous methods, the MKF method can obtain a tighter estimation interval as shown by the solid magenta lines, but it cannot guarantee that the estimation interval contains the true state value at all times. Fig. 5.14 shows a comparison of the state estimation interval sampled within time instant $k = 268 \sim 276$. It can be seen clearly that the estimation interval of the MKF method cannot always contain the true state value represented by the solid point. To sum up, the IOF method is the most effective approach.

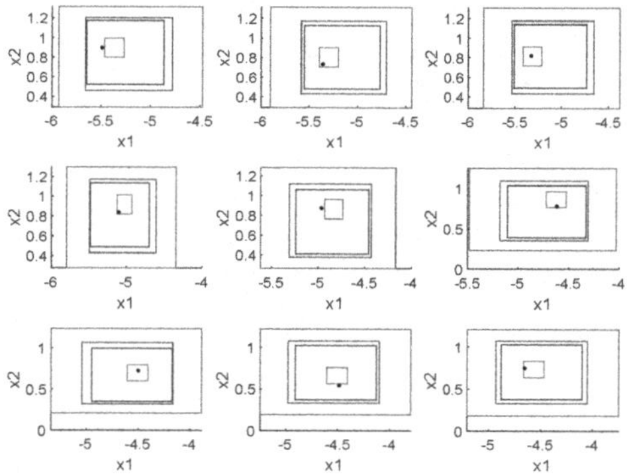

Figure 5.14: Partial state estimation interval comparison of different methods.

5.2.3.3 Fault detection, isolation and identification

Consider two types of additive actuator faults that occur separately. Fault type 1 is

$$f_{1,k} = \begin{cases} [0.46,\ 0]^{\mathrm{T}}, & k > 100, \\ [0,\ 0]^{\mathrm{T}}, & 0 < k \leqslant 100, \end{cases}$$

fault type 2 is

$$f_{2,k} = \begin{cases} [0,\ 3.9]^{\mathrm{T}}, & k > 100, \\ [0,\ 0]^{\mathrm{T}}, & 0 < k \leqslant 100, \end{cases}$$

and fault type 3 is

$$f_{3,k} = \begin{cases} [-\sin(0.012(k-70)),\ 0]^{\mathrm{T}}, & k > 100, \\ [0,\ 0]^{\mathrm{T}}, & 0 < k \leqslant 100. \end{cases}$$

The results are shown in Figs. 5.15 and 5.16. The detection results for the three types of faults are shown in Fig. 5.15. The brown line represents the fault detection signal obtained by the IOF method in this section, and the blue solid line is obtained based on the EZGKF method. For fault type 2, both method can detect the occurrence of the fault at time instant $k = 102$, but the latter has the higher false alarm rate and missing detection rate because it detects the fault when the fault does not occur and cannot guarantee that all fault are detected when the fault occurs. From the detection result of fault type 1 and 3, we can see that the time instant when the fault detection signal obtained based on the EZGKF method jumps from 0 to 1 is later than that based on the IOF method. When the fault always exists, the detection signal changes from 1 to 0 occasionally. Owing to the existence of Gaussian disturbance and noise,

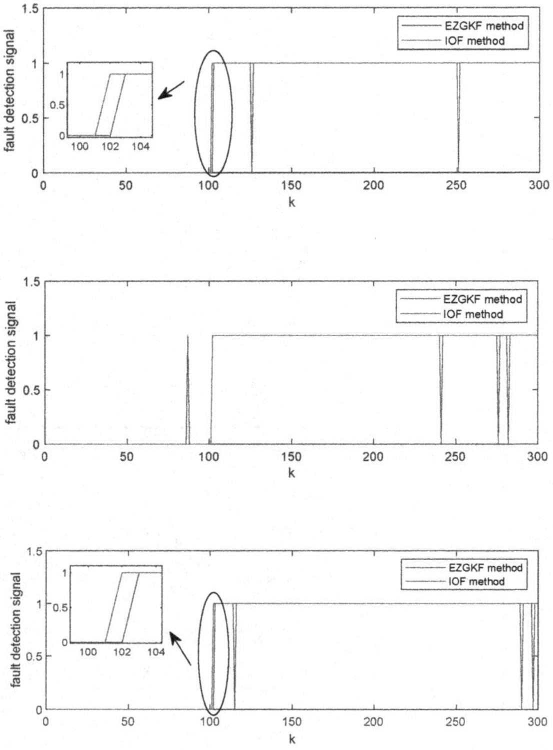

Figure 5.15: Fault detection results of faults type 1, 2 and 3.

choosing different probability values and only using the residual at the current moment when using the EZGKF method for simulations results in a corresponding false alarm rate in the fault detection result, which cannot be avoided. The IOF method determines that the fault has occurred (or not occurred) when the center of the fault detection interval indicator is greater than (or less than) the threshold value within a certain length of time L.

As shown in Fig. 5.16, black dot, red dots, blue dots and green dots indicate no fault condition and the fault values of fault types 1, 2 and 3 in all three subgraphs. In this section, no occurrence means that this type of fault does not occur in the simulation, because only one type of fault occurs at the same time. The difference is that they use different colored rectangles to indicate the fault interval estimation when different types of faults occur in the system. The initial fault estimation interval contains the origin representing no fault. In the process of implementing fault isolation and identification according to the fault estimation results, the fault estimation

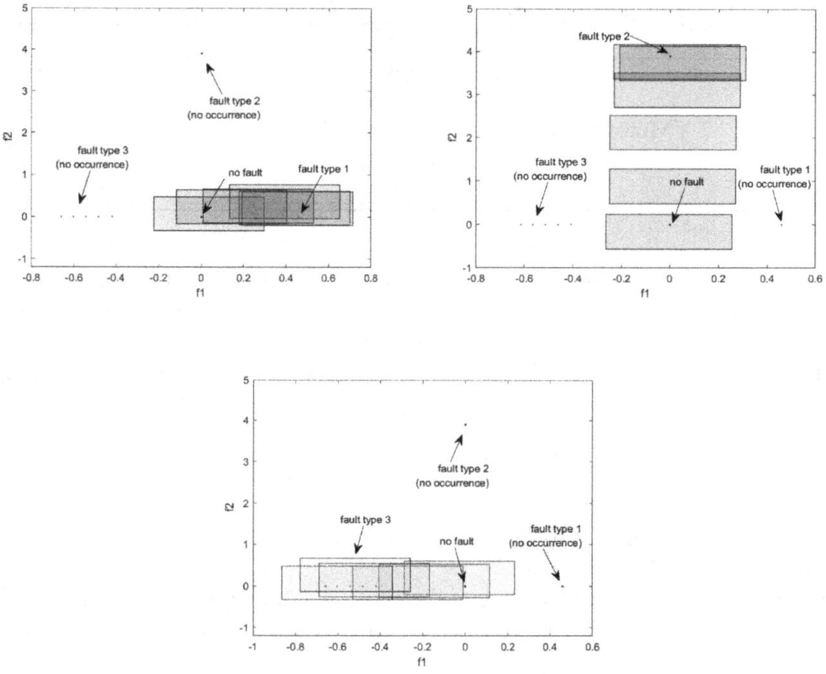

Figure 5.16: Fault isolation and identification results of fault type 1 (top left), 2 (top right) and 3 (bottom).

interval approaches the actual fault type. In the MATLAB R2018b simulation platform with system specification Intel Corei7-6500U CPU, the time for fault detection of the three types of faults is 0.0513s, 0.0712s and 0.0673s, respectively, and the time for fault isolation and identification is 0.0882s, 0.0833s and 0.0872s, respectively.

5.2.3.4 DC motor system simulation

To further verify the effectiveness and feasibility of the IOF method, fault diagnosis is performed on a DC servomotor dynamic system. With the help of a speed sensor, position sensor and current sensor, the dynamic model of a DC motor is described as [16]

$$
\begin{aligned}
J\frac{d^2\theta}{dt^2}+b\frac{d\theta}{dt} &= Ki(t),\\
L\frac{di}{dt}+Ri &= u-K\frac{d\theta}{dt},
\end{aligned}
$$

Table 5.1: DC motor parameters.

Parameters	Meanings	Values
J	Armature moment of inertia	$0.0985\ \mathrm{kg\,m^2}$
b	Frictional coefficient	0.1482 N m s
K	Motor torque constant and back EMF constant	0.4901 V s/rad
L	Inductance	1.3726 H
R	Resistance	$0.0062\ \Omega$

where θ, i and u represent the angular position, armature current of the motor, and voltage source, respectively, applied to the motor armature. The meanings and values of the model parameters are shown in Table 5.1.

By defining the state as $x = [\theta,\ v,\ i]^{\mathrm{T}}$ and the input as u, the DC system can be transformed as

$$\begin{bmatrix} \frac{d\theta}{dt} \\ \frac{dv}{dt} \\ \frac{di}{dt} \end{bmatrix} = \begin{bmatrix} 0 & 1 & 0 \\ 0 & -\frac{b}{J} & \frac{K}{J} \\ 0 & -\frac{K}{L} & -\frac{R}{L} \end{bmatrix} \begin{bmatrix} \theta \\ v \\ i \end{bmatrix} + \begin{bmatrix} 0 \\ 0 \\ \frac{1}{L} \end{bmatrix} u.$$

Substituting the model parameters into the above and taking the sampling time $T_s = 0.1$s, the DC model is discretized into the form of system (5.45a) with the matrix parameters as

$$A = \begin{bmatrix} 1 & 0.1 & 0 \\ 0 & 0.8495 & 0.4977 \\ 0 & -0.0357 & 0.9995 \end{bmatrix}, B = \begin{bmatrix} 0 \\ 0 \\ 0.729 \end{bmatrix},$$

Meanwhile, measurement matrix $C = \boldsymbol{I}_3$, which means that the entire state is measured directly that and fault matrix $F = \boldsymbol{I}_3$.

In the simulation, we set the input signal $u_k = 2$V, UBB disturbance signal $|w_k| \leqslant [0.0125,\ 0.0125]^{\mathrm{T}}$, UBB noise signal $|v_k| \leqslant [0.0345,\ 0.0345]^{\mathrm{T}}$, Gaussian disturbance signal $d_k \sim \mathcal{N}(0,R)$, $R = 0.305^2\boldsymbol{I}_3$, Gaussian noise signal $e_k \sim \mathcal{N}(0,Q)$, $Q = 0.48^2\boldsymbol{I}_3$, the initial state $\hat{x}_0 = [0.0015,\ 0.0148,\ 0]^{\mathrm{T}}$, $H_0 = 0.2\boldsymbol{I}_3$ and $P_0 = 0.02\boldsymbol{I}_3$ and initial actuator fault $f_k = \boldsymbol{0}$.

5.2.3.5 DC motor working normally

When there is no fault, the Ff_k item in the DC motor system is omitted. From the estimation results (dark red lines) obtained by IOF method in Fig. 5.17, we can see that the interval estimation of the motor angular position, motor speed and armature current can effectively track the true state (black line) of the system. In order to further illustrate the superiority of the proposed method, IOF method is compared with different methods. In Fig. 5.17, it can be clearly seen that the interval state estimation of DC motor obtained by IOF method is more tighter than those obtained by EIE method (brown lines) and EZGKF method (blue lines), although their upper and

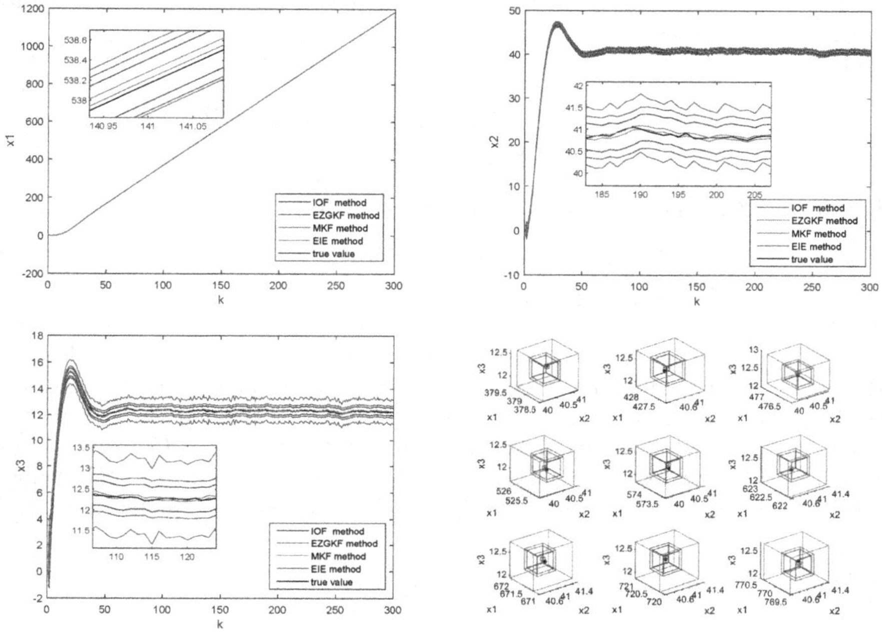

Figure 5.17: State estimation interval and comparison of different methods.

lower bounds can also wrap the true value all the time. MFK method cannot guarantee to obtain the accurate estimation interval (magenta lines) that always surrounds the true value compared with IOF method, so the estimation results with higher accuracy obtained by it have no higher application value. The results sampled every 12 time instant within $k = 102 \sim 198$ are shown in the lower right of Fig. 5.17, which shows the comparison of the state estimation interval of the above several methods more clearly. In general, IOF method can not only attenuate the effect of the UBB and Gaussian disturbances and noises and track the actual state value of the DC motor quickly, but also handle the balance between the precision and accuracy of the state estimation results very well.

5.2.3.6 DC motor fails

In this part, three types of DC actuator faults are considered. Fault type 1 is an abrupt fault with 20% and 50% performance decreases of the actuator:

$$f_{1,k} = \begin{cases} [-0.2u_k,\ 0,\ 0]^{\mathrm{T}}, 100 < k \leqslant 150, \\ [-0.5u_k,\ 0,\ 0]^{\mathrm{T}}, k > 150, \\ [0,\ 0,\ 0]^{\mathrm{T}}, 0 < k \leqslant 100. \end{cases}$$

Fault type 2 is a time-varying actuator fault in which the performance of the actuator decreases exponentially:

$$f_{2,k} = \begin{cases} [0,\ -(1-\exp(-0.05(k-100))),\ 0]^{\mathrm{T}}, k > 100, \\ [0,\ 0,\ 0]^{\mathrm{T}}, 0 < k \leqslant 100. \end{cases}$$

Fault type 3 is a sinusoidal time-varying actuator fault:

$$f_{3,k} = \begin{cases} [0,\ 0,\ -0.9\sin(0.01(k-70))]^{\mathrm{T}}, k > 100, \\ [0,\ 0,\ 0]^{\mathrm{T}}, 0 < k \leqslant 100. \end{cases}$$

The three types of faults occur separately in the simulation. From the fault signal (brown line) obtained by the proposed fault detection method in Fig. 5.18, we can see that faults are detected at time instants $k = 102$, $k = 106$ and $k = 103$, respectively. Because the fault value is small when the actuator of the DC motor has just failed, the IOF method can detect the fault more quickly than the EZGKF method and has higher detection accuracy during a continuous fault time. Similar to the fault

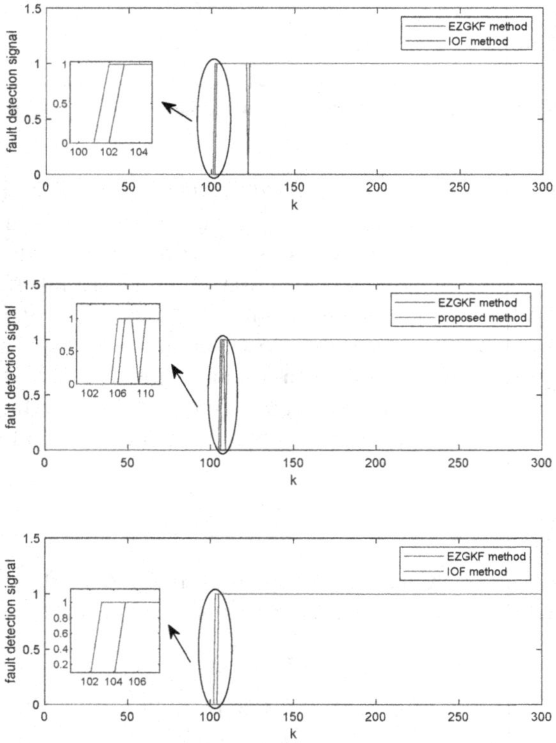

Figure 5.18: Fault detection results of faults type 1, 2 and 3.

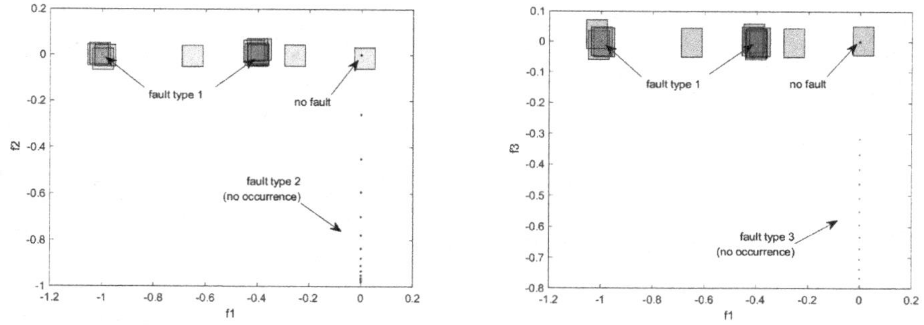

Figure 5.19: Fault isolation and identification result of fault type 1.

detection results of the numerical example in Section 5.1.3.1, the EZGKF method may lead to false detection, but this situation does not exist in the IOF method. The simulation results indicate that the IOF method can effectively detect both abrupt and time-varying faults in real-time.

Fault isolation is required after a fault is detected. To ensure the readability of the simulation graph, we sampled some of the fault estimation results during the isolation process and expressed them in pairs with two-dimensional graphs. As shown in Fig. 5.19, the red box representing the fault estimation interval includes a black dot representing no fault when the DC motor is working normally. Once the motor fails, the red box gradually approaches the red dots representing fault type 1 rather than the blue dots representing fault type 2 or the green dots representing fault type 3. In Fig. 5.20, the blue fault estimation interval just encloses the non-fault origin initially and surrounds the blue point following the change in the fault value finally. This indicates that a type-2 fault has occurred. The position change of the green box in

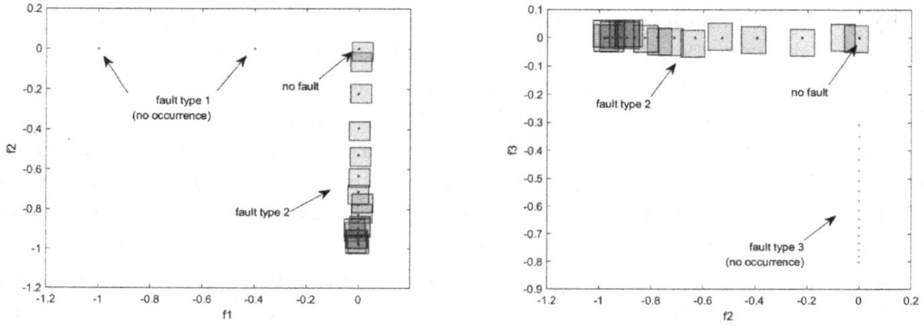

Figure 5.20: Fault isolation and identification result of fault type 2.

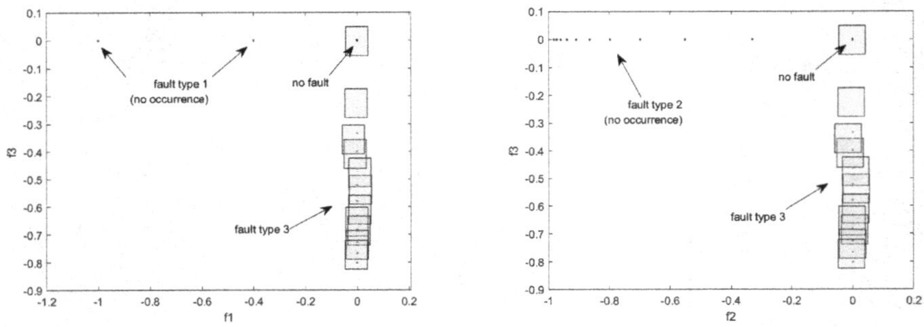

Figure 5.21: Fault isolation and identification result of fault type 3.

Fig. 5.21 indicates that the fault isolation of fault type 3 has been achieved. It can be seen from the above content that the fault isolation method is applicable to both time-varying and time-invariant faults. In the same simulation environment as the numerical example simulation, the detection time is 0.0933s, 0.0889s and 0.0578s, respectively, for the three types of actuator faults of the DC motor. The time for isolation and identification is 0.100s, 0.105s and 0.114s, respectively.

5.2.4 Conclusions

A fault diagnosis algorithm for dual uncertain systems combining UBB and stochastic uncertainties is presented in this section. First, a dynamic observer designed by determining the relationship between bounded and Gaussian signals according to the signal power ratio ensure a more accurate state estimation interval. Second, a fault detection indicator is designed based on the probability ratio of the residual of the Gaussian part. This indicator not only performs real-time system fault detection tests but also determines different types of faults based on the fault estimation interval. Finally, through simulations of a numerical system and a DC motor system, the given method is shown to be effective. The proposed interval observer filtering-based fault diagnosis method not only can be extended to DC motor systems with uncertain parameters [96] or continuous-time [10], but also is applicable to the fault diagnosis of other dual uncertain systems in the industrial field, such as heat exchanger systems [12], wind turbine systems [38], servo actuator systems [69] and robot systems [60,73]. Moreover, the goal of fault-tolerant control is to make the system maintain the desired performance and stable conditions when system fails. As the basis of fault tolerance control, fault diagnosis provides information for it to better compensate the impact of fault on the system [48,51,116,128]. The sliding mode observer is inherent robustness to matched uncertainties and it can be introduced into the design of interval observer to obtain more accurate interval estimation [121, 122]. Further research on fault-tolerant control and sliding mode control can be carried out on the basis of this section.

5.3 Orthometric hyperparallel spatial directional expansion filtering based fault diagnosis method

5.3.1 Pre-knowledge

Define the strip space of n dimension as

$$S(\boldsymbol{p},c)=\left\{\boldsymbol{\theta}:|\boldsymbol{p}^{\mathrm{T}}\boldsymbol{\theta}-c|\leqslant 1\right\},\quad \boldsymbol{p}\in R^{n},\tag{5.108}$$

in fact, any polytope $\mathcal{V}$ that contains a close combination of truth values can be formed by the intersection of m non-parallel n dimensional strip spaces, where $m\geqslant n$. When $m=n$, the polytope is a hyperparallel space, namely

$$\mathcal{P}=\bigcap_{i=1}^{n}S(\boldsymbol{p}_i,c_i).\tag{5.109}$$

For n dimensional hyperparallel space $\mathcal{P}$, $\boldsymbol{P}=[\boldsymbol{p}_1,\ldots,\boldsymbol{p}_n]^{\mathrm{T}}\in R^{n\times n}$, where $\boldsymbol{P}$ is reversible. Another expression of $\mathcal{P}$ is [85]:

$$\mathcal{P}=\{\boldsymbol{\theta}:\|\boldsymbol{P}(\boldsymbol{\theta}-\boldsymbol{\theta}_c)\|_{\infty}\leqslant 1\},\tag{5.110}$$

where

$$\boldsymbol{P}=[\boldsymbol{p}_1,\ldots,\boldsymbol{p}_n]^{\mathrm{T}}\in R^{n\times n},\tag{5.111}$$

$$\boldsymbol{\theta}_c=\boldsymbol{P}^{-1}[c_1,\ldots,c_n]^{\mathrm{T}}\in R^{n}.\tag{5.112}$$

Define $\boldsymbol{T}=\boldsymbol{P}^{-1}$, the Eq. (5.110) can be transformed into [85]:

$$\mathcal{P}(\boldsymbol{T},\boldsymbol{\theta}_c)=\left\{\boldsymbol{\theta}:\boldsymbol{\theta}=\boldsymbol{\theta}_c+\boldsymbol{T}\tilde{\boldsymbol{\theta}},\ \|\tilde{\boldsymbol{\theta}}\|_{\infty}\leqslant 1\right\},\tag{5.113}$$

where $\boldsymbol{\theta}_c$ is the center of the hyperparallel space, $\boldsymbol{T}$ is the generator matrix of the hyperparallel space.

5.3.2 Problem description

Consider the following linear recursive model:

$$y(k)=\boldsymbol{\varphi}^{\mathrm{T}}(k)(\boldsymbol{\theta}_o+\Delta\boldsymbol{\theta}_{f_a}(k))+e(k),\tag{5.114}$$

where $k=1,2,\ \ldots,\ N$. N is the total sample length, $\boldsymbol{\varphi}(k)\in R^{n}$ is an observable data vector, $y(k)$ is the output data for the system, $\boldsymbol{\theta}_o$ is the true value of the system parameter in the fault-free state, $\Delta\boldsymbol{\theta}_{f_a}(k)$ is the parameter change vector when the fault occurs at time k, f_a is the fault type of the parameter. Where, assuming $k=0$, $f_a=f_0$, $\Delta\boldsymbol{\theta}_{f_0}(k)=\boldsymbol{0}_n$, that is, the system is in a normal state, $\boldsymbol{\theta}_o+\Delta\boldsymbol{\theta}_{f_a}(k)\in R^{n}$ is the parameter vector to be estimated. For the convenience of the following description, let $\boldsymbol{\theta}_{f_a}(k)=\boldsymbol{\theta}_o+\Delta\boldsymbol{\theta}_{f_a}(k)$. $e(k)$ is defined as unknown but bounded noise at time k, the amplitude of the noise signal is $\delta(k)$, that is $|e(k)|\leqslant\delta(k)$, $\delta(k)\in R^{+}$.

When the system fails at time k, $\Delta\boldsymbol{\theta}_{fa}(k)$ will change, at this time, the unknown change is bounded, given $\boldsymbol{\gamma}^{\max} = [\gamma_1^{\max}, \ldots, \gamma_n^{\max}]^{\mathrm{T}}$, that is [61]

$$\left|\Delta\boldsymbol{\theta}_{fa}(k)\right| \leqslant \boldsymbol{\gamma}^{\max}, \quad \forall k. \tag{5.115}$$

The measurement set $s(k)$ at time k can be expressed as:

$$S(k) = \left\{\boldsymbol{\theta} : \left|y(k) - \boldsymbol{\varphi}^{\mathrm{T}}(k)\boldsymbol{\theta}\right| \leqslant \delta(k)\right\}, \tag{5.116}$$

the Eq. (5.116) can be converted to the strip space as shown in the Eq. (5.108):

$$S(\boldsymbol{p}_k, c_k) = \left\{\boldsymbol{\theta} : |\boldsymbol{p}_k^{\mathrm{T}}\boldsymbol{\theta} - c_k| \leqslant 1\right\}, \tag{5.117}$$

where

$$\boldsymbol{p}_k = \frac{\boldsymbol{\varphi}(k)}{\delta(k)}, \quad c_k = \frac{y(k)}{\delta(k)}. \tag{5.118}$$

From the Eq. (5.116), it can be seen that the parameter set of the system at time k is between two parallel hyperplanes. Therefore, with the increase of sampling time, in a given time, all possible parameters are finally included in the convex space $\theta(k)$, that is

$$\begin{aligned}\Theta(k) &= \left\{\boldsymbol{\theta} : \left|y(k) - \boldsymbol{\varphi}^{\mathrm{T}}(k)\boldsymbol{\theta}\right| \leqslant \delta(k), k = 1, 2, \ldots, N\right\} \\ &= \bigcap_{k=1}^{N} S(k) \\ &= \Theta(k-1) \cap S(k).\end{aligned} \tag{5.119}$$

As the sampling time increases, the complexity of the convex space structure shown in Eq. (5.119) continues to increase. The purpose of this section is to solve the fault diagnosis problem of linear recursive systems (5.114), considering the influence of unknown but bounded noise, and propose a filter fault diagnosis method based on the orthometric hyperparallel spatial directional expansion.

5.3.3 Orthometric hyperparallel spatial directional expansion filtering based fault diagnosis method

Lemma 5.1

Given a n dimensional hyperparallel space $\mathcal{P}(\boldsymbol{T}_{k-1}, \boldsymbol{\theta}_{c,k-1})$ intersects with the measurement strip $S(\boldsymbol{p}_k, c_k)$, let each column vector in the hyperparallel space generator matrix be $\boldsymbol{t}_{k-1,i}$, where $i = 1, \ldots, n$. If $\boldsymbol{p}_k^{\mathrm{T}}\boldsymbol{t}_{k-1,i} \geqslant 0$ all hold, otherwise let $\boldsymbol{t}_{k-1,i} = -\boldsymbol{t}_{k-1,i}$, then there is a n dimensional hyperparallel space $\mathcal{P}(\boldsymbol{T}_k, \boldsymbol{\theta}_{c,k})$ [85] with the smallest volume that wraps the intersecting area,

$$\mathcal{P}(\boldsymbol{T}_k, \boldsymbol{\theta}_{c,k}) = \begin{cases} \bar{\mathcal{P}}(\bar{\boldsymbol{T}}, \bar{\boldsymbol{\theta}}_c), & i^* = 0, \\ \mathcal{P}^*(\boldsymbol{T}^*, \boldsymbol{\theta}_c^*), & i^* \neq 0, \end{cases} \tag{5.120}$$

where

$$i^* = \arg \max_{b=0,1,\ldots,n} \bar{\boldsymbol{p}}_k^{\mathrm{T}} \bar{\boldsymbol{t}}_b, \quad \bar{\boldsymbol{t}}_0 = \bar{\boldsymbol{p}}_k / |\bar{\boldsymbol{p}}_k|^2, \tag{5.121}$$

$$\bar{\boldsymbol{\theta}}_c = \boldsymbol{\theta}_{c,k-1} + \sum_{i=1}^{n} \frac{r_i^+ - r_i^-}{2} \boldsymbol{t}_{k-1,i}, \tag{5.122}$$

$$\bar{\boldsymbol{t}}_i = \frac{r_i^+ + r_i^-}{2} \boldsymbol{t}_{k-1,i}. \tag{5.123}$$

$$\boldsymbol{\theta}_c^* = \bar{\boldsymbol{\theta}}_c + \frac{1}{\bar{\boldsymbol{p}}_k^{\mathrm{T}} \bar{\boldsymbol{t}}_{i^*}} \bar{\boldsymbol{t}}_{i^*} \left(\bar{c}_k - \bar{\boldsymbol{p}}_k^{\mathrm{T}} \bar{\boldsymbol{\theta}}_c \right), \tag{5.124}$$

$$\boldsymbol{t}_i^* = \begin{cases} \bar{\boldsymbol{t}}_i - \frac{\bar{\boldsymbol{p}}_k^{\mathrm{T}} \bar{\boldsymbol{t}}_i}{\bar{\boldsymbol{p}}_k^{\mathrm{T}} \bar{\boldsymbol{t}}_{i^*}} \bar{\boldsymbol{t}}_{i^*}, & i \neq i^*, \\ \frac{1}{\bar{\boldsymbol{p}}_k^{\mathrm{T}} \bar{\boldsymbol{t}}_{i^*}} \bar{\boldsymbol{t}}_{i^*}, & i = i^*. \end{cases} \tag{5.125}$$

In Eqs. (5.121)–(5.125),

$$\bar{\boldsymbol{p}}_k = \frac{2}{r_0^+ + r_0^-} \boldsymbol{p}_k, \tag{5.126}$$

$$\bar{c}_k = \frac{2}{r_0^+ + r_0^-} \left(c_k + \frac{r_0^+ - r_0^-}{2} \right), \tag{5.127}$$

$$r_i^{\pm} = \begin{cases} \min\left(1, \frac{1 \mp \varepsilon_0^{\pm}}{\boldsymbol{p}_k^{\mathrm{T}} \boldsymbol{t}_{k-1,i}} - 1\right), & \boldsymbol{p}_k^{\mathrm{T}} \boldsymbol{t}_{k-1,i} \neq 0, \\ 1, & \boldsymbol{p}_k^{\mathrm{T}} \boldsymbol{t}_{k-1,i} = 0, \end{cases} \tag{5.128}$$

where $r_0^+ = \min(1, \varepsilon_0^+)$, $r_0^- = \min(1, -\varepsilon_0^-)$, ε_0^+ *and* ε_0^- *are scalar:*

$$\begin{cases} \varepsilon_0^+ = (\boldsymbol{p}_k^{\mathrm{T}} \boldsymbol{\theta}_{c,k-1} - c_k) + \sum_{i=1}^{n} \boldsymbol{p}_k^{\mathrm{T}} \boldsymbol{t}_{k-1,i}, \\ \varepsilon_0^- = (\boldsymbol{p}_k^{\mathrm{T}} \boldsymbol{\theta}_{c,k-1} - c_k) - \sum_{i=1}^{n} \boldsymbol{p}_k^{\mathrm{T}} \boldsymbol{t}_{k-1,i}. \end{cases} \tag{5.129}$$

According to Lemma 5.1, the minimum volume hyperparallel space $\mathcal{P}(\boldsymbol{T}_k, \boldsymbol{\theta}_{c,k})$ of the feasible set of wrapping parameters at time k can be obtained. Define each column of the n dimensional hyperparallel space generation matrix $\boldsymbol{T}$ as $\boldsymbol{t}_i$, where $i = 1, \ldots, n$, the vertex $\boldsymbol{V}_l$ of hyperparallel space can be calculated from the expression of Eq. (5.113):

$$\boldsymbol{V}_l = \boldsymbol{\theta}_c + \sum_{i=1}^{n} \alpha_{l,i} \boldsymbol{t}_i \in R^n, \tag{5.130}$$

where $l = 1, 2, \ldots, 2^n$, $\alpha_{l,i} \in \{-1, 1\}$.

Define $\boldsymbol{V}_k = [\boldsymbol{V}_1, \boldsymbol{V}_2, \ldots, \boldsymbol{V}_{2^n}] \in R^{n\times 2^n}$ represents the vertex matrix of n dimensional hyperparallel space at time k, $\boldsymbol{V}_k(c,l)$ represents the cth element of the lth column of the $\boldsymbol{V}_k$ matrix, $c = 1, \ldots, n$. The maximum values $\theta_c^+(k)$ and minimum values $\theta_c^-(k)$ of each parameter at time k can be obtained from Eq. (5.130),

$$\begin{cases} \theta_c^+(k) = \max(\boldsymbol{V}_k(c,l)), \\ \theta_c^-(k) = \min(\boldsymbol{V}_k(c,l)), \end{cases} \tag{5.131}$$

however, with the increase of sampling time, it is difficult to ensure the monotonic convergence of its parameter boundary values. In order to solve this problem, this section defines the orthometric hyperparallel space $\mathcal{P}_o(\boldsymbol{T}_k, \boldsymbol{\theta}_{c,k})$ at time k:

$$\mathcal{P}_o(\boldsymbol{T}_k, \boldsymbol{\theta}_{c,k}) = \left\{ \boldsymbol{\theta} : \boldsymbol{\theta} = \boldsymbol{\theta}_{c,k} + \boldsymbol{T}_k \tilde{\boldsymbol{\theta}},\ \|\tilde{\boldsymbol{\theta}}\|_\infty \leqslant 1 \right\}, \tag{5.132}$$

where

$$\boldsymbol{\theta}_{c,k} = \begin{bmatrix} \frac{\theta_{o_1}^+(k)+\theta_{o_1}^-(k)}{2} \\ \vdots \\ \frac{\theta_{o_n}^+(k)+\theta_{o_n}^-(k)}{2} \end{bmatrix}, \quad \boldsymbol{T}_k = \mathrm{diag}\left\{ \begin{matrix} \frac{\theta_{o_1}^+(k)-\theta_{o_1}^-(k)}{2} \\ \vdots \\ \frac{\theta_{o_n}^+(k)-\theta_{o_n}^-(k)}{2} \end{matrix} \right\}, \tag{5.133}$$

$$\begin{cases} \theta_{o_u}^+(k) = \min(\theta_{o_u}^+(k-1), \theta_u^+(k)), \\ \theta_{o_u}^-(k) = \max(\theta_{o_u}^-(k-1), \theta_u^-(k)), \end{cases} \tag{5.134}$$

where $u = 1, \ldots, n$. From Eqs. (5.132)–(5.134), the orthometric hyperparallel space at time k is to compare the maximum value of parameter boundary between the hyperparallel space at time k and the orthometric hyperparallel space at time $k-1$, and take the minimum value of parameter boundary as the maximum, which guarantees $\mathcal{P}_o(\boldsymbol{T}_k, \boldsymbol{\theta}_{c,k}) \subseteq \mathcal{P}_o(\boldsymbol{T}_{k-1}, \boldsymbol{\theta}_{c,k-1})$. In this section, the minimum volume hyperparallel space of the feasible set of the wrapped parameters at each time is obtained by Eq. (5.120), and then the orthometric hyperparallel space at each time is obtained by Eqs. (5.130)–(5.134). The boundary value of the orthometric hyperparallel space is used as the upper and lower bounds of the wrapped true value, which ensures the monotonic convergence of the boundary value of the parameters and makes the feasible set of the wrapped true value more compact. At the same time, for the fault state, A fault diagnosis strategy based on directional expansion of orthometric hyperparallel space is designed.

5.3.3.1 Fault detection

Lemma 5.2
When scalar $\varepsilon_0^+ < -1$ or $\varepsilon_0^- > 1$, the hyperparallel space $\mathcal{P}(\boldsymbol{T}_{k-1}, \boldsymbol{\theta}_{c,k-1})$ does not intersect with the measuring strip $S(\boldsymbol{p}_k, c_k)$ [85].

Fault detection is based on the consistency analysis of hyperparallel space $\mathcal{P}(\boldsymbol{T}_{k-1}, \boldsymbol{\theta}_{c,k-1})$ and measuring strip $S(\boldsymbol{p}_k, c_k)$, that is, when $\mathcal{P}(\boldsymbol{T}_{k-1}, \boldsymbol{\theta}_{c,k-1}) \cap S(\boldsymbol{p}_k, c_k) = \varnothing$, the system fails. Therefore, according to Lemma 5.2, when the scalar in Eq. (5.129) satisfies $\varepsilon_0^+ < -1$ or $\varepsilon_0^- > 1$, it can be determined that the fault occurs.

5.3.3.2 Fault isolation and identification

Assuming that at time k, the system detects a fault and directionally expands $\mathcal{P}_o(\boldsymbol{T}_{k-1}, \boldsymbol{\theta}_{c,k-1})$ to obtain n orthometric hyperparallel space test sets $\mathcal{P}^t_{o,d}(\boldsymbol{T}_{k-1}, \boldsymbol{\theta}_{c,k-1})$, where $d = 1, \ldots, n$.

Define the n dimensional orthometric hyperparallel space test set $\mathcal{P}^t_{o,d}(\boldsymbol{T}_{k-1}, \boldsymbol{\theta}_{c,k-1})$ is:

$$\mathcal{P}^t_{o,d}(\boldsymbol{T}_{k-1}, \boldsymbol{\theta}_{c,k-1}) = \{\boldsymbol{\theta} : \boldsymbol{\theta} = \boldsymbol{\theta}_{c,k-1} + \boldsymbol{T}_{k-1}\tilde{\boldsymbol{\theta}}, \|\tilde{\boldsymbol{\theta}}\|_\infty \leqslant 1\}, \tag{5.135}$$

where

$$\boldsymbol{\theta}_{c,k-1} = \begin{bmatrix} \frac{\theta^{t+}_{(o,d)_1}(k-1)+\theta^{t-}_{(o,d)_1}(k-1)}{2} \\ \vdots \\ \frac{\theta^{t+}_{(o,d)_n}(k-1)+\theta^{t-}_{(o,d)_n}(k-1)}{2} \end{bmatrix}, \tag{5.136}$$

$$\boldsymbol{T}_{k-1} = \operatorname{diag}\left\{ \begin{matrix} \frac{\theta^{t+}_{(o,d)_1}(k-1)-\theta^{t-}_{(o,d)_1}(k-1)}{2} \\ \vdots \\ \frac{\theta^{t+}_{(o,d)_n}(k-1)-\theta^{t-}_{(o,d)_n}(k-1)}{2} \end{matrix} \right\}, \tag{5.137}$$

(1) when $p \neq d$ and $1 \leqslant p \leqslant n$,

$$\begin{cases} \theta^{t+}_{(o,d)_p}(k-1) = \theta^{+}_{op}(k-1) + 2\gamma_p^{\max}, \\ \theta^{t-}_{(o,d)_p}(k-1) = \theta^{-}_{op}(k-1) - 2\gamma_p^{\max}, \end{cases} \tag{5.138}$$

(2) when $p = d$,

$$\begin{cases} \theta^{t+}_{(o,d)_p}(k-1) = \theta^{+}_{op}(k-1), \\ \theta^{t-}_{(o,d)_p}(k-1) = \theta^{-}_{op}(k-1). \end{cases} \tag{5.139}$$

From Eqs. (5.135) to (5.139), $\mathcal{P}^t_{o,d}(\boldsymbol{T}_{k-1}, \boldsymbol{\theta}_{c,k-1})$ is an orthometric hyperparallel space test set extending in all $n-1$ parameter directions except the d parameter direction. Given the time length L, assuming that no new fault occurs in L time after the fault is detected at time k, the measuring strips $S(\boldsymbol{p}_k, c_k), \ldots, S(\boldsymbol{p}_{k+L}, c_{k+L})$ can be obtained. Using Eqs. (5.120), the n orthometric hyperparallel space test sets obtained by directional expansion are calculated with L measuring strips to obtain the n hyperparallel space test sets $\mathcal{P}^t_d(\boldsymbol{T}_{k+L}, \boldsymbol{\theta}_{c,k+L})$.

Theorem 5.7
If $\mathcal{P}_j^t(\boldsymbol{T}_{k+L},\boldsymbol{\theta}_{c,k+L})=\varnothing$, where $j\in\{1,\ldots,n\}$, it means that the jth component of the parameter vector fails. Then the parameter boundary of the jth component in the initialized orthometric hyperparallel space $\mathcal{P}_o^{in}(\boldsymbol{T}_k,\boldsymbol{\theta}_{c,k})$ is directionally expanded as

$$\begin{cases}\theta_{o_j}^{in+}(k)=\theta_{o_j}^{+}(k-1)+2\gamma_j^{\max},\\ \theta_{o_j}^{in-}(k)=\theta_{o_j}^{-}(k-1)-2\gamma_j^{\max},\end{cases}\tag{5.140}$$

otherwise, if $\mathcal{P}_j^t(\boldsymbol{T}_{k+L},\boldsymbol{\theta}_{c,k+L})\neq\varnothing$, it means that the jth component of the parameter vector has not failed. Then the parameter boundary of jth component of the initialized orthometric hyperparallel space $\mathcal{P}_o^{in}(\boldsymbol{T}_k,\boldsymbol{\theta}_{c,k})$ does not need to be extended:

$$\begin{cases}\theta_{o_j}^{in+}(k)=\theta_{o_j}^{+}(k-1),\\ \theta_{o_j}^{in-}(k)=\theta_{o_j}^{-}(k-1).\end{cases}\tag{5.141}$$

Proof 5.7 Assuming that the jth component in the parameter vector fails, the parameter vector $\boldsymbol{\theta}_{fa}(k)$ at this time satisfies:

$$\begin{cases}\theta_{fa_u}(k)=\theta_{fa-1_u}(k), & u\neq j,\\ \theta_{fa_u}(k)=\theta_{fa-1_u}(k)+\Delta\theta_{fa_u}(k), & u=j,\end{cases}\tag{5.142}$$

where $u=1,\ldots,n$. According to Eqs. (5.138) and (5.139), $\mathcal{P}_{o,d}^t(\boldsymbol{T}_{k-1},\boldsymbol{\theta}_{c,k-1})$ is an orthometric hyperparallel space test set whose $n-1$ parameter directions are extended except the d parameter direction, So $\mathcal{P}_{o,j}^t(\boldsymbol{T}_{k-1},\boldsymbol{\theta}_{c,k-1})$ is the orthometric hyperparallel space test set without expanding the jth component, that is

$$\begin{cases}\theta_{fa_j}(k)\notin\mathcal{P}_{o,j}^t(\boldsymbol{T}_{k-1},\boldsymbol{\theta}_{c,k-1}), & u=j,\\ \theta_{fa_u}(k)\in\mathcal{P}_{o,j}^t(\boldsymbol{T}_{k-1}.\boldsymbol{\theta}_{c,k-1}), & u\neq j.\end{cases}\tag{5.143}$$

It can be seen from the intersection of Eq. (5.120) and L measuring strips,

$$\mathcal{P}_j^t(\boldsymbol{T}_{k+L},\boldsymbol{\theta}_{c,k+L})\subseteq\mathcal{P}_{o,j}^t(\boldsymbol{T}_{k-1},\boldsymbol{\theta}_{c,k-1}),$$

due to

$$\theta_{fa_j}(k)\notin\mathcal{P}_{o,j}^t(\boldsymbol{T}_{k-1},\boldsymbol{\theta}_{c,k-1}),\tag{5.144}$$

so

$$\theta_{fa_j}(k)\notin\mathcal{P}_j^t(\boldsymbol{T}_{k+L},\boldsymbol{\theta}_{c,k+L}),\tag{5.145}$$

thus

$$\mathcal{P}_j^t(\boldsymbol{T}_{k+L},\boldsymbol{\theta}_{c,k+L})=\varnothing.\tag{5.146}$$

At this time, it indicates that the j-component of the parameter vector fails, then its parameter boundary is directionally extended to

$$\begin{cases} \theta_{o_j}^{in+}(k) = \theta_{o_j}^{+}(k-1) + 2\gamma_j^{\max}, \\ \theta_{o_j}^{in-}(k) = \theta_{o_j}^{-}(k-1) - 2\gamma_j^{\max}, \end{cases}$$

where $\gamma_j^{\max}$ is the maximum change value of the jth component of the parameter vector. On the contrary, if $\mathcal{P}_j^t(\boldsymbol{T}_{k+L}, \boldsymbol{\theta}_{c,k+L}) \neq \varnothing$, which indicates $\theta_{fa_j}(k) \in \mathcal{P}_j^t(\boldsymbol{T}_{k+L}, \boldsymbol{\theta}_{c,k+L})$, then $\theta_{fa_j}(k) \in \mathcal{P}_{o,j}^t(\boldsymbol{T}_{k-1}, \boldsymbol{\theta}_{c,k-1})$, indicating that the jth component of the parameter vector does not fail, its parameter boundary will not be extended in a directional way, that is

$$\begin{cases} \theta_{o_j}^{in+}(k) = \theta_{o_j}^{+}(k-1), \\ \theta_{o_j}^{in-}(k) = \theta_{o_j}^{-}(k-1). \end{cases}$$

Taking two-dimensional space as an example, Fig. 5.22 describes the process of fault isolation when a fault occurs at time k. at time $k+1$, $\mathcal{P}_1^t(k+1) \neq \varnothing$, while $\mathcal{P}_2^t(k+1) \neq \varnothing$, fault isolation cannot be achieved at this time; at time $k+2$, $\mathcal{P}_1^t(k+2) = \varnothing$, ¶ø$\mathcal{P}_2^t(k+2) \neq \varnothing$, it can be determined that the fault occurred in the parameter θ_1, fault isolation completed.

After realizing system fault isolation, expansion of $\mathcal{P}_o(\boldsymbol{T}_{k-1}, \boldsymbol{\theta}_{c,k-1})$ directionally to obtain the initialized orthometric hyperparallel space $\mathcal{P}_o^{in}(\boldsymbol{T}_k, \boldsymbol{\theta}_{c,k})$.

Theorem 5.8
If $\boldsymbol{\theta}_{fa-1}(k)$ is in the orthometric hyperparallel space $\mathcal{P}_o(\boldsymbol{T}_{k-1}, \boldsymbol{\theta}_{c,k-1})$, at this time, the jth component of the parameter vector fails at time k, then the new fault parameter value $\boldsymbol{\theta}_{fa}(k)$ will be in the initialized orthometric hyperparallel space

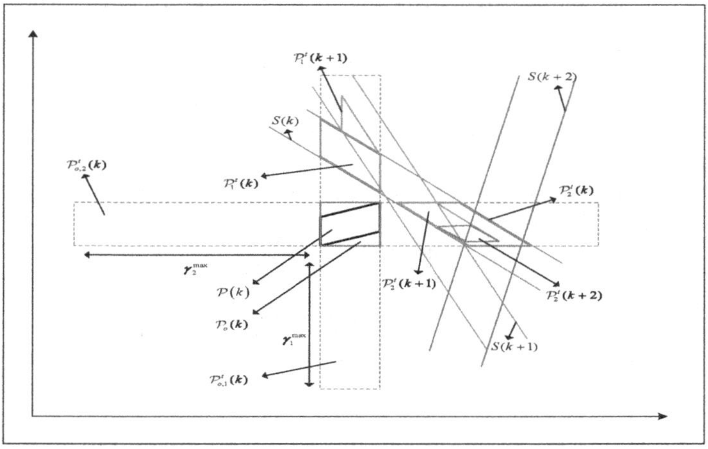

Figure 5.22: Process of fault isolation.

$\mathcal{P}_o^{in}(\boldsymbol{T}_k, \boldsymbol{\theta}_{c,k})$,

$$\mathcal{P}_o^{in}(\boldsymbol{T}_k, \boldsymbol{\theta}_{c,k}) = \left\{\boldsymbol{\theta} : \boldsymbol{\theta} = \boldsymbol{\theta}_{c,k} + \boldsymbol{T}_k\tilde{\boldsymbol{\theta}},\ \|\tilde{\boldsymbol{\theta}}\|_\infty \leqslant 1\right\}, \tag{5.147}$$

where

$$\boldsymbol{\theta}_{c,k} = \begin{bmatrix} \frac{\theta_{o_1}^{in+}(k)+\theta_{o_1}^{in-}(k)}{2} \\ \vdots \\ \frac{\theta_{o_n}^{in+}(k)+\theta_{o_n}^{in-}(k)}{2} \end{bmatrix}, \tag{5.148}$$

$$\boldsymbol{T}_k = \operatorname{diag}\left\{ \begin{matrix} \frac{\theta_{o_1}^{in+}(k)-\theta_{o_1}^{in-}(k)}{2} \\ \vdots \\ \frac{\theta_{o_1}^{in+}(k)-\theta_{o_1}^{in-}(k)}{2} \end{matrix} \right\}, \tag{5.149}$$

(1) when $q = j$,

$$\begin{cases} \theta_{o_q}^{in+}(k) &= \theta_{o_q}^{+}(k-1) + 2\gamma_q^{\max}, \\ \theta_{o_q}^{in-}(k) &= \theta_{o_q}^{-}(k-1) - 2\gamma_q^{\max}, \end{cases} \tag{5.150}$$

(2) when $q \neq j$ and $1 \leqslant q \leqslant n$,

$$\begin{cases} \theta_{o_q}^{in+}(k) &= \theta_{o_q}^{+}(k-1), \\ \theta_{o_q}^{in-}(k) &= \theta_{o_q}^{-}(k-1). \end{cases} \tag{5.151}$$

Proof 5.8 From Eq. (5.115), it can be seen that for a fault at any time, the change of any component of the parameter vector is within a certain range, which is

$$\begin{bmatrix} |\Delta\theta_{fa_1}(k)| \\ \vdots \\ |\Delta\theta_{fa_n}(k)| \end{bmatrix} \leqslant \begin{bmatrix} \gamma_1^{\max} \\ \vdots \\ \gamma_n^{\max}. \end{bmatrix}, \qquad \forall k, \tag{5.152}$$

if the parameter vector fails at the jth component at the time k, that is

$$|\Delta\theta_{fa_j}(k)| \leqslant \gamma_j^{\max}, \tag{5.153}$$

because of

$$|\Delta\theta_{fa_{-1_j}}(k)| \leqslant \gamma_j^{\max}, \tag{5.154}$$

therefore:

$$|\Delta\theta_{fa_j}(k) - \Delta\theta_{fa_{-1_j}}(k)| \leqslant |\Delta\theta_{fa_j}(k)| + |\Delta\theta_{fa_{-1_j}}(k)| \leqslant 2\gamma_j^{\max}, \tag{5.155}$$

due to

$$\begin{aligned}\theta_{fa_j}(k) &= \theta_{o_j} + \Delta\theta_{fa_j}(k) \\ &= \theta_{o_j} + \Delta\theta_{fa-1_j}(k) + (\Delta\theta_{fa_j}(k) - \Delta\theta_{fa-1_j}(k)) \\ &= \theta_{fa-1_j}(k) + \Delta\theta_{fa_j}(k) - \Delta\theta_{fa-1_j}(k),\end{aligned} \tag{5.156}$$

Use Eq. (5.155) to get Eq. (5.155)

$$-2\gamma_j^{\max} \leqslant \Delta\theta_{fa_j}(k) - \Delta\theta_{fa-1_j}(k) \leqslant 2\gamma_j^{\max}, \tag{5.157}$$

because of

$$\theta_{o_j}^-(k-1) \leqslant \theta_{fa-1_j}(k) \leqslant \theta_{o_j}^+(k-1), \tag{5.158}$$

from Eqs. (5.156) to (5.158), $\theta_{fa_j}(k)$ must be expanded in the following direction inside:

$$\begin{cases}\theta_{o_j}^{in+}(k) = \theta_{o_j}^+(k-1) + 2\gamma_j^{\max}, \\ \theta_{o_j}^{in-}(k) = \theta_{o_j}^-(k-1) - 2\gamma_j^{\max}.\end{cases}$$

Since the other components of the parameter vector have not failed, the parameter value has not changed, that is, $q \neq j$, so the boundary value does not need to be expanded, that is

$$\begin{cases}\theta_{o_q}^{in+}(k) = \theta_{o_q}^+(k-1), \\ \theta_{o_q}^{in-}(k) = \theta_{o_q}^-(k-1).\end{cases}$$

In summary, the steps of the filtering fault diagnosis method (Orthometric hyper-parallel spatial directional expansion filtering based fault diagnosis, OHSDEF-FD) proposed in this section are as follows:

Step 1: Let $k = 1$, initialize $\mathcal{P}(\boldsymbol{T}_1, \boldsymbol{\theta}_{c,1})$ and $\mathcal{P}_o(\boldsymbol{T}_1, \boldsymbol{\theta}_{c,1})$, given length of time L, maximum value of parameter change $\boldsymbol{\gamma}^{\max}$ and sampling time N.

Step 2: Obtain measuring strip $S(\boldsymbol{p}_k, c_k)$, use Eq. (5.129) to calculate scalar ε_0^+ and ε_0^-, if $\varepsilon_0^+ < -1$ or $\varepsilon_0^- > 1$, a system failure is detected, jump to step 4; on the contrary, judge that the system has no fault and continue with step 3.

Step 3: From Eq. (5.120) to calculate $\mathcal{P}(\boldsymbol{T}_k, \boldsymbol{\theta}_{c,k})$, then, $\mathcal{P}_o(\boldsymbol{T}_k, \boldsymbol{\theta}_{c,k})$ is solved by Eqs. (5.130)–(5.134), jump to step 6.

Step 4: Obtain the measuring strips $S(\boldsymbol{p}_k, c_k), \ldots, S(\boldsymbol{p}_{k+L}, c_{k+L})$, n orthometric hyperparallel space test sets $\mathcal{P}_{o,d}^t(\boldsymbol{T}_{k-1}, \boldsymbol{\theta}_{c,k-1})$ are obtained from Eqs. (5.138) and (5.139), and then use Eq. (5.120) to calculate L times to obtain n hyperparallel space test sets $\mathcal{P}_d^t(\boldsymbol{T}_{k+L}, \boldsymbol{\theta}_{c,k+L})$.

Step 5: According to $\mathcal{P}_d^t(\boldsymbol{T}_{k+L}, \boldsymbol{\theta}_{c,k+L})$, the specific fault parameters are isolated by Theorem 5.7 and the initialized orthometric hyperparallel space $\mathcal{P}_o^{in}(\boldsymbol{T}_k, \boldsymbol{\theta}_{c,k})$ is obtained by directional expansion of Eq. (5.140).

Step 6: Let $k = k+1$, return to Step 2, until $k = n$, the algorithm ends and the fault diagnosis result is output.

5.3.4 Simulation

Example 1: For the following linear dynamic system models:

$$y(k) = u(k-1) + 0.8u(k-2) + e(k), \tag{5.159}$$

where $\boldsymbol{\theta} = [1, 0.8]^{\mathrm{T}}$ is the system truth value, $e(k)$ is unknown but bounded noise and $|e(t)| \leqslant 0.1$, given length of time $L = 5$, $\gamma_1^{\max} = \gamma_2^{\max} = 1$. Set fault 1 at $k = 501$ as: $\boldsymbol{\theta}_{f_1}(k) = [1.5, 0.8]^{\mathrm{T}}$; fault 2 at $k = 1001$ as: $\boldsymbol{\theta}_{f_2}(k) = [1.5, 1.2]^{\mathrm{T}}$. The algorithm (OHSDEF-FD) proposed in this section is used for fault diagnosis, at the same time, it is compared with the fault diagnosis based on set-membership identification using output-error models(SMIOE-FD) in literature [79], and the results are shown in Figs. 5.23–5.25.

In Fig. 5.23, blue represents the area where the ellipsoid wraps the feasible set in the SMIOE-FD algorithm, rose red represents the area where the hyperparallel space wraps the feasible set in the OHSDEF-FD algorithm, and red star points represent the system truth value. It can be seen from Fig. 5.23 that the OHSDEF-FD algorithm can tightly wrap the truth value at different sampling points, and compared with the SMIOE-FD algorithm, the parameter feasible set area wrapped in the OHSDEF-FD algorithm is smaller, it means that the proposed algorithm is less conservative and has lower spatial redundancy.

As can be seen from Figs. 5.24 and 5.25, when $k = 501$, θ_2 directional expansion but θ_1 does not directional expansion, indicates that the fault is detected and the fault parameter θ_2 is isolated. Similarly, when the OHSDEF-FD algorithm detects $k = 1001$, the system fails and isolates the fault parameter θ_1. Compared with SMIOE-FD algorithm, OHSDEF-FD algorithm converges faster, the upper and lower

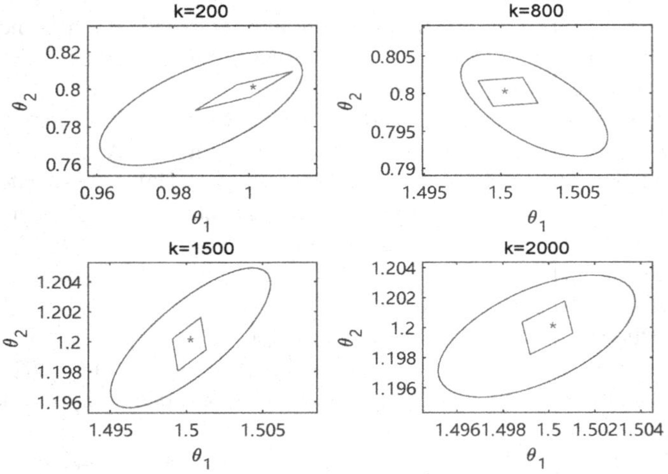

Figure 5.23: Comparison between OHSDEF-FD algorithm and SMIOE-FD algorithm.

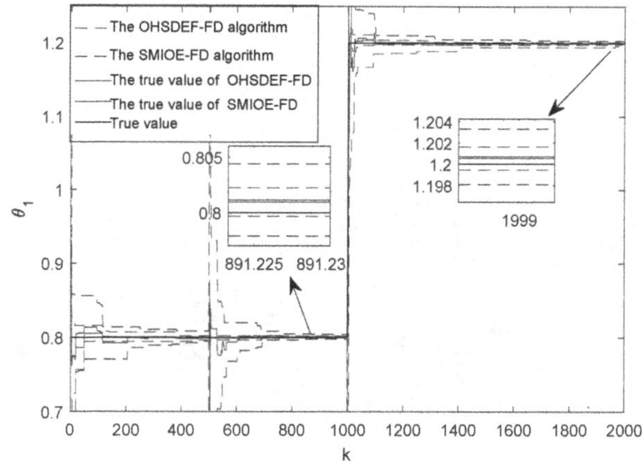

Figure 5.24: Upper and lower bounds of $\boldsymbol{\theta}_1$.

bounds are more densely wrapped with the truth value, and the truth value estimated by OHSDEF-FD is closer to the system truth value, which further reflects that the proposed algorithm has less conservatism and lower spatial redundancy.

Example 2: The pitch subsystem is an important part of the control of the blade and pitch angle in the wind turbine system. In order to further test the effectiveness and feasibility of the fault diagnosis algorithm proposed in this section, this section uses

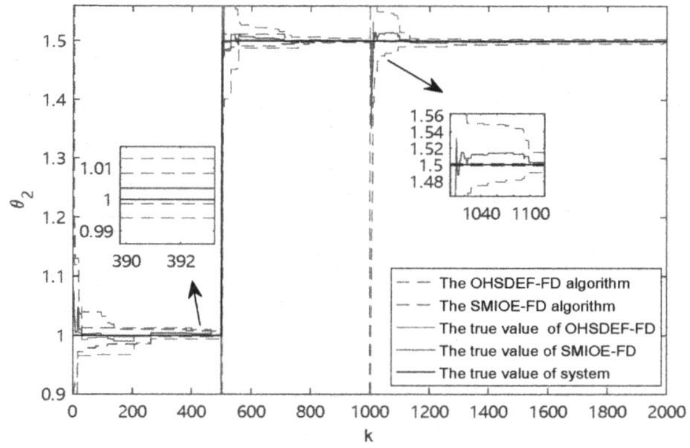

Figure 5.25: Upper and lower bounds of $\boldsymbol{\theta}_2$.

Figure 5.26: The physical map of pitch subsystem.

the OHSDEF-FD algorithm to simulate the pitch subsystem of the wind turbine. The physical map and diagram of the pitch subsystem is shown in Figs. 5.26 and 5.27.

The mathematical model of the pitch subsystem can be expressed as [19, 118]:

$$\begin{bmatrix} \dot{\beta} \\ \dot{\beta}_a \end{bmatrix} = \begin{bmatrix} 0 & 1 \\ -\omega_n^2 & -2\zeta\omega_n \end{bmatrix} \begin{bmatrix} \beta \\ \beta_a \end{bmatrix} + \begin{bmatrix} 0 \\ \omega_n^2 \end{bmatrix} \beta_r, \tag{5.160}$$

where β and β_a are the pitch angle and angular velocity, respectively, β_r is the literature value of the pitch and ζ and ω_n are the damping coefficient and the natural

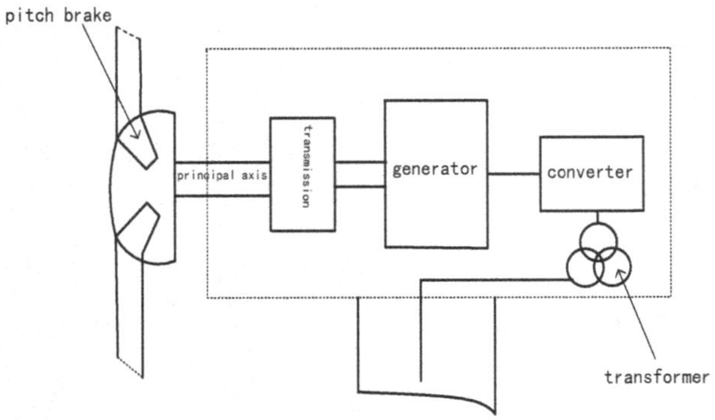

Figure 5.27: The diagram of pitch subsystem.

frequency of the system, respectively. From formula (5.160), we can get:

$$\begin{cases} \dot{\beta} = \beta_a, \\ \dot{\beta}_a = -\omega_n^2\beta - 2\zeta\omega_n\beta_a + \omega_n^2\beta_r, \end{cases}$$

it can be seen that β_a is the first derivative of β, and β_a is eliminated to obtain:

$$\ddot{\beta} = -\omega_n^2\beta - 2\zeta\omega_n\dot{\beta} + \omega_n^2\beta_r,$$

by Laplace transformation:

$$s^2\beta = -\omega_n^2\beta - 2\zeta\omega_n s\beta + \omega_n^2\beta_r,$$

after finishing,

$$(s^2 + 2\zeta\omega_n s + \omega_n^2)\beta = \omega_n^2\beta_r,$$

that is

$$\frac{y}{u} = \frac{\omega_n^2}{s^2 + 2\zeta\omega_n s + \omega_n^2}, \tag{5.161}$$

where $y = \beta, u = \beta_r$. According to literature [19], take $\zeta = 0.6, \omega_n = 11.11$ rad/s, given the sampling time $T_s = 0.01$s, the system (5.161) can be discretized as:

$$A(z)y(k) = B(z)u(k) + e(k), \tag{5.162}$$

where

$$A(z) = 1 - 1.864z^{-1} + 0.8752z^{-2},$$
$$B(z) = 0.0059z^{-1} + 0.0056z^{-2},$$

It can be seen that the number of parameter of the system (5.162) is $n = 4$, and the true value of the parameter is $\boldsymbol{\theta} = [-1.864, 0.8752, 0.0059, 0.0056]^{\mathrm{T}}$. Set input signal $|u(k)| \leqslant 20$, noise amplitude$|e(k)| \leqslant 0.2$, given time length $L = 5$, $\gamma_1^{\max} = \gamma_2^{\max} = \gamma_3^{\max} = \gamma_4^{\max} = 2$. According to the literature [19] add three fault states as shown in below Table 5.2 to the system.

Within the predetermined time range, add fault 1, fault 2 and fault 3 to the system at $k = 501$, $k = 1001$ and $k = 1501$ respectively. In the process of parameter change, the OHSDEF-FD algorithm mentioned in this section is used to calculate the upper

Table 5.2: Operation states and parameters of the system

System state	Cause of failure	θ_1	θ_2	θ_3	θ_4
No fault	No fault	−1.864	0.8752	0.0059	0.0056
Fault 1	Hydraulic leakage	−1.520	0.8752	0.0059	0.0056
Fault 2	Pump wear	−1.520	1.020	0.4038	0.0056
Fault 3	Increased air content	−1.520	1.020	0.2038	0.3028

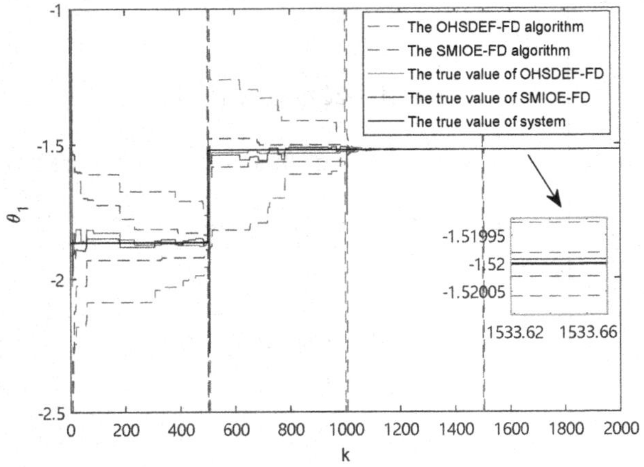

Figure 5.28: Upper and lower bounds of θ_1.

and lower bounds of each parameter, and compared with the SMIOE-FD algorithm in literature [79]. The results are shown in Figs. 5.28–5.31.

As can be seen from Figs. 5.28–5.31, at $k = 501$, θ_1 directional expansion but θ_2, θ_3, θ_4 does not directional expansion indicates that the fault is detected and the fault parameter θ_1 is isolated. Similarly, at $k = 1001$, the fault is detected and the

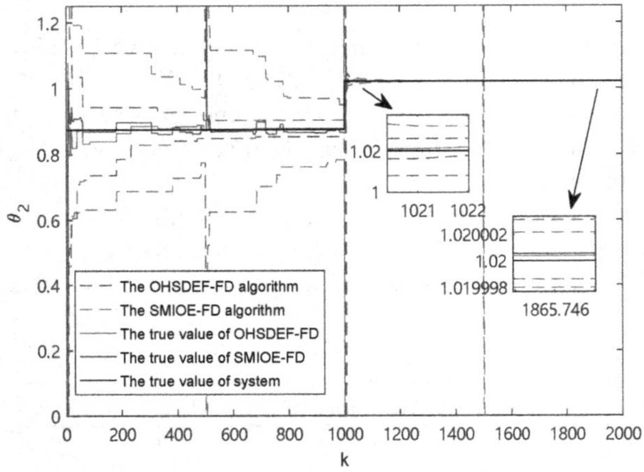

Figure 5.29: Upper and lower bounds of θ_2.

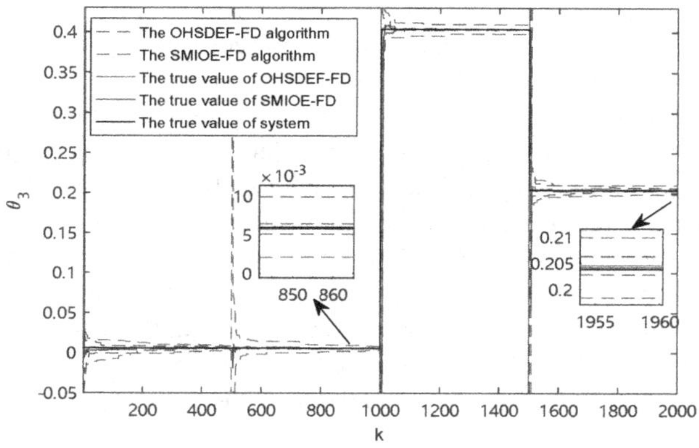

Figure 5.30: Upper and lower bounds of θ_3.

fault parameters θ_2, θ_3 are isolated. At $k = 1501$, the fault is detected and the fault parameters θ_3, θ_4 are isolated. It shows that this algorithm can not only quickly detect faults, but also quickly isolate fault parameters. At the same time, it can be seen from Figs. 5.28–5.31 that compared with SMIOE-FD algorithm, OHSDEF-FD algorithm only directionally expands fault parameters without affecting the filter estimation of other parameters, and the filter estimation curve of OHSDEF-FD algorithm converges faster regardless of system fault, The range of the upper and lower bounds of the given parameters is smaller, which shows that the OHSDEF-FD algorithm proposed in this section has lower conservatism and spatial redundancy.

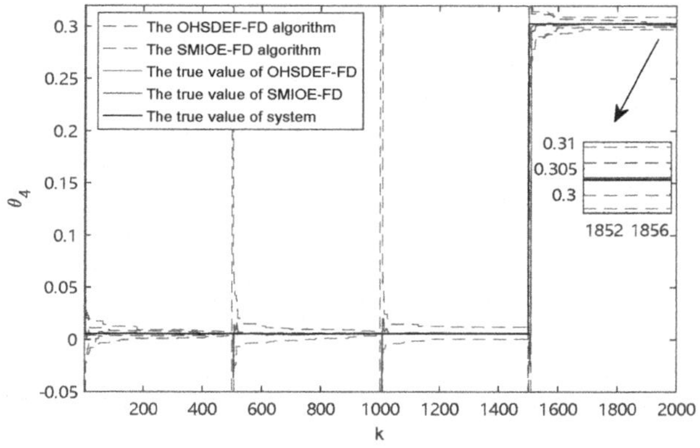

Figure 5.31: Upper and lower bounds of θ_4.

5.3.5 Conclusion

Aiming at the problem of fault diagnosis of linear recursive systems with unknown but bounded noise interference, a filter fault diagnosis method based on directional expansion of orthometric hyperparallel space is proposed in this section. First, the feasible set of parameter is wrapped in hyperparallel space to complete the parameter identification based on filter in the fault-free state, at the same time, the orthometric hyperparallel space is set to ensure the monotonic convergence of the boundary value of the parameter. Whether the intersection of the current hyperparallel space and the measuring strip is empty is used to detect whether the system has a fault. After the fault is detected, the fault parameters are isolated by using the state of the hyperparallel space test set, and then the directional expansion strategy is designed to obtain the initialization orthometric hyperparallel space of the fault parameters.

Chapter 6

Fault diagnosis method based on zonotopic Kalman filtering

6.1 Zonotopic Kalman filtering-based fault diagnosis algorithm for linear system with state constraints

6.1.1 Problem formulation and preliminaries

6.1.1.1 Problem formulation

Consider the following discrete-time linear system

$$\begin{cases} x_{k+1} = Ax_k + Bu_k + D_1 w_k, \\ y_k = Cx_k + f_k + D_2 v_k, \end{cases} \tag{6.1}$$

where $x_k \in \mathbb{R}^n$,$u_k \in \mathbb{R}^r$,$y_k \in \mathbb{R}^p$ and $f_k \in \mathbb{R}^p$ are the state, input, output and sensor fault vectors, respectively. A, B, D_1, C, D_2 are the parameter matrices with appropriate dimensions. $w_k \in \mathbb{R}^w$ and $v_k \in \mathbb{R}^v$ are the unknown but bounded disturbances and measurement noises.

Without loss of generality, given $\tilde{x}_0$, $\tilde{w}$ and $\tilde{v}$, , we assume that the initial state x_0, the UBB disturbances w_k and v_k as follows: $|x_0 - p_0| \leq \tilde{x}_0, |w_k| \leq \tilde{w}, |v_k| \leq \tilde{v}$. They can also be wrapped by the zonotopes: $x_0 \in \mathcal{X}_0 = \langle p_0, G_0 \rangle, w_k \in \mathcal{W} = \langle 0, W \rangle, v_k \in \mathcal{V} = \langle 0, V \rangle$, where $G_0 = \text{diag}(\tilde{x}_0), W = \text{diag}(\tilde{w})$ and $V = \text{diag}(\tilde{v})$, respectively.

In this section, the state of the system satisfies the constraint in the fault-free situation.

$$\|h(x_k)\|_2 \leq l_k, \; h(x_k) = D_k - H_k x_k, \tag{6.2}$$

DOI: 10.1201/b23146-6

Therefore, considering the constraint of the system when a sensor fault occurs, the system in Eq. (6.1) can be extended as:

$$\begin{cases} x_{k+1} = Ax_k + Bu_k + D_1 w_k, \\ \bar{y}_k = \bar{C}x_k + \bar{D}_2\bar{v}_k + \bar{f}_k, \end{cases} \tag{6.3}$$

where $\bar{y}_k = \begin{bmatrix} y_k \\ \gamma_k \end{bmatrix}$, $\bar{C} = \begin{bmatrix} C \\ H_k \end{bmatrix}$, $\bar{D}_2 = \begin{bmatrix} D_2 & 0_{p\times\gamma} \\ 0_{p\times v} & I_{p\times\gamma} \end{bmatrix}$, $\bar{v}_k = \begin{bmatrix} v_k \\ l_k \end{bmatrix}$, $\bar{f}_k = \begin{bmatrix} f_k \\ 0_\gamma \end{bmatrix}$, $|\bar{v}_k| \leq \bar{\bar{v}}, \bar{V} = \text{diag}(\bar{\bar{v}}), \gamma_k = D_k + l_k, l_k \in \langle 0, l_k \cdot I_{p\times\gamma}\rangle, \gamma_k \in \mathbb{R}^\gamma$.

In this section, we aim to design a zonotopic Kalman filter by integrating Kalman filter structure with zonotope techniques. To achieve interval estimation of sensor fault, the state vector is augmented as follows [124]:

$$\bar{x}_k = \begin{bmatrix} x_k \\ \bar{f}_k \end{bmatrix}, \tag{6.4}$$

then the system in Eq. (6.3) can be rewritten as

$$\begin{cases} E\bar{x}_{k+1} = \bar{A}\bar{x}_k + \bar{B}u_k + \bar{D}_1 w_k, \\ \bar{y}_k = C_1\bar{x}_k + \bar{D}_2\bar{v}_k, \end{cases} \tag{6.5}$$

where

$$E = \begin{bmatrix} I_n & 0_{n\times(p+\gamma)} \\ 0_{(p+\gamma)\times n} & 0_{p+\gamma} \end{bmatrix}, \bar{A} = \begin{bmatrix} A & 0_{n\times(p+\gamma)} \\ 0_{(p+\gamma)\times n} & 0_{p+\gamma} \end{bmatrix},$$

$$\bar{B} = \begin{bmatrix} B \\ 0_{(p+\gamma)\times r} \end{bmatrix}, C_1 = \begin{bmatrix} \bar{C} & I \end{bmatrix}, \bar{D}_1 = \begin{bmatrix} D_1 \\ 0_{(p+\gamma)\times w} \end{bmatrix}.$$

6.1.1.2 Preliminaries

Some definitions and properties are adopted in this section.

Definition 6.1 An s-order zonotope $\mathcal{Z} \subset \mathbb{R}^n (n \leq s)$ is an affine transformation of a hypercube $\mathbf{B}^s = [-1, +1]^s$ as follows [82]:

$$\mathcal{Z} = \boldsymbol{p} + \sum_{i=1}^{s} \alpha_i \boldsymbol{g}_i \triangleq \langle \boldsymbol{p}, G\rangle = \boldsymbol{p} \oplus G\mathbf{B}^s, \tag{6.6}$$

where $-1 \leq \alpha_i \leq 1$, $\boldsymbol{p} \in \mathbf{B}^s$ is the center of $\mathcal{Z}$, $G = \{\boldsymbol{g}_1, \boldsymbol{g}_2, \cdots, \boldsymbol{g}_s\} \in \mathbb{R}^{n\times s}$ is named the generator matrix.

The zonotopes have the following properties:

$$\langle \boldsymbol{p}_1, G_1\rangle \oplus \langle \boldsymbol{p}_2, G_2\rangle = \langle \boldsymbol{p}_1 + \boldsymbol{p}_2, [G_1 \;\; G_2]\rangle, \tag{6.7}$$

$$M \otimes \langle \boldsymbol{p}, G\rangle = \langle M\boldsymbol{p}, MG\rangle, \tag{6.8}$$

where M is a matrix with appropriate dimension.

Lemma 6.1
A zonotope can be contained by an interval hull box [101]

$$\mathcal{Z} \subset \square\mathcal{Z} = \boldsymbol{p} \oplus rs(G)\mathbf{B}^n, \tag{6.9}$$

where $rs(G) = \mathrm{diag}([G_1, \ldots, G_S])$ *with*

$$G_i = \sum_{j=1}^{s} |G(i,j)|, \quad i = 1, \ldots, n.$$

Lemma 6.2
A reduction operator can be used to bound a zonotope by a lower dimensional one [27]:

$$\mathcal{Z} = \langle \boldsymbol{p}, G \rangle \subseteq \langle \boldsymbol{p}, \downarrow_{re}(G) \rangle,$$

where $\downarrow_{re}(G) \in \mathbb{R}^{n \times q}$ *denotes the complexity reduction operator with* q $(n \leq q \leq s)$ *is the maximum number of columns of the generated matrix.*

Define $\tilde{G}$ as the matrix obtained by recording the column of matrix G in decreasing Euclidean norm. The last $s-q+n$ smallest column of $\tilde{G}$ can be replaced by a diagonal matrix $rs(G) \in \mathbb{R}^{n \times n}$, since the zonotope generated by these columns are enclosed in a box.

Lemma 6.3
Consider a zonotope $\mathcal{Z} = \langle \boldsymbol{p}, G \rangle$ *and a strip* $\mathcal{S} = \{x \in \mathbb{R}^n : |Kx - c| \leq \delta\}$, *then a group of zonotope containing the intersection* $\mathcal{Z} \cap \mathcal{S}$ *between* $\mathcal{Z}$ *and* $\mathcal{S}$ *can be obtained by* $\mathcal{Z}(\Lambda) = \langle \boldsymbol{p}(\Lambda), G(\Lambda) \rangle$ *with*

$$\boldsymbol{p}(\Lambda) = \boldsymbol{p} + \Lambda(c - K\boldsymbol{p}), \quad G(\Lambda) = [(I_n - \Lambda K)G \quad \Lambda\Delta],$$

where G *and* K *are known matrices,* $\boldsymbol{p}, c, \delta$ *are known vectors,* $\Lambda = \mathrm{diag}(\delta)$ *and* Δ *is a matrix to be specified [20].*

6.1.2 Main results

The ZKF algorithm is divided into two steps: the prediction step and the update step. The prediction step uses the previous state zonotope, the zonotopes of disturbances and measurement noises, the input and output of the system to produce the prediction zonotope; the update step combines the prediction zonotope with a strip and the state zonotope at current time instant.

6.1.2.1 Prediction step

Define parameter matrices T and N to decouple the state $\bar{x}_{k+1}$ to the specified fault. T and N satisfy $TE + NC_1 = I_{n+p+\gamma}$. Denote $S \in \mathbb{R}^{(n+p+\gamma)\times(n+2p+2\gamma)}$ an arbitrary matrix.

Theorem 6.1
Consider the augmented system in Eq. (6.5), if the system's state $\bar{x}_k \in \bar{\mathcal{X}}_k = \langle \bar{\boldsymbol{p}}_k, \bar{G}_k \rangle$ is obtained, the state prediction at time $k+1$, that is, $\bar{x}_{k+1}$ can be bounded in a prediction zonotope

$$\hat{\bar{\mathcal{X}}}_{k+1} = \left\langle \hat{\bar{\boldsymbol{p}}}_{k+1}, \hat{\bar{G}}_{k+1} \right\rangle,$$

where

$$\hat{\bar{\boldsymbol{p}}}_{k+1} = T\bar{A}\bar{\boldsymbol{p}}_k + T\bar{B}u_k + N\bar{y}_{k+1}, \tag{6.10}$$

$$\hat{\bar{G}}_{k+1} = \begin{bmatrix} T\bar{A} \downarrow_{re} \bar{G}_k & T\bar{D}_1 W & -N\bar{D}_2\bar{V} \end{bmatrix}, \tag{6.11}$$

$$T = \Theta^{\dagger}\alpha_1 + S\Psi\alpha_1, N = \Theta^{\dagger}\alpha_2 + S\Psi\alpha_2, \tag{6.12}$$

$$\Psi = I_{n+2p+2\gamma} - \Theta\Theta^{\dagger}, \Theta = \begin{bmatrix} E \\ C_1 \end{bmatrix}, \tag{6.13}$$

$$\alpha_1 = \begin{bmatrix} I_{n+p+\gamma} \\ 0_{(p+\gamma)\times(n+p+\gamma)} \end{bmatrix}, \alpha_2 = \begin{bmatrix} 0_{(n+p+\gamma)\times(p+\gamma)} \\ I_{p+\gamma} \end{bmatrix}. \tag{6.14}$$

Proof 6.1 The augmented system in Eq. (6.5) can be written as

$$\begin{cases} TE\bar{x}_{k+1} = T\bar{A}\bar{x}_k + T\bar{B}u_k + T\bar{D}_1 w_k, \\ NC_1\bar{x}_{k+1} = N\bar{y}_{k+1} - N\bar{D}_2\bar{v}_{k+1}. \end{cases}$$

According to $TE + NC_1 = I_{n+p+\gamma}$ and zonotope properties, we have

$$\begin{aligned} \bar{x}_{k+1} &= (TE + NC_1)\bar{x}_{k+1} \\ &= TE\bar{x}_{k+1} + NC_1\bar{x}_{k+1} \\ &= T\bar{A}\bar{x}_k + T\bar{B}u_k + N\bar{y}_{k+1} + T\bar{D}_1 w_k - N\bar{D}_2\bar{v}_{k+1} \\ &\in [T\bar{A} \otimes \langle \bar{\boldsymbol{p}}_k, \bar{G}_k \rangle] \oplus T\bar{B}u_k \oplus N\bar{y}_{k+1} \\ &\oplus [T\bar{D}_1 \otimes \langle 0, W \rangle] \oplus [-N\bar{D}_2 \otimes \langle 0, \bar{V} \rangle] \\ &= \langle T\bar{A}\bar{\boldsymbol{p}}_k + T\bar{B}u_k + N\bar{y}_{k+1}, \\ &\begin{bmatrix} T\bar{A}\bar{G}_k & T\bar{D}_1 W & -N\bar{D}_2\bar{V} \end{bmatrix} \rangle = \langle \hat{\bar{\boldsymbol{p}}}_k, \hat{\bar{G}}_k \rangle. \end{aligned}$$

6.1.2.2 Update step

In the update step, we combine a strip $\mathcal{S}$ with the perdition zonotope. Then, the state $\bar{x}_{k+1}$ is bounded in the strip $\mathcal{S}$ as

$$\mathcal{S} = \left\{ \bar{x}_{k+1} \in \mathbb{R}^{n+p+\gamma} : |C_1\bar{x}_{k+1} - \bar{y}_{k+1}| \leq \bar{D}_2\bar{\bar{v}} \right\}.$$

Since $\bar{x}_{k+1} \in \hat{\bar{\mathcal{X}}}_{k+1}$ and $\bar{x}_{k+1} \in \mathcal{S}$, $\bar{x}_{k+1}$ is determined by the intersection of $\hat{\bar{\mathcal{X}}}_{k+1}$ and $\mathcal{S}$. According to Lemma 6.3, the intersection $\hat{\bar{\mathcal{X}}}_{k+1} \cap \mathcal{S}$ can be contained in a zonotope $\bar{\mathcal{X}}_{k+1}(L_{k+1}) = \langle \bar{\boldsymbol{p}}_{k+1}(L_{k+1}), \bar{G}_{k+1}(L_{k+1}) \rangle$, which is also called the update zonotopes of $\bar{x}_{k+1}$.

The updated zonotope can be calculated as follows:

$$\bar{p}_{k+1}(L_{k+1}) = \hat{\bar{p}}_{k+1} + L_{k+1}\left(\bar{y}_{k+1} - C_1\hat{\bar{p}}_{k+1}\right), \tag{6.15}$$

$$\bar{G}_{k+1}(L_{k+1}) = \left[(I_{n+s+\gamma} - L_{k+1}C_1)\hat{\bar{G}}_{k+1} \quad L_{k+1}D_v\right], \tag{6.16}$$

where $D_v = \text{diag}(\bar{D}_2\bar{\tilde{v}}), L_{k+1}$ is the filter gain matrix for ZKF.

In order to obtain a more accurate state range estimation, the use of filter gain matrix L_{k+1} should minimize the volume of $\bar{\mathcal{X}}_{k+1}(L_{k+1})$. We take the Frobenius norm of the generator matrix $\bar{G}_{k+1}(L_{k+1})$ as optimality criterion to design the optimal filter gain matrix L_{k+1}. To solve the optimal filter gain matrix L_{k+1}, the following theorem is given.

Theorem 6.2
Given $\bar{x}_{k+1} \in \bar{\mathcal{X}}_{k+1}(L_{k+1}) = \langle \bar{p}_{k+1}(L_{k+1}), \bar{G}_{k+1}(L_{k+1})\rangle$, the optimal filter gain matrix L_{k+1} can be calculated by

$$L_{k+1} = \hat{\bar{G}}_{k+1}\hat{\bar{G}}_{k+1}^{\mathrm{T}}C_1^{\mathrm{T}}\left(C_1\hat{\bar{G}}_{k+1}\hat{\bar{G}}_{k+1}^{\mathrm{T}}C_1^{\mathrm{T}} + D_vD_v^{\mathrm{T}}\right)^{-1}. \tag{6.17}$$

Proof 6.2 To find $L_{k+1} = L_{k+1}^*$ that minimizes the Frobenius norm of $\bar{G}_{k+1}(L_{k+1})$, we decompose $\bar{G}_{k+1}(L_{k+1})$ as

$$\bar{G}_{k+1}(L_{k+1}) = M + L_{k+1}a^{\mathrm{T}}, \tag{6.18}$$

where

$$M = \left[\hat{\bar{G}}_{k+1} \quad 0_{(n+p+\gamma)\times(v+\gamma)}\right], \; a^{\mathrm{T}} = \left[-C_1\hat{\bar{G}}_{k+1} \quad D_v\right].$$

Then,

$$\begin{aligned}
&\|\bar{G}_{k+1}(L_{k+1})\|_F^2 = \|M + L_{k+1}a^{\mathrm{T}}\|_F^2 \\
&= tr\left((M^{\mathrm{T}} + aL_{k+1}^{\mathrm{T}})(M + L_{k+1}a^{\mathrm{T}})\right) \\
&= tr(M^{\mathrm{T}}M) + tr(aL_{k+1}^{\mathrm{T}}M) + tr(M^{\mathrm{T}}L_{k+1}a^{\mathrm{T}}) + tr(aL_{k+1}^{\mathrm{T}}L_{k+1}a^{\mathrm{T}}) \\
&= tr(M^{\mathrm{T}}M) + 2\times tr(M^{\mathrm{T}}L_{k+1}a^{\mathrm{T}}) + tr(aL_{k+1}^{\mathrm{T}}L_{k+1}a^{\mathrm{T}}),
\end{aligned}$$

where $tr(\cdot)$ represents the trace of the matrix. For matrix X and Y, $tr(XY) = \text{tr}(Y^{\mathrm{T}}X^{\mathrm{T}})$. Then, the value of L_{k+1} that minimizes $\bar{G}_{k+1}$ can be obtained by

$$\begin{aligned}
L_{k+1}^* &= -Ma(a^{\mathrm{T}}a)^{-1} \\
&= \hat{\bar{G}}_{k+1}\hat{\bar{G}}_{k+1}^{\mathrm{T}}C_1^{\mathrm{T}} \times \left(C_1\hat{\bar{G}}_{k+1}\hat{\bar{G}}_{k+1}^{\mathrm{T}}C_1^{\mathrm{T}} + D_vD_v^{\mathrm{T}}\right)^{-1}.
\end{aligned}$$

For the augmented system in Eq. (6.5), the filtering process of the proposed ZKF method can be summarized as follows:

$$\begin{aligned}
&\bar{x}_k \in \bar{\mathcal{X}}_k = \langle \bar{\boldsymbol{p}}_k, \bar{G}_k \rangle, \hat{\bar{\mathcal{X}}}_{k+1} = \left\langle \hat{\bar{\boldsymbol{p}}}_{k+1}, \hat{\bar{G}}_{k+1} \right\rangle, \\
&\hat{\bar{\boldsymbol{p}}}_{k+1} = T\bar{A}\bar{\boldsymbol{p}}_k + T\bar{B}u_k + N\bar{y}_{k+1}, \\
&\hat{\bar{G}}_{k+1} = \left[T\bar{A} \downarrow_{re} (\bar{G}_k) \quad T\bar{D}_1 W \quad -N\bar{D}_2\bar{V}\right], \\
&\bar{\boldsymbol{p}}_{k+1} = \hat{\bar{\boldsymbol{p}}}_{k+1} + L_{k+1}\left(\bar{y}_{k+1} - C_1\hat{\bar{\boldsymbol{p}}}_{k+1}\right), \\
&\bar{G}_{k+1} = \left[(I_{n+s+\gamma} - L_{k+1}C_1)\hat{\bar{G}}_{k+1} \quad L_{k+1}D_v\right], \\
&L_{k+1} = \hat{\bar{G}}_{k+1}\hat{\bar{G}}_{k+1}^{\mathrm{T}}C_1^{\mathrm{T}}\left(C_1\hat{\bar{G}}_{k+1}\hat{\bar{G}}_{k+1}^{\mathrm{T}}C_1^{\mathrm{T}} + D_vD_v^{\mathrm{T}}\right)^{-1}.
\end{aligned}$$

6.1.2.3 Fault diagnosis

The zonotope $\mathcal{Z} = \langle \boldsymbol{p}, G \rangle \subset \mathbb{R}^n$ can be bounded by an interval hull $\mathrm{Box}(\mathcal{Z}) = [z^-, z^+]$. where z^- and z^+ denote the lower and upper bounds of z. They can be calculated as follows :

$$\begin{cases} z^-(i) = \boldsymbol{p}(i) - \sum_{j=1}^{s} |G(i,j)|, & i = 1,\dots,n, \\ z^+(i) = \boldsymbol{p}(i) - \sum_{j=1}^{s} |G(i,j)|, & i = 1,\dots,n. \end{cases}$$

Based on the proposed ZKF, we can obtain the interval estimation of $\bar{x}_k$. $\bar{x}_k$ can be bounded by the interval hull $\mathrm{Box}(X_k) = [\bar{x}_k^-, \bar{x}_k^+]$.

$$\begin{cases} \bar{x}_k^+(i) = \bar{\boldsymbol{p}}_k(i) + \sum\limits_{j=1}^{q} \left|\bar{G}_k(i,j)\right|, i = 1,\dots,n+s+\gamma, \\ \bar{x}_k^-(i) = \bar{\boldsymbol{p}}_k(i) - \sum\limits_{j=1}^{q} \left|\bar{G}_k(i,j)\right|, i = 1,\dots,n+s+\gamma, \end{cases}$$

where $\bar{p}_k(i), \bar{x}_k^+(i)$ and $\bar{x}_k^-(i)$ denote the ith components of $\bar{p}_k, \bar{x}_k^+$ and $\bar{x}_k^-$. $\bar{G}_k(i,j)$ denotes the element of $\bar{G}_k$ in the ith row and the jth column, l is the column of number of $\bar{G}_k$.

Further, the interval estimation of f_k can be calculated by

$$\begin{aligned} f_k^+ &= \left[0_{p\times n} \quad I_p \quad 0_{p\times\gamma}\right]\bar{x}_k^+, \\ f_k^- &= \left[0_{p\times n} \quad I_p \quad 0_{p\times\gamma}\right]\bar{x}_k^-. \end{aligned}$$

Based on the minimum upper and maximum lower bounds of the fault estimation, it can be judged whether the system has a fault and then find the fault type.

6.1.3 Simulation

To verify the effectiveness of the proposed method, we use the electrothermal-coupled model of Li-ion battery for verification. The electrothermal-coupled model of Li-ion battery is formed by coupling the equivalent circuit model of Li-ion battery and the heat generation model [99, 108].

As shown in Fig. 6.1, R_1 and C_{p1} are electrochemical polarization internal resistance and capacitance respectively, R_2 and C_{p2} are concentration polarization internal

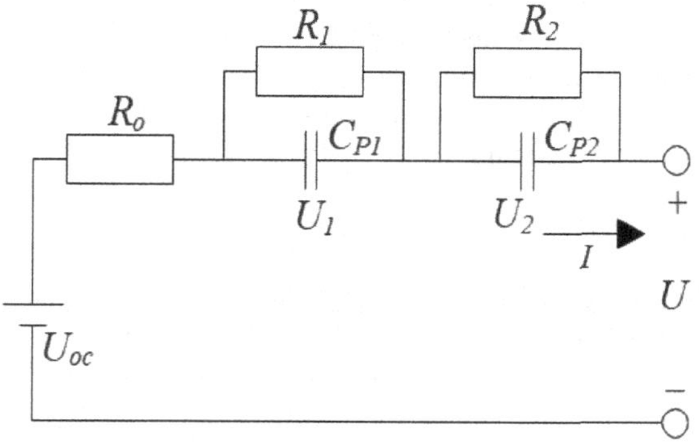

Figure 6.1: Schematic diagram of the second-order RC circuit model.

resistance and capacitance. U_1 and U_2 are the voltage drop of the two RC circuits, U_{oc} represents the open circuit voltage of the battery, U is the terminal voltage of the battery, I is the battery current, R_0 is the ohmic internal resistance. The mathematical expression of the model is as follows:

$$U = U_{oc} - R_o I - U_1 - U_2,$$

$$\begin{cases} \dot{U}_1 = -\frac{1}{R_1 C_{p1}} U_1 + \frac{1}{C_{p1}} I, \\ \dot{U}_2 = -\frac{1}{R_2 C_{p2}} U_2 + \frac{1}{C_{p2}} I, \end{cases}$$

where $\dot{U}_1$ and $\dot{U}_2$ represent changing rates of voltage U_1 and U_2, respectively. The parameters of equivalent circuit model in Fig. 6.1 is listed in Table 6.1.

When the battery is working at high power, it will release a lot of heat, causing the internal temperature of the battery to rise, making the core temperature higher than the surface temperature. To better maintain the battery performance, a heat generation model needs to be established to estimate the battery core temperature to ensure the normal operation of battery. According to the energy conservation equation inside and outside the battery, the Fourier's law of heat and the Newton's law of heat dissipation, the heat generation model is defined as

$$\begin{cases} C_c \dot{T}_c = Q_{gen} + \frac{T_s - T_c}{R_c}, \\ C_s \dot{T}_s = \frac{T_e - T_s}{R_u} - \frac{T_s - T_c}{R_c}, \end{cases}$$

where C_s and C_c represent the heat capacity of internal battery material and the heat capacity at battery surface, respectively. R_c represents the thermal resistance between

Table 6.1: Parameters of equivalent circuit model

R_0/Ω	R_1/Ω	R_2/Ω	C_{p1}/F	C_{p2}/F	I/A
0.0501	0.0298	0.03819	1789.7	5.26	5

battery core and surface, while R_u is the convective resistance between battery surface and cooling air. Q_{gen} represents the heat generation rate at the core of the battery. T_e represents the ambient temperature for convection. $\dot{T}_c$ and $\dot{T}_s$ represent changing rates of the core and surface temperature (T_c and T_s), respectively.

According to Bernardi formula,

$$Q_{gen} = I(U_{oc} - U) = I(R_o I + U_1 + U_2),$$

the heat generation rate Q_{gen} and the ambient temperature T_e can be regarded as the inputs, while the surface temperature measurement $\bar{T}_s$ and the core temperature measurement $\bar{T}_c$ is regarded as the output of the electrothermal-coupled model.

After the discretization, the state-space form of the electrothermal-coupled model can be given as

$$\begin{cases} x_{k+1} = Ax_k + Bu_k + w_k, \\ y_k = Cx_k + v_k, \end{cases}$$

where $x_k = \left[T_{c,k}, T_{s,k}\right]^{\mathrm{T}}$, $u_k = \left[Q_{\mathrm{gen},k}, T_{e,k}\right]^{\mathrm{T}}$, $y_k = \left[\bar{T}_{c,k}, \bar{T}_{s,k}\right]^{\mathrm{T}}$, w_k and v_k are the unknown but bounded disturbance and measurement noises of the system. Δt is the sampling interval, and

$$A = \begin{bmatrix} 1 - \frac{\Delta t}{R_c C_c} & \frac{\Delta t}{R_c C_c} \\ \frac{\Delta t}{R_c C_s} & 1 - \frac{\Delta t}{R_c C_s} - \frac{\Delta t}{R_u C_s} \end{bmatrix},$$

$$B = \begin{bmatrix} \frac{\Delta t}{C_c} & 0 \\ 0 & \frac{\Delta t}{R_u C_s} \end{bmatrix}, \quad C = \begin{bmatrix} 1 & 0 \\ 0 & 1 \end{bmatrix}.$$

The system parameters of the battery under normal conditions in the simulation is shown in Table 6.2.

6.1.3.1 Fault-free system simulation

When the Li-ion battery is working under normal conditions, the core temperature and surface temperature meet specific differential conditions and the core temperature fluctuates in a small range, that is, it generally restricted to remain within a certain range when the system is stable. In this case, when the system works normally at $T_e = 25°\mathrm{C}$, the difference between the core temperature and the surface temperature is $5.82°\mathrm{C} \pm 0.05°\mathrm{C}$ and the core temperature $T_c \in [35.8°\mathrm{C}, 35.9°\mathrm{C}]$. Therefore, we set

$$l = \begin{bmatrix} 0.05 \\ 0.05 \end{bmatrix}, \quad D_k = \begin{bmatrix} -5.82 \\ -35.85 \end{bmatrix}, \quad H_k = \begin{bmatrix} -1 & 1 \\ -1 & 0 \end{bmatrix}.$$

Table 6.2: Parameters of heat generation model

C_c/J K^{-1}	C_s/J K^{-1}	R_u/K W^{-1}	R_c/K W^{-1}
63.5	4.5	1.718	1.98

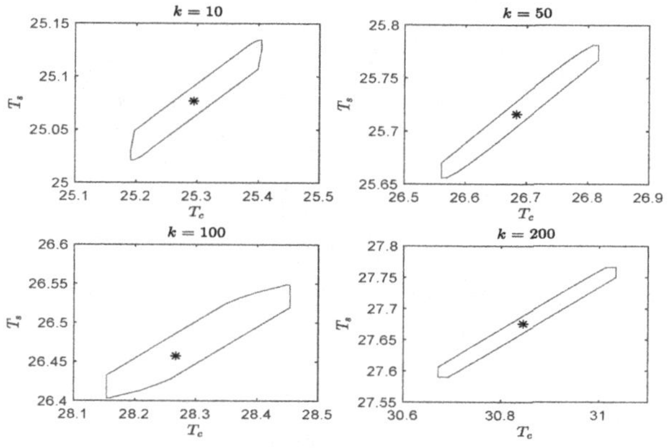

Figure 6.2: Comparison of actual states and estimated states.

In the simulation, we set the initial state vector $x_0 = [25,25]^{\mathrm{T}}$, $p_0 = [25,25]^{\mathrm{T}}$ and $G_0 = 0.1I_4$. At the same time, the reduction order of the matrix $\downarrow_{re} \bar{G}_k$ is regarded as $l = 20$, to limit the column number of zonotopes generator matrix. The disturbances and noises are bounded by $|w_k|_\infty \leq 0.001$ and $|v_k|_\infty \leq 0.1$. The simulation sampling period Δt is set to 1s and the system parameter matrices are

$$A = \begin{bmatrix} 0.992 & 0.0080 \\ 0.112 & 0.758 \end{bmatrix}, \ B = \begin{bmatrix} 0.0157 & 0 \\ 0 & 0.1293 \end{bmatrix},$$

$$\bar{C} = \begin{bmatrix} 1 & 0 \\ 0 & 1 \\ -1 & 1 \\ -1 & 0 \end{bmatrix}, \ \bar{D}_2 = \begin{bmatrix} 1 & 0 & 0 & 0 \\ 0 & 1 & 0 & 0 \\ 0 & 0 & 1 & 0 \\ 0 & 0 & 0 & 1 \end{bmatrix}.$$

Fig. 6.2 shows the comparison between the estimated states and the actual states of the system at different sampling times. The zonotope is the feasible state set, and the black dot is the actual state of the system. As shown in Fig. 6.2, the proposed ZKF estimates the states with state constraints well. And it can be seen from the scale of the coordinate axis that the volume of the zonotope is getting smaller, which means the accuracy of the state estimation is higher with sampling increases. From Fig. 6.3, the actual state in each sampling time is in the feasible set, which show the proposed method has a good robust performance.

6.1.3.2 Faulty system simulation

According to the state space expression of the battery core and surface temperature, the fault model can be written as

$$\begin{cases} x_{k+1} = Ax_k + Bu_k + w_k, \\ y_k = Cx_k + v_k + f_k, \end{cases}$$

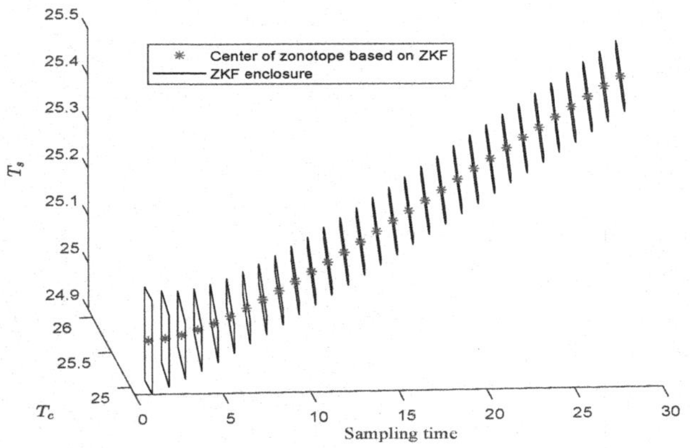

Figure 6.3: Recursive evolution process of zonotopic set.

where $f_k = \begin{bmatrix} f_{1,k} & f_{2,k} \end{bmatrix}^{\mathrm{T}}$ is a sensor fault vector. When the core temperature sensor fails, $f_{1,k} \neq 0$. When the surface temperature sensor fails, $f_{2,k} \neq 0$.

In this simulation, the sensor fault has the following form:

$$f_k = \begin{bmatrix} f_{1,k} & f_{2,k} \end{bmatrix}^{\mathrm{T}} = \begin{cases} \begin{bmatrix} 0 & 0 \end{bmatrix}^{\mathrm{T}}, & 0 \leq k < 400, \\ \begin{bmatrix} 5 & 0 \end{bmatrix}^{\mathrm{T}}, & 400 \leq k < 700, \\ \begin{bmatrix} 5 & 10 \end{bmatrix}^{\mathrm{T}}, & k \geq 700. \end{cases}$$

The parameter matrices of the augmented systems are

$$\bar{A} = \begin{bmatrix} 0.992 & 0.0080 & 0 & 0 & 0 & 0 \\ 0.112 & 0.758 & 0 & 0 & 0 & 0 \\ 0 & 0 & 0 & 0 & 0 & 0 \\ 0 & 0 & 0 & 0 & 0 & 0 \\ 0 & 0 & 0 & 0 & 0 & 0 \\ 0 & 0 & 0 & 0 & 0 & 0 \end{bmatrix}, \bar{D}_1 = \begin{bmatrix} 1 & 0 \\ 0 & 1 \\ 0 & 0 \\ 0 & 0 \\ 0 & 0 \\ 0 & 0 \end{bmatrix},$$

$$\bar{B} = \begin{bmatrix} 0.0157 & 0 \\ 0 & 0.1293 \\ 0 & 0 \\ 0 & 0 \\ 0 & 0 \\ 0 & 0 \end{bmatrix}, C_1 = \begin{bmatrix} 1 & 0 & 1 & 0 & 0 & 0 \\ 0 & 1 & 0 & 1 & 0 & 0 \\ -1 & 1 & 0 & 0 & 1 & 0 \\ -1 & 0 & 0 & 0 & 0 & 1 \end{bmatrix},$$

$$E = \begin{bmatrix} 1 & 0 & 0 & 0 & 0 & 0 \\ 0 & 1 & 0 & 0 & 0 & 0 \\ 0 & 0 & 0 & 0 & 0 & 0 \\ 0 & 0 & 0 & 0 & 0 & 0 \\ 0 & 0 & 0 & 0 & 0 & 0 \\ 0 & 0 & 0 & 0 & 0 & 0 \end{bmatrix}.$$

By setting

$$S = \begin{bmatrix} 1 & 0 & 0 & 0 & 0 & 0 & 0 & 0 & 0 & 0 \\ 0 & 1 & 0 & 0 & 0 & 0 & 0 & 0 & 0 & 0 \\ 0 & 0 & 1 & 0 & 0 & 0 & 0 & 0 & 0 & 0 \\ 0 & 0 & 0 & 1 & 0 & 0 & 0 & 0 & 0 & 0 \\ 0 & 0 & 0 & 0 & 1 & 0 & 0 & 0 & 0 & 0 \\ 0 & 0 & 0 & 0 & 0 & 1 & 0 & 0 & 0 & 0 \end{bmatrix},$$

we can compute the matrices T and N as follows:

$$T = \begin{bmatrix} 1 & 0 & 0 & 0 & 0 & 0 \\ 0 & 1 & 0 & 0 & 0 & 0 \\ -1 & 0 & 1 & 0 & 0 & 0 \\ 0 & -1 & 0 & 1 & 0 & 0 \\ 1 & -1 & 0 & 0 & 1 & 0 \\ 1 & 0 & 0 & 0 & 0 & 1 \end{bmatrix},$$

$$N = \begin{bmatrix} 0 & 0 & 0 & 0 \\ 0 & 0 & 0 & 0 \\ 1 & 0 & 0 & 0 \\ 0 & 1 & 0 & 0 \\ 0 & 0 & 1 & 0 \\ 0 & 0 & 0 & 1 \end{bmatrix}.$$

Sensor faults and their interval estimation results are shown in Fig. 6.4, the actual faults are depicted by black lines. As shown in Fig. 6.4, we can quickly find the time when the sensor faults occurs, that is, the ZKF can quickly track the actual sensor faults. As far as the system is concerned, when the system fails, the failure can be detected and the type of sensor failure can be judged fastly and accurately. At the same time, according to the upper and lower bounds of the fault estimates, we can obtain the sensor fault estimates within a certain error range. Fig. 6.5 shows the comparison between the fault estimates and the actual fault at different sampling times. The zonotope is the fault estimate set, and the circle is the actual fault of the system. It can be seen from Fig. 6.5 that the actual faults are in the accurate but uncertain fault set. It also shows the accuracy of the given method in fault diagnosis.

6.1.4 Conclusion

This section presents a fault diagnosis method based on zonotopic Kalman filter for linear systems with state constraints. First, the state constraints of the system are extended to the output vector of the system and faults are extended to the state vector to construct an augmented system. The designed filter is used to estimate the upper and lower boundaries of faults to detect and diagnose faults. Finally, the sensor faults of the electrothermal-coupled model for the Li-ion battery is taken as an example for simulation verification, which illustrates the effectiveness and accuracy of the method proposed in this section. The fault diagnosis method proposed in this section

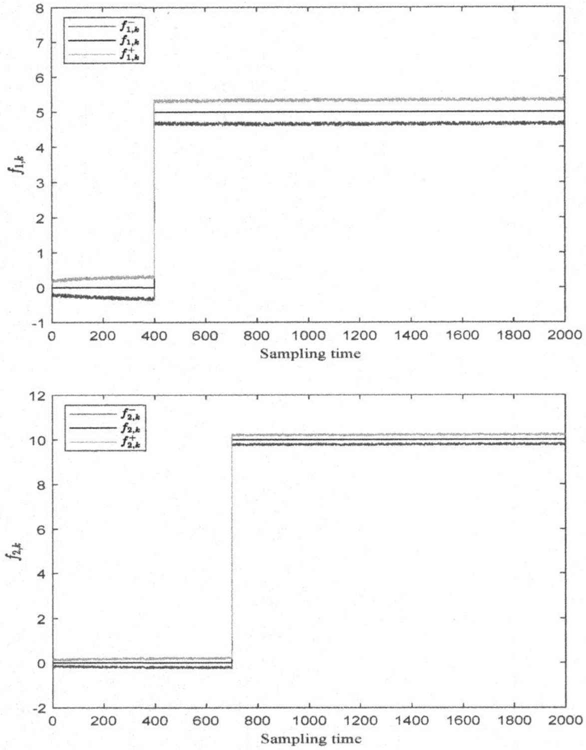

Figure 6.4: Sensor faults and their upper/lower diagnosis bounds.

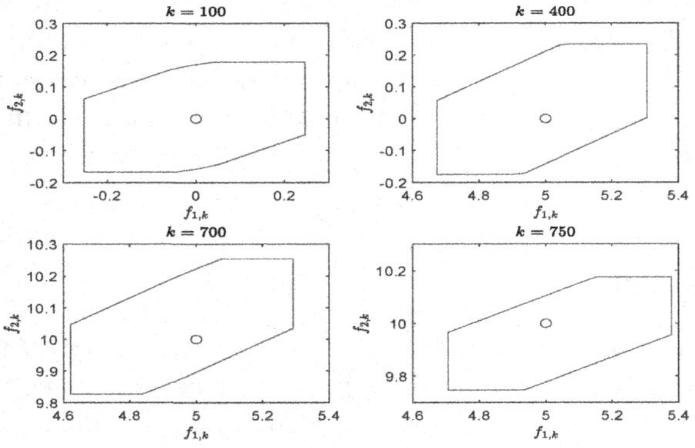

Figure 6.5: Comparison of actual faults and feasible set of faults.

has the characteristics of high fault detection efficiency, accurate fault diagnose and low computational complexity. It can be also applied to the field of fault diagnosis of nonlinear system [59,77] and fault-tolerant control [41,67].

6.2 Sensor fault estimation based on the constrained zonotopic Kalman filter

6.2.1 Preliminaries

Definition 6.2 For matrices A, B, C and X with appropriate dimensions, the correlation operations of the matrix traces are as follows [27,103]:

$$\begin{aligned} tr(A) &= tr(A^{\mathrm{T}}), && (6.19)\\ tr(AB) &= tr(BA), && (6.20)\\ \frac{\partial tr(AX^{\mathrm{T}}B)}{\partial X} &= A^{\mathrm{T}}B^{\mathrm{T}}, && (6.21)\\ \frac{\partial tr(AXBX^{\mathrm{T}}C)}{\partial X} &= BX^{\mathrm{T}}CA + B^{\mathrm{T}}X^{\mathrm{T}}A^{\mathrm{T}}C^{\mathrm{T}}. && (6.22) \end{aligned}$$

Definition 6.3 $I_a = [a^-, a^+] = \{a | a^- \leq a \leq a^+, a^-, a^+ \in \mathbb{R}\}$ and $I_b = [b^-, b^+] = \{b | b^- \leq b \leq b^+, b^-, b^+ \in \mathbb{R}\}$ denote two intervals, and the basic arithmetic operations are defined as [87]:

$$\begin{aligned} I_a + I_b &= [a^- + b^-, a^+ + b^+], && (6.23)\\ I_a - I_b &= [a^- - b^+, a^+ - b^-], && (6.24)\\ I_a \times I_b &= [\min\{a^-b^-, a^-b^+, a^+b^-, a^+b^+\}, \max\{a^-b^-, a^-b^+, a^+b^-, a^+b^+\}], && (6.25)\\ I_a \div I_b &= \left[\min\left\{\frac{a^-}{b^+}, \frac{a^-}{b^-}, \frac{a^+}{b^+}, \frac{a^+}{b^-}\right\}, \max\left\{\frac{a^-}{b^+}, \frac{a^-}{b^-}, \frac{a^+}{b^+}, \frac{a^+}{b^-}\right\}\right], \ (0 \notin I_b) && (6.26) \end{aligned}$$

Definition 6.4 A r-order zonotope $\mathcal{Z} = \langle p, G \rangle \subset \mathbb{R}^n$ $(n \leq r)$ is defined as [27,124]:

$$\mathcal{Z} = p \oplus G\mathbf{B}^r = \{z \in \mathbb{R}^n : z = p + G\boldsymbol{b}, \mathbf{B} \in \mathbf{B}^r\}, \tag{6.27}$$

where $\|\boldsymbol{b}\|_\infty \leq 1$, $\|\cdot\|_\infty$ is the infinity norm, hypercube $\boldsymbol{B}^r = [-1,1]^r$, $p \in \mathbb{R}^n$ is the center of $\mathcal{Z}$ and $G \in \mathbb{R}^{n \times r}$ is the generator matrix of $\mathcal{Z}$.

For zonotope $\mathcal{Z}$, there are some basic properties as follows:

Property 8 *The Minkowski sum of two zonotopes $\mathcal{Z}_1 = \langle p_1, G_1 \rangle \subset \mathbb{R}^n$ and $\mathcal{Z}_2 = \langle p_2, G_2 \rangle \subset \mathbb{R}^n$ is also a zonotope, and its defined by:*

$$\langle p_1, G_1 \rangle \oplus \langle p_2, G_2 \rangle = \langle p_1 + p_2, [G_1\ G_2] \rangle. \tag{6.28}$$

Property 9 *The linear image of the zonotope $\mathcal{Z} = \langle p, G\rangle \subset \mathbb{R}^n$ by $L \in \mathbb{R}^{l\times n}$ is computed by a matrix product:*

$$L \odot \langle p, G\rangle = \langle Lp, LG\rangle. \tag{6.29}$$

Property 10 *A zonotope $\mathcal{Z} = \langle p, G\rangle \subset \mathbb{R}^n$ can be contained in an interval hull* $\text{Box}(\mathcal{Z}) = [\text{z}^-, \text{z}^+]$*:*

$$\begin{cases} z_i^- = p_i - \sum_{j=1}^{r} |G_{i,j}|, i = 1, \ldots, n, \\ z_i^+ = p_i + \sum_{j=1}^{r} |G_{i,j}|, i = 1, \ldots, n, \end{cases} \tag{6.30}$$

where z_i^-, z_i^+ and p_i are the ith *components of z^-,z^+ and p, respectively. $G_{i,j}$ represents the element in the ith row and jth column of the generator matrix G. The interval hull* $\text{Box}(\mathcal{Z}) = [\text{z}^-, \text{z}^+] = \{\text{z} : \text{z} \in \mathcal{Z}, \text{z}^- \leq \text{z} \leq \text{z}^+\}$ *is the smallest box containing the set $\mathcal{Z}$, z^- and z^+ are the lower and upper bounds of z, respectively. And according to the above definitions, the interval hull of $\mathcal{Z}$ can also be expressed as $\langle p, rs(G)\rangle$, where $rs(G)$ is a diagonal matrix and satisfying:*

$$rs(G) = \begin{bmatrix} \sum_{j=1}^{r} |G_{1,j}| & 0 & \cdots & 0 \\ 0 & \sum_{j=1}^{r} |G_{2,j}| & \cdots & 0 \\ \vdots & \vdots & \ddots & \vdots \\ 0 & 0 & \cdots & \sum_{j=1}^{r} |G_{n,j}| \end{bmatrix}. \tag{6.31}$$

Lemma 6.4

For a r-order zonotope $\mathcal{Z} = p \oplus G\mathbf{B}^r \subset \mathbb{R}^n$, it can be bounded by a more conservative zonotope with lower dimensions [27]:

$$\mathcal{Z} = \langle p, G\rangle \subseteq \langle p, \downarrow_s (G)\rangle, \tag{6.32}$$

where $\downarrow_s (\cdot)$, $(n \leq s \leq r)$ is the reduction operator, and s specifies the maximum number of column of the generator matrix after reduction. Thus, the implement procedure of $\downarrow_s (G)$ can be summarized as:

- *Rearranged the column in the matrix G in the decreasing order of the Euclidean norm to generate the norm decreasing matrix $\downarrow (G)$:*

$$\downarrow (G) = [g_1, g_2, \ldots, g_r], \ \|g_j\|_2 \geq \|g_{j+1}\|_2, \ j = 1, \ldots, r, \tag{6.33}$$

where $\|\cdot\|_2$ denotes the Euclidean norm.

- *Divide* $\downarrow(G)$ *into two part:* $G_>$ *and* $G_<$*, and replace* $G_<$ *by a diagonal matrix* $rs(G_<) \in \mathbb{R}^{n\times n}$*:*

  If $r \leq s$ ***then*** $\downarrow_s(G) = \downarrow(G)$.
  Else $\downarrow_s(G) = [G_>, rs(G_<)] \in \mathbb{R}^{n\times s}$, $G_> = [g_1,\ldots,g_{s-n}]$, $G_< = [g_{s-n+1},\ldots,g_r]$.

Definition 6.5 The Frobenius radius can be used as an index to measure the size of the zonotope $\mathcal{Z} = \langle p, G\rangle$, which is generally expressed by the Frobenius norm of the generator matrix G [27]:

$$||\langle p,G\rangle||_F = ||G||_F = \sqrt{tr(G^{\mathrm{T}}G)} = \sqrt{tr(GG^{\mathrm{T}})}. \tag{6.34}$$

By analogy, in the weighted case, $||G||_{F,M} = \sqrt{tr(G^{\mathrm{T}}MG)}$, and $M \in \mathbb{R}^{n\times n}$ is a weighting symmetric positive definite matrix.

Definition 6.6 The covariation of a zonotope $\langle p,G\rangle$ is defined as $cov(\langle p,G\rangle) = GG^{\mathrm{T}}$ [27].

The minimum zonotope can be obtained by solving the minimization problem of Frobenius radius $||G||_F$, and it is equivalent to minimizing the trace of its covariation $cov(\langle p,G\rangle) = GG^{\mathrm{T}}$. Thus, $J = tr(GG^{\mathrm{T}}) = ||G||_F^2$ can be defined as the judgment criteria for the size of the zonotope, and $J_M = tr(MGG^{\mathrm{T}}) = tr(G^{\mathrm{T}}MG) = ||G||_{F,M}^2$ is the judgment criteria in the weighted case.

6.2.2 Problem formulation

Considering the following linear discrete system:

$$\begin{cases} x(k+1) = Ax(k) + Bu(k) + Ew(k), \\ y(k) = Cx(k) + Fv(k), \end{cases} \tag{6.35}$$

where $x(k) \in \mathbb{R}^{n_x}$, $u(k) \in \mathbb{R}^{n_u}$, $y(k) \in \mathbb{R}^{n_y}$ refer to the unknown state, the known input, the known output, respectively. A, B, C, E, F denote the parameter matrices with appropriately fixed size. $w(k) \in \mathbb{R}^{n_w}$ and $v(k) \in \mathbb{R}^{n_v}$ are the unknown but bounded process interference and measurement noise of the system.

Without loss of generality, the initial state, process interference and measurement noise of system (6.35) are confined to the zonotopic sets as follows:

$$x(0) \in X(0) = \langle p(0), G(0)\rangle, w(k) \in W = \langle 0, G_w\rangle, v(k) \in V = \langle 0, G_v\rangle, \tag{6.36}$$

where $|w(k)| \leq \tilde{w}, |v(k)| \leq \tilde{v}$, $\tilde{w}$ and $\tilde{v}$ are the limit values of $w(k)$ and $v(k)$, respectively. $X(0)$, W and V are the zonotopic sets of initial state, process interference and measurement noise, respectively, $p(0)$ is the center of $X(0)$, $G(0)$ is the generator matrix of $X(0)$ and G_w and G_v are the generator matrices of W and V, respectively.

The stable state of the system satisfies the constraint $|h(x)| \leq \tilde{h}$, $h(x) = \gamma(k) - H(k)x(k)$ in the fault-free working state, where $\tilde{h}$ is the limit value of $h(k)$. Therefore, considering the constraint of the system, the state space equation of system (6.35) can be extended as:

$$\begin{cases} x(k+1) = Ax(k) + Bu(k) + Ew(k), \\ \bar{y}(k) = \bar{C}x(k) + \bar{F}\bar{v}(k). \end{cases} \tag{6.37}$$

where $\bar{y}(k) = \begin{bmatrix} y(k) \\ \gamma(k) \end{bmatrix} \in \mathbb{R}^{n_y+n_\gamma}, \bar{C} = \begin{bmatrix} C \\ H(k) \end{bmatrix} \in \mathbb{R}^{(n_y+n_\gamma)\times n_x}, \bar{F} = \begin{bmatrix} F & 0_{n_y\times n_\gamma} \\ 0_{n_\gamma\times n_v} & I_{n_\gamma\times n_\gamma} \end{bmatrix}$ $\in \mathbb{R}^{(n_y+n_\gamma)\times(n_v+n_\gamma)}$, $\bar{v}(k) = \begin{bmatrix} v(k) \\ h(k) \end{bmatrix} \in \mathbb{R}^{n_v+n_\gamma}$, $|\bar{v}(k)| \leq \tilde{\bar{v}}$, $\tilde{\bar{v}}$ is the limit value of $\bar{v}(k)$, $h(k) \in \langle 0, \mathrm{diag}(\tilde{h})\rangle$, $\gamma(k) \in \mathbb{R}^{n_\gamma}$.

For the system with constraint mentioned above, two possible sensor faults are considered: additive sensor fault and multiplicative sensor fault.

1. Additive sensor fault:

$$\begin{cases} x(k+1) = Ax(k) + Bu(k) + Ew(k), \\ \bar{y}(k) = \bar{C}x(k) + \bar{F}\bar{v}(k) + \bar{f}(k), \end{cases} \tag{6.38}$$

where $\bar{f}(k) = \begin{bmatrix} f(k) & 0_{n_\gamma} \end{bmatrix}^{\mathrm{T}} \in \mathbb{R}^{n_y+n_\gamma}$, $f(k) \in \mathbb{R}^{n_y}$ is the additive sensor fault in the system.

2. Multiplicative sensor fault:

$$\begin{cases} x(k+1) = Ax(k) + Bu(k) + Ew(k), \\ \bar{y}(k) = \bar{f}^{-1}(k)\bar{C}x(k) + \bar{F}\bar{v}(k), \end{cases} \tag{6.39}$$

where $\bar{f}(k) = \begin{bmatrix} f(k) & 0_{n_y\times n_\gamma} \\ 0_{n_\gamma\times n_y} & I_{n_\gamma\times n_\gamma} \end{bmatrix}$, $f(k) = \begin{bmatrix} f_1(k) & 0 & 0 \\ 0 & \ddots & 0 \\ 0 & 0 & f_{n_y}(k) \end{bmatrix}$ ($f_i(k) \neq 0$, $i = 1, \ldots, n_y$) is the multiplicative sensor fault in the system, and $\bar{f}(k) \in \mathbb{R}^{(n_y+n_\gamma)\times(n_y+n_\gamma)}$, $f(k) \in \mathbb{R}^{n_y\times n_y}$

Remark 6.1 In the actual system, the output of the system will change due to the multiplicative sensor fault, which will make it deviate from the true output of the system. Affected by the multiplicative sensor fault, the output of the system will become larger or smaller. Generally, it is necessary to assume $C_i x(k) \neq 0$ when studying multiplicative sensor fault. Once $C_i x(k) = 0$, the multiplicative sensor fault will have no effect on the system. Therefore, the research of the multiplicative sensor fault in this section is based on the condition of $C_i x(k) \neq 0$. Meanwhile, since the output of the sensor is bounded by a certain limit in the practical engineering system, so the infinity situation will not appear in output $y(k)$. Therefore, $f_i(k) = \frac{C_i x(k)}{y_i(k) - F_i v(k)} \neq 0$, where C_i, $y_i(k)$ and F_i is the ith row of C, $y(k)$ and F, respectively. Thus, $f(k)$ is invertible.

For system affected by multiplicative sensor fault, there is another expression: $y(k) = m(k)Cx(k) + Fv(k)$, where $m(k) = \begin{bmatrix} m_1(k) & 0 & 0 \\ 0 & \ddots & 0 \\ 0 & 0 & m_{n_y}(k) \end{bmatrix}$ is the multiplicative sensor fault of the system. However, the main purpose of this section is to obtain the minimum estimation interval of the multiplicative sensor fault by minimizing the zonotopic set of estimated state $\hat{x}(k)$, so the minimum fault estimation interval cannot be reasonably derived according to the calculation rule $\hat{m}_i(k) = \frac{y_i(k)-F_i v(k)}{C_i\hat{x}(k)} (i = 1,\dots,n_y)$ when using this fault expression. But according to $\hat{f}_i(k) = \frac{C_i\hat{x}(k)}{y_i(k)-F_i v(k)} (i = 1,\dots,n_y)$, the interval of the denominator is only related to G_v, and G_v remains unchanged in the whole fault estimation process. Therefore, an appropriate estimated interval of multiplicative sensor fault can be obtained by minimizing the zonotopic set of estimated state $\hat{x}(k)$. However, this method needs to assume that $y_i(k) - F_i v(k) \neq 0$. Compared with the former method, it has one more restriction. But $y_i(k) - F_i v(k) = 0 (i = 1,\dots,n_y)$ is a special case, and when $y_i(k) - F_i v(k) = 0$, $C_i x(k) \neq 0$, then $f_i^{-1}(k) = 0$ can be easily obtained. Thus, the expression form of multiplicative sensor fault is selected as $y(k) = f^{-1}(k)Cx(k) + Fv(k)$ after the comprehensive consideration. ■

6.2.3 Main results

6.2.3.1 *Design of the constrained zonotopic Kalman filter*

For the system (6.35) with constraint, the constraint works after the system reaches stability, and forming the state-space equation (6.37). The stable condition of the system is judged based on whether $\frac{\|\hat{p}(k+l)-\hat{p}(k+l-1)\|}{\|\hat{p}(k+l-1)\|} \leq \delta$ $(l = 1,\dots,L)$ holds or not, L is the selectable length and δ is the selectable threshold. The selection of the values affect the specific adding time of the system's constraint. For different systems, different values can be selected according to the actual situation.

When the system is stable, the design of CZKF is as follows:

$$\hat{x}(k+1) = A\hat{x}(k) + Bu(k) + Ew(k) + L(k)(\bar{y}(k) - \bar{C}\hat{x}(k) - \bar{F}\bar{v}(k)), \tag{6.40}$$

where $L(k)$ is the estimated optimal gain matrix of the constrained zonotopic Kalman filter and $\hat{x}(k)$ is the estimated state at time k.

Theorem 6.3

Assuming the estimated state $\hat{x}(k) \in \hat{X}(k) = \langle \hat{p}(k), \hat{G}(k) \rangle$ at time k is known, so the estimated state at time $k+1$ is

$$\hat{x}(k+1) \in \hat{X}(k+1) = \langle \hat{p}(k+1), \hat{G}(k+1) \rangle, \tag{6.41}$$

where

$$\begin{aligned}\hat{p}(k+1) &= (A-L(k)\bar{C})\hat{p}(k)+Bu(k)+L(k)\bar{y}(k), &(6.42)\\ \hat{G}(k+1) &= [(A-L(k)\bar{C})\tilde{\hat{G}}(k)\ EG_w\ -L(k)\bar{F}\bar{G}_v], \tilde{\hat{G}}(k)=\downarrow_s \hat{G}(k). &(6.43)\end{aligned}$$

Proof 6.3 From Eq. (6.40), we can obtain

$$\begin{aligned}\hat{x}(k+1) &= A\hat{x}(k)+Bu(k)+Ew(k)+L(k)(\bar{y}(k)-\bar{C}\hat{x}(k)-\bar{F}\bar{v}(k))\\ &= (A-L(k)\bar{C})\hat{x}(k)+Bu(k)+Ew(k)+L(k)\bar{y}(k)-L(k)\bar{F}\bar{v}(k)\\ &\in ((A-L(k)\bar{C})\odot\langle\hat{p}(k),\hat{G}(k)\rangle)\oplus Bu(k)\oplus(E\odot\langle 0,G_w\rangle)\\ &\oplus L(k)\bar{y}(k)\oplus(-L(k)\bar{F}\odot\langle 0,\bar{G}_v\rangle)\\ &= \langle(A-L(k)\bar{C})\hat{p}(k)+Bu(k)+L(k)\bar{y}(k),\\ &\quad [(A-L(k)\bar{C})\hat{G}(k)\ EG_w\ -L(k)\bar{F}\bar{G}_v]\rangle. &(6.44)\end{aligned}$$

Applying Lemma 6.4, the complexity reduction matrix $\downarrow_s \hat{G}(k)$ corresponding to $\hat{G}(k)$ can be obtained, so the estimated zonotopic set $\hat{X}(k+1)=\langle\hat{p}(k+1),\hat{G}(k+1)\rangle=\langle(A-L(k)\bar{C})\hat{p}(k)+Bu(k)+L(k)\bar{y}(k),[(A-L(k)\bar{C})\tilde{\hat{G}}(k)\ EG_w\ -L(k)\bar{F}\bar{G}_v]\rangle,(\tilde{\hat{G}}(k)=\downarrow_s\hat{G}(k))$ at time $k+1$ can be obtained.

Theorem 6.4

For the zonotope of Eq. (6.41), the optimal gain matrix $L(k)$ is designed by minimizing $J_M(k+1)=\|\hat{G}(k+1)\|_{F,M}^2=tr(\hat{G}^{\mathrm{T}}(k+1)M\hat{G}(k+1))$. Define $\tilde{P}(k)=\tilde{\hat{G}}(k)\tilde{\hat{G}}^{\mathrm{T}}(k)$, $\bar{Q}_v=\bar{F}\bar{G}_v\bar{G}_v^{\mathrm{T}}\bar{F}^{\mathrm{T}}$, $Q_w=EG_wG_w^{\mathrm{T}}E^{\mathrm{T}}$, it can be obtained that:

$$L(k)=AK(k),\ K(k)=R(k)S^{-1}(k),\ R(k)=\tilde{P}(k)\bar{C}^{\mathrm{T}},\ S(k)=\bar{C}\tilde{P}(k)\bar{C}^{\mathrm{T}}+\bar{Q}_v. \quad (6.45)$$

Proof 6.4 According to Definitions 6.5 and 6.6, the judgment criteria $J_M(k+1)=\|\hat{G}(k+1)\|_{F,M}^2=tr(\hat{G}^{\mathrm{T}}(k+1)M\hat{G}(k+1))$ can be used to determine the size of the zonotope $\hat{X}(k+1)=\langle\hat{p}(k+1),\hat{G}(k+1)\rangle$, the minimum zonotope can be obtained by minimizing $J_M(k+1)$. Based on Eq. (6.43), there is only one unknown matrix $L(k)$ when calculating $G(k+1)$. According to the calculation rules, when $\partial_{L(k)}J_M(k+1)=0$, $J_M(k+1)$ is the minimum, and the size of the zonotope is the smallest, $L(k)$ is called the optimal observer gain at this time.

From Eq. (6.43), we can obtain

$$\begin{aligned}\partial_{L(k)}J_M(k+1) &= \partial_{L(k)}tr(\hat{G}^{\mathrm{T}}(k+1)M\hat{G}(k+1))\\ &= \partial_{L(k)}tr(M(A-L(k)\bar{C})\tilde{P}(k)(A-L(k)\bar{C})^{\mathrm{T}}\\ &+ MQ_w+ML(k)\bar{Q}_vL^{\mathrm{T}}(k))\\ &= \partial_{L(k)}tr(M(A\tilde{P}(k)A^{\mathrm{T}}-A\tilde{P}(k)\bar{C}^{\mathrm{T}}L^{\mathrm{T}}(k)-L(k)\bar{C}\tilde{P}(k)A^{\mathrm{T}}\\ &+ L(k)\bar{C}\tilde{P}(k)\bar{C}^{\mathrm{T}}L^{\mathrm{T}}(k)+L(k)\bar{Q}_vL^{\mathrm{T}}(k)))\\ &= \partial_{L(k)}tr(M(L(k)\bar{C}\tilde{P}(k)\bar{C}^{\mathrm{T}}L^{\mathrm{T}}(k)+L(k)\bar{Q}_vL^{\mathrm{T}}(k)))\\ &- 2\partial_{L(k)}tr(MA\tilde{P}(k)\bar{C}^{\mathrm{T}}L^{\mathrm{T}}(k)). &(6.46)\end{aligned}$$

Let $\partial_{L(k)}J_M(k+1)=0$, then $\partial_{L(k)}tr(ML(k)\bar{C}\tilde{P}(k)\bar{C}^{\mathrm{T}}L^{\mathrm{T}}(k))+\partial_{L(k)}tr(ML(k)\bar{Q}_vL^{\mathrm{T}}(k)) = 2\partial_{L(k)}tr(MA\tilde{P}(k)\bar{C}^{\mathrm{T}}L^{\mathrm{T}}(k))$. Considering $M=M^{\mathrm{T}}>0$ and the operations of the matrix traces defined in Definition 6.2, the following equations hold:

$$ML(k)(\bar{C}\tilde{P}(k)\bar{C}^{\mathrm{T}}+\bar{Q}_v) = MA\tilde{P}(k)\bar{C}^{\mathrm{T}} \tag{6.47}$$

$$L(k)S(k) = AR(k). \tag{6.48}$$

Thus,

$$L(k)=AK(k), \tag{6.49}$$

where $K(k)=R(k)S^{-1}(k)$, $R(k)=\tilde{P}(k)\bar{C}^{\mathrm{T}}$, $S(k)=\bar{C}\tilde{P}(k)\bar{C}^{\mathrm{T}}+\bar{Q}_v$.

6.2.3.2 Fault detection

According to the state-space equation of the system (6.35), based on the measurement output and the known input, the state $\hat{x}(k)$ can be determined, which is contained by a zonotopic set $\hat{X}(k)$ in this section. When the system is fault-free, the estimated output set is

$$\begin{aligned}\hat{y}(k) &= C\hat{x}(k)+Fv(k)\\ &\in (C\odot\langle\hat{p}(k),\hat{G}(k)\rangle)\oplus(F\odot\langle 0,G_v\rangle)\\ &= \langle C\hat{p}(k),[C\hat{G}(k)\ FG_v]\rangle\\ &= \langle\hat{p}_y(k),\hat{G}_y(k)\rangle.\end{aligned} \tag{6.50}$$

Based on the estimated output set $\hat{Y}(k)=\langle\hat{p}_y(k),\hat{G}_y(k)\rangle$, the estimated upper bound $\hat{y}_u(k)$ and lower bound $\hat{y}_l(k)$ satisfy:

$$\begin{cases}\hat{y}_u(k)=\hat{p}_y(k)+\begin{bmatrix}\sum\limits_{j=1}^{r_{\hat{G}_y}}|\hat{G}_{y1j}(k)|\\ \vdots\\ \sum\limits_{j=1}^{r_{\hat{G}_y}}|\hat{G}_{yn_yj}(k)|\end{bmatrix},\\ \hat{y}_l(k)=\hat{p}_y(k)-\begin{bmatrix}\sum\limits_{j=1}^{r_{\hat{G}_y}}|\hat{G}_{y1j}(k)|\\ \vdots\\ \sum\limits_{j=1}^{r_{\hat{G}_y}}|\hat{G}_{yn_yj}(k)|\end{bmatrix}.\end{cases} \tag{6.51}$$

If $\hat{y}_{l1}(k)\le y_1(k)\le\hat{y}_{u1}(k),\ldots,\hat{y}_{ln_y}(k)\le y_{n_y}(k)\le\hat{y}_{un_y}(k)$, the system's true output $y(k)\in\hat{Y}(k)$, which indicates the system is fault-free and the fault detection signal $f_s(k)=0$, otherwise, the system is in the faulty state and $f_s(k)=1$. When a sensor fault occurs in the system, the true output of the system will change suddenly due to the effect of the sensor fault, and it is no longer included in the estimated zonotopic set $\hat{Y}(k)$ calculated by the CZKF-FE algorithm. And in order to facilitate the development of this section, it is assumed that only one type of sensor fault occurs in the system at one time, the sensor fault does not change during a time length and the sensor fault will occur in the system after the system is stable for a period of time.

The sensor fault detection method based on the CZKF-FE algorithm proposed in this section is implemented according to whether the true output of the system is within the upper and lower bounds of the estimated zonotopic set $\hat{Y}(k)$. If the true output of the system is within the upper and lower bounds of $\hat{Y}(k)$, it is determined that the system has no sensor fault, otherwise, it is judged that there is a sensor fault occurred in the system. In each iteration of the algorithm, the upper and lower bounds of the estimated output $\hat{Y}(k)$ can be calculated according to the estimated state set $\hat{X}(k)$. When the system does not fail, the true output of the system must be included in $\hat{Y}(k)$, which can be equivalent to the true output of the system is included in the upper and lower bounds of $\hat{Y}(k)$. Once the true output is not within the estimated upper and lower bounds, it indicates that the system has failed. Therefore, based on the given noise bounds and according to the CZKF-FE algorithm proposed in this section, there will be no false alarms, that is, the false alarm rate is zero.

6.2.3.3 *Design of constrained zonotopic Kalman filter based fault estimator*

For the system with additive sensor fault (system (6.38)), a fault estimation method based on the constrained zonotopic Kalman filter is proposed in this section to estimate the sensor fault of the system, and the zonotopic set is used to contain the estimated additive sensor fault value. When an additive sensor fault is detected in the system at time $k+1$, the design process of the additive sensor fault estimator is as follows:

- State estimation:

$$\hat{x}(k+1) = A\hat{x}(k) + Bu(k) + Ew(k) + L(k)(\bar{y}(k) - \bar{C}\hat{x}(k) - \bar{F}\bar{v}(k) - \hat{\bar{f}}_a(k)), \tag{6.52}$$

  where $L(k)$ is the estimated optimal gain matrix of the additive sensor fault estimator based on the constrained zonotopic Kalman filter, $\hat{x}(k)$ is the estimated state at time k, $\hat{\bar{f}}_a(k) = [\ \hat{f}_a(k) \quad 0_{n_\gamma}\ ]^{\mathrm{T}}$ and $\hat{f}_a(k)$ is an arbitrary value in the zonotopic set $\hat{X}_f(k)$, $\hat{X}_f(k)$ is the estimated zonotopic set of the additive sensor fault at time k.

- Fault estimation:

$$\hat{f}(k+1) \quad = \quad y(k+1) - C\hat{x}(k+1) - Fv(k+1). \tag{6.53}$$

In the design process of the additive sensor fault estimator, according to Eq. (6.53), the estimated value of $f(k+1)$ mainly depends on $\hat{x}(k+1)$, $y(k+1)$ and $v(k+1)$.

Remark 6.2 In Eq. (6.52), the main reasons for using the arbitrary value $\hat{\bar{f}}_a(k)$ instead of $\hat{\bar{f}}(k)$ in the calculation of $\hat{x}(k+1)$ are as follows: (1) If $\hat{\bar{f}}(k)$ is used in Eq. (6.52), then according to Eq. (6.53), $\hat{\bar{f}}(k) = \bar{y}(k) - \bar{C}\hat{x}(k) - \bar{F}\bar{v}(k)$ can be substituted into the calculation, and a calculation equation of $\hat{x}(k+1)$ independent of $\bar{y}(k)$ can be obtained after simplification. However, for the system with constraints which extend the constraints in $\bar{y}(k)$, it is necessary to retain the constraint conditions in order to smaller the size of the zonotopic set $\hat{X}(k+1)$ corresponding to $\hat{x}(k+1)$. (2) According to the calculation rules, the calculation equation of $\hat{x}(k+1)$ is still valid by substituting the arbitrary value $\hat{\bar{f}}_a(k)$ for the operation. Although due to the arbitrariness of $\hat{\bar{f}}_a(k)$, a certain uncertainty will be introduced in the calculation of $\hat{x}(k+1)$, but the conservativeness of $\hat{X}(k+1)$ and $\hat{X}_f(k+1)$ is greatly reduced under the constraints. Through the comprehensive consideration, the small uncertainty can be negligible. ▮

Theorem 6.5

Consider the additive sensor fault system in (6.38), given the system's state $\hat{x}(k) \in \hat{X}(k) = \langle \hat{p}(k), \hat{G}(k) \rangle$ and the estimated zonotopic set $\hat{X}_f(k)$ of the additive sensor fault, $\hat{f}_a(k)$ is an arbitrary value in $\hat{X}_f(k)$ and $\hat{\bar{f}}_a(k) = [\ \hat{f}_a(k) \quad 0_{n_\gamma}\ ]^{\mathrm{T}}$, then the estimated zonotopic set of the additive fault at time $k+1$ is

$$\hat{X}_f(k+1) = \langle \hat{p}_f(k+1), \hat{G}_f(k+1) \rangle, \tag{6.54}$$

where

$$\begin{aligned}
\hat{p}_f(k+1) &= y(k+1) - C\hat{p}(k+1), &(6.55)\\
\hat{G}_f(k+1) &= [\ -C\hat{G}(k+1) \quad -FG_v\ ], &(6.56)\\
\hat{p}(k+1) &= (A - L(k)\bar{C})\hat{p}(k) + Bu(k) + L(k)(\bar{y}(k) - \hat{\bar{f}}_a(k)), &(6.57)\\
\hat{G}(k+1) &= [\ (A - L(k)\bar{C})\tilde{\hat{G}}(k) \quad EG_w \quad -L(k)\bar{F}\bar{G}_v\ ], &\\
\tilde{\hat{G}}(k) &= \downarrow_s \hat{G}(k). &(6.58)
\end{aligned}$$

Proof 6.5 From Eq. (6.52), we can obtain:

$$\begin{aligned}
\hat{x}(k+1) &= A\hat{x}(k) + Bu(k) + Ew(k) + L(k)(\bar{y}(k) - \bar{C}\hat{x}(k) - \bar{F}\bar{v}(k) - \hat{\bar{f}}_a(k))\\
&= (A - L(k)\bar{C})\hat{x}(k) + Bu(k) + Ew(k) + L(k)\bar{y}(k)\\
&- L(k)\bar{F}\bar{v}(k) - L(k)\hat{\bar{f}}_a(k)\\
&\in ((A - L(k)\bar{C}) \odot \langle \hat{p}(k), \hat{G}(k) \rangle) \oplus Bu(k) \oplus (E \odot \langle 0, G_w \rangle) \oplus L(k)\bar{y}(k)\\
&\oplus (-L(k)\bar{F} \odot \langle 0, \bar{G}_v \rangle) \oplus (-L(k)\hat{\bar{f}}_a(k))\\
&= \langle (A - L(k)\bar{C})\hat{p}(k) + Bu(k) + L(k)(\bar{y}(k) - \hat{\bar{f}}_a(k)),\\
&\quad [(A - L(k)\bar{C})\hat{G}(k)\ EG_w\ -L(k)\bar{F}\bar{G}_v] \rangle.
\end{aligned} \tag{6.59}$$

Taking into account the dimensionality of the generator matrix, the dimensionality reduction method in Lemma 6.4 is used to reduce the dimensionality of $\hat{G}(k)$, so the estimated zonotopic set at time $k+1$ can be obtained as $\hat{X}(k+1) = \langle \hat{p}(k+1), \hat{G}(k+1)\rangle = \langle (A-L(k)\bar{C})\hat{p}(k) + Bu(k) + L(k)(\bar{y}(k) - \hat{\bar{f}}_a(k)), [(A-L(k)\bar{C})\tilde{\hat{G}}(k)\ EG_w\ -L(k)\bar{F}\bar{G}_v]\rangle, (\tilde{\hat{G}}(k) = \downarrow_s \hat{G}(k))$.

According to Eq. (6.53), it easy to derive the zonotopic set $\hat{X}_f(k+1)$ of the additive sensor fault:

$$\begin{aligned} \hat{f}(k+1) &= y(k+1) - C\hat{x}(k+1) - Fv(k+1) \\ &\in y(k+1) \oplus (-C \odot \langle \hat{p}(k+1), \hat{G}(k+1)\rangle) \oplus (-F \odot \langle 0, G_v\rangle) \\ &= \langle y(k+1) - C\hat{p}(k+1), [-C\hat{G}(k+1)\ \ -FG_v]\rangle \\ &= \langle \hat{p}_f(k+1), \hat{G}_f(k+1)\rangle, \end{aligned} \tag{6.60}$$

where

$$\begin{aligned} \hat{p}_f(k+1) &= y(k+1) - C\hat{p}(k+1), \\ \hat{G}_f(k+1) &= [\ -C\hat{G}(k+1) \quad -FG_v\ ]. \end{aligned}$$

Remark 6.3 From Eq. (6.56), it can be obtained that the generator matrix $\hat{G}_f(k+1)$ of the zonotopic set $\hat{X}_f(k+1)$ corresponding to the additive sensor fault $f(k+1)$ depends on $\hat{G}(k+1)$ and G_v. Since the G_v is fixed throughout the working process, the size of the zonotopic set $\hat{X}_f(k+1)$ is directly determined by $\hat{G}(k+1)$. Therefore, it is easy to get that when the Frobenius norm of $\hat{G}(k+1)$ is smallest, we can get a smallest zonotopic set $\hat{X}_f(k+1)$. Since the equation of $\hat{G}(k+1)$ is the same in the fault-free state and the additive sensor fault state, so we can obtain the optimal gain matrix $L(k)$ in the additive sensor fault state the same as Theorem 6.4. Thus, the whole design process of additive sensor fault estimator is completed. ■

For the system with multiplicative sensor fault, as in (6.39), the multiplicative sensor fault value is contained in the interval box estimated by the fault estimation method based on the constrained zonotopic Kalman filter. When a multiplicative sensor fault is detected in the system at time $k+1$, the design process of the multiplicative sensor fault estimator is as follows:

- State estimation:

$$\hat{x}(k+1) = A\hat{x}(k) + Bu(k) + Ew(k) + L(k)(\bar{y}(k) - \hat{\bar{f}}_a^{-1}(k)\bar{C}\hat{x}(k) - \bar{F}\bar{v}(k)), \tag{6.61}$$

where $L(k)$ is the estimated optimal gain matrix of the multiplicative sensor fault estimator based on the constrained zonotopic Kalman filter, $\hat{x}(k)$ is the estimated state at time k, $\hat{\bar{f}}_a(k) = \begin{bmatrix} \hat{f}_a(k) & 0_{n_y \times n_\gamma} \\ 0_{n_\gamma \times n_y} & I_{n_\gamma \times n_\gamma} \end{bmatrix}$, $\hat{f}_a(k) = \begin{bmatrix} \hat{f}_{a1}(k) & 0 & 0 \\ 0 & \ddots & 0 \\ 0 & 0 & \hat{f}_{an_y}(k) \end{bmatrix}$, $\hat{f}_{ai}(k) \neq 0$ is an arbitrary value in the interval

$\hat{I}_{fi}(k) = [\hat{f}_i^-(k), \hat{f}_i^+(k)](i = 1, \dots, n_y)$ and $\hat{I}_f(k) = [\hat{I}_{f1}, \dots, \hat{I}_{fn_y}]$ is the estimated interval box of the multiplicative sensor fault at time k.

- Fault estimation:

$$\hat{f}(k+1) = \begin{bmatrix} \hat{f}_1(k+1) & 0 & 0 \\ 0 & \ddots & 0 \\ 0 & 0 & \hat{f}_{n_y}(k+1) \end{bmatrix}, \tag{6.62}$$

$$\begin{aligned} \hat{f}_i(k+1) &= \frac{C_i\hat{x}(k+1)}{y_i(k+1) - F_i v(k+1)}, \\ (C_i\hat{x}(k+1) &\neq 0,\ y_i(k+1) - F_i v(k+1) \neq 0), (i = 1, \dots, n_y), \end{aligned} \tag{6.63}$$

where C_i, $\hat{y}_i(k+1)$ and F_i is the ith row of C, $\hat{y}(k+1)$ and F, respectively.

From the two equations above, the value of $\hat{f}(k+1)$ is determined by $\hat{x}(k+1)$, $y(k+1)$ and $v(k+1)$. Because of the fixed size of the sets of $v(k+1)$, so the estimated interval box $\hat{I}_f(k+1)$ containing the multiplicative sensor fault $f(k+1)$ can be computed from the zonotopic set corresponding to $\hat{x}(k+1)$.

Theorem 6.6
Consider the multiplicative sensor fault system in (6.39), given the system's state $\hat{x}(k) \in \hat{X}(k) = \langle \hat{p}(k), \hat{G}(k) \rangle$ and the estimated interval box of the multiplicative sensor fault $\hat{I}_f(k) = [\hat{I}_{f1}, \dots, \hat{I}_{fn_y}]$, $\hat{f}_{ai}(k) \neq 0$ is an arbitrary value in $\hat{I}_{fi}(k)$ $(i = 1, \dots, n_y)$, $\hat{f}_a(k) = \begin{bmatrix} \hat{f}_{a1}(k) & 0 & 0 \\ 0 & \ddots & 0 \\ 0 & 0 & \hat{f}_{an_y}(k) \end{bmatrix}$ *and* $\hat{\hat{f}}_a(k) = \begin{bmatrix} \hat{f}_a(k) & 0_{n_y \times n_\gamma} \\ 0_{n_\gamma \times n_y} & I_{n_\gamma \times n_\gamma} \end{bmatrix}$, *then the estimated interval box of the multiplicative sensor fault at time $k+1$ is*

$$\hat{I}_f(k+1) = [\hat{I}_{f1}(k+1), \dots, \hat{I}_{fi}(k+1), \dots, \hat{I}_{fn_y}(k+1)]\ (i = 1, \dots, n_y), \tag{6.64}$$

where

$$\hat{I}_{fi}(k+1) = \left[\hat{f}_i^-(k+1), \hat{f}_i^+(k+1)\right], \tag{6.65}$$

$$\begin{aligned} \hat{f}_i^-(k+1) &= \min\left\{ \frac{\hat{I}_{Gi}^-(k+1)}{\hat{I}_{vi}^+(k+1)}, \frac{\hat{I}_{Gi}^-(k+1)}{\hat{I}_{vi}^-(k+1)}, \frac{\hat{I}_{Gi}^+(k+1)}{\hat{I}_{vi}^+(k+1)}, \frac{\hat{I}_{Gi}^+(k+1)}{\hat{I}_{vi}^-(k+1)} \right\}, \\ &\quad (0 \notin [\hat{I}_{vi}^-(k+1), \hat{I}_{vi}^+(k+1)]), \end{aligned} \tag{6.66}$$

$$\begin{aligned} \hat{f}_i^+(k+1) &= \max\left\{ \frac{\hat{I}_{Gi}^-(k+1)}{\hat{I}_{vi}^+(k+1)}, \frac{\hat{I}_{Gi}^-(k+1)}{\hat{I}_{vi}^-(k+1)}, \frac{\hat{I}_{Gi}^+(k+1)}{\hat{I}_{vi}^+(k+1)}, \frac{\hat{I}_{Gi}^+(k+1)}{\hat{I}_{vi}^-(k+1)} \right\}, \\ &\quad (0 \notin [\hat{I}_{vi}^-(k+1), \hat{I}_{vi}^+(k+1)]), \end{aligned} \tag{6.67}$$

$$
\begin{cases}
\hat{I}_{vi}^{-}(k+1) = y_i(k+1) - F_i \begin{bmatrix} \sum_{j=1}^{r_{G_v}} |G_{v1j}(k+1)| \\ \vdots \\ \sum_{j=1}^{r_{G_v}} |G_{vn_vj}(k+1)| \end{bmatrix}, \ i = 1,\ldots,n_y, \\
\hat{I}_{vi}^{+}(k+1) = y_i(k+1) + F_i \begin{bmatrix} \sum_{j=1}^{r_{G_v}} |G_{v1j}(k+1)| \\ \vdots \\ \sum_{j=1}^{r_{G_v}} |G_{vn_vj}(k+1)| \end{bmatrix}, \ i = 1,\ldots,n_y,
\end{cases} \tag{6.68}
$$

$$
\begin{cases}
\hat{I}_{Gi}^{-}(k+1) = C_i\hat{p}(k+1) - C_i \begin{bmatrix} \sum_{j=1}^{r_{\hat{G}}} |\hat{G}_{1j}(k+1)| \\ \vdots \\ \sum_{j=1}^{r_{\hat{G}}} |\hat{G}_{n_xj}(k+1)| \end{bmatrix}, \ i = 1,\ldots,n_y, \\
\hat{I}_{Gi}^{+}(k+1) = C_i\hat{p}(k+1) + C_i \begin{bmatrix} \sum_{j=1}^{r_{\hat{G}}} |\hat{G}_{1j}(k+1)| \\ \vdots \\ \sum_{j=1}^{r_{\hat{G}}} |\hat{G}_{n_xj}(k+1)| \end{bmatrix}, \ i = 1,\ldots,n_y,
\end{cases} \tag{6.69}
$$

$$
\hat{p}(k+1) = (A - L(k)\hat{\bar{f}}_a^{-1}(k)\bar{C})\hat{p}(k) + Bu(k) + L(k)\bar{y}(k), \tag{6.70}
$$

$$
\begin{aligned}
\hat{G}(k+1) &= \begin{bmatrix} (A - L(k)\hat{\bar{f}}_a^{-1}(k)\bar{C})\tilde{\hat{G}}(k) & EG_w & -L(k)\bar{F}\bar{G}_v \end{bmatrix}, \\
\hat{G}(k) &= \downarrow_s \hat{G}(k).
\end{aligned} \tag{6.71}
$$

Proof 6.6 From Eq. (6.61), we can obtain

$$
\begin{aligned}
\hat{x}(k+1) &= A\hat{x}(k) + Bu(k) + Ew(k) + L(k)(\bar{y}(k) - \hat{\bar{f}}_a^{-1}(k)\bar{C}\hat{x}(k) - \bar{F}\bar{v}(k)), \\
&= (A - L(k)\hat{\bar{f}}_a^{-1}(k)\bar{C})\hat{x}(k) + Bu(k) + Ew(k) + L(k)\bar{y}(k) - L(k)\bar{F}\bar{v}(k) \\
&\in ((A - L(k)\hat{\bar{f}}_a^{-1}(k)\bar{C}) \odot \langle \hat{p}(k), \hat{G}(k)\rangle) \oplus Bu(k) \oplus (E \odot \langle 0, G_w\rangle) \\
&\oplus L(k)\bar{y}(k) \oplus (-L(k)\bar{F} \odot \langle 0, \bar{G}_v\rangle) \\
&= \langle (A - L(k)\hat{\bar{f}}_a^{-1}(k)\bar{C})\hat{p}(k) + Bu(k) + L(k)\bar{y}(k), \\
&\quad [(A - L(k)\hat{\bar{f}}_a^{-1}(k)\bar{C})\hat{G}(k)\ EG_w\ -L(k)\bar{F}\bar{G}_v]\rangle.
\end{aligned} \tag{6.72}
$$

Considering the complexity reduction matrix $\hat{\tilde{G}}(k)$ of $\hat{G}(k)$, the estimated zonotopic set $k+1$ can be obtained as $\hat{X}(k+1) = \langle \hat{p}(k+1), \hat{G}(k+1) \rangle = \langle (A - L(k)\hat{\bar{f}}_a^{-1}(k)\bar{C})\hat{p}(k) + Bu(k) + L(k)\bar{y}(k), [(A - L(k)\hat{\bar{f}}_a^{-1}(k)\bar{C})\hat{\tilde{G}}(k) \ EG_w \ - L(k)\bar{F}\bar{G}_v] \rangle, (\hat{\tilde{G}}(k) =\downarrow_s \hat{G}(k))$.

According to Eq. (6.63) and Property 10, the numerator of $\hat{f}_i(k+1)$ can be further expanded as:

$$C_i\hat{x}(k+1) \in C_i \odot \langle \hat{p}(k+1), \hat{G}(k+1) \rangle \subset [\hat{I}_{Gi}^-(k+1), \hat{I}_{Gi}^+(k+1)], \tag{6.73}$$

$$\begin{cases} \hat{I}_{Gi}^-(k+1) = C_i\hat{p}(k+1) - C_i \begin{bmatrix} \sum_{j=1}^{r_{\hat{G}}} |\hat{G}_{1j}(k+1)| \\ \vdots \\ \sum_{j=1}^{r_{\hat{G}}} |\hat{G}_{n_xj}(k+1)| \end{bmatrix}, \ i = 1, \ldots, n_y, \\ \hat{I}_{Gi}^+(k+1) = C_i\hat{p}(k+1) + C_i \begin{bmatrix} \sum_{j=1}^{r_{\hat{G}}} |\hat{G}_{1j}(k+1)| \\ \vdots \\ \sum_{j=1}^{r_{\hat{G}}} |\hat{G}_{n_xj}(k+1)| \end{bmatrix}, \ i = 1, \ldots, n_y. \end{cases} \tag{6.74}$$

Based on the same method above, we can also extend the denominator to an interval:

$$y_i(k+1) - F_iv(k+1) \in y_i(k+1) \oplus [-F_i \odot \langle 0, G_v \rangle] \subset [\hat{I}_{vi}^-(k+1), \hat{I}_{vi}^+(k+1)], \tag{6.75}$$

$$\begin{cases} \hat{I}_{vi}^-(k+1) = y_i(k+1) - F_i \begin{bmatrix} \sum_{j=1}^{r_{G_v}} |G_{v1j}(k)| \\ \vdots \\ \sum_{j=1}^{r_{G_v}} |G_{vn_vj}(k)| \end{bmatrix}, \ i = 1, \ldots, n_y, \\ \hat{I}_{vi}^+(k+1) = y_i(k+1) + F_i \begin{bmatrix} \sum_{j=1}^{r_{G_v}} |G_{v1j}(k)| \\ \vdots \\ \sum_{j=1}^{r_{G_v}} |G_{vn_vj}(k)| \end{bmatrix}, \ i = 1, \ldots, n_y. \end{cases} \tag{6.76}$$

Thus, according to Definition 6.3, the estimated interval of the ith multiplicative sensor fault can be obtained as:

$$\begin{aligned} \hat{I}_{fi}(k+1) &= [\hat{I}_{Gi}^-(k+1), \hat{I}_{Gi}^+(k+1)] \div [\hat{I}_{vi}^-(k+1), \hat{I}_{vi}^+(k+1)] \\ &= [\hat{f}_i^-(k+1), \hat{f}_i^+(k+1)], \ (0 \notin [\hat{I}_{vi}^-(k+1), \hat{I}_{vi}^+(k+1)]), \end{aligned} \tag{6.77}$$

where

$$\hat{f}_i^-(k+1) = \min\left\{\frac{\hat{I}_{Gi}^-(k+1)}{\hat{I}_{vi}^+(k+1)}, \frac{\hat{I}_{Gi}^-(k+1)}{\hat{I}_{vi}^-(k+1)}, \frac{\hat{I}_{Gi}^+(k+1)}{\hat{I}_{vi}^+(k+1)}, \frac{\hat{I}_{Gi}^+(k+1)}{\hat{I}_{vi}^-(k+1)}\right\},$$

$$\hat{f}_i^+(k+1) = \max\left\{\frac{\hat{I}_{Gi}^-(k+1)}{\hat{I}_{vi}^+(k+1)}, \frac{\hat{I}_{Gi}^-(k+1)}{\hat{I}_{vi}^-(k+1)}, \frac{\hat{I}_{Gi}^+(k+1)}{\hat{I}_{vi}^+(k+1)}, \frac{\hat{I}_{Gi}^+(k+1)}{\hat{I}_{vi}^-(k+1)}\right\}.$$

Subsequently, $\hat{I}_f(k+1) = [\hat{I}_{f1}(k+1), \ldots, \hat{I}_{fi}(k+1), \ldots, \hat{I}_{fn_y}(k+1)]$ $(i = 1, \ldots, n_y)$ can be obtained.

Theorem 6.7

For the interval box of Eq. (6.62), the optimal gain matrix $L(k)$ for the multiplicative sensor fault estimator is designed by minimizing $J_M(k+1) = \|\hat{G}(k+1)\|_{F,M}^2 = tr(\hat{G}^{\mathrm{T}}(k+1)M\hat{G}(k+1))$. Define $\tilde{P}(k) = \tilde{G}(k)\tilde{G}^{\mathrm{T}}(k)$, $Q_w = EG_wG_w^{\mathrm{T}}E^{\mathrm{T}}$, $\bar{Q}_v = \bar{F}\bar{G}_v\bar{G}_v^{\mathrm{T}}\bar{F}^{\mathrm{T}}$, and the expression of the optimal gain matrix is:

$$L(k) = AK(k),\ K(k) = R(k)S^{-1}(k),\ R(k) = \tilde{P}(k)\bar{C}^{\mathrm{T}}\hat{\bar{f}}_a^{-1}(k), \tag{6.78}$$

$$S(k) = \hat{\bar{f}}_a^{-1}(k)\bar{C}\tilde{P}(k)\bar{C}^{\mathrm{T}}\hat{\bar{f}}_a^{-1}(k) + \bar{Q}_v. \tag{6.79}$$

Proof 6.7 From Theorem 6.6, the size of the interval $\hat{I}_{fi}(k+1)(i = 1, \ldots, n_y)$ is related to $\hat{X}(k+1)$ and V. Since V is fixed in the system, so the size of $\hat{I}_{fi}(k+1)$ mainly depends on $\hat{X}(k+1)$. For the zonotope $\hat{X}(k+1)$, $J_M(k+1) = \|\hat{G}(k+1)\|_{F,M}^2 = tr(\hat{G}^{\mathrm{T}}(k+1)M\hat{G}(k+1))$ is the judgment criteria using to determine its size according to Definitions 6.5 and 6.6. Thus, the optimal gain matrix $L(k)$ for the multiplicative sensor fault estimator is designed by making $\partial_{L(k)}J_M(k+1) = 0$ to minimize $J_M(k+1)$.

From Eq. (6.71), it can be obtained that:

$$\begin{aligned}
\partial_{L(k)}J_M(k+1) &= \partial_{L(k)}tr(\hat{G}^{\mathrm{T}}(k+1)M\hat{G}(k+1)) \\
&= \partial_{L(k)}tr(M(A - L(k)\hat{\bar{f}}_a^{-1}(k)\bar{C})\tilde{P}(k)(A - L(k)\hat{\bar{f}}_a^{-1}(k)\bar{C})^{\mathrm{T}} \\
&+ MQ_w + ML(k)\bar{Q}_vL^{\mathrm{T}}(k)) \\
&= \partial_{L(k)}tr(M(A\tilde{P}(k)A^{\mathrm{T}} - A\tilde{P}(k)\bar{C}^{\mathrm{T}}\hat{\bar{f}}_a^{-1}(k)L^{\mathrm{T}}(k) \\
&- L(k)\hat{\bar{f}}_a^{-1}(k)\bar{C}\tilde{P}(k)A^{\mathrm{T}} \\
&+ L(k)\hat{\bar{f}}_a^{-1}(k)\bar{C}\tilde{P}(k)\bar{C}^{\mathrm{T}}\hat{\bar{f}}_a^{-1}(k)L^{\mathrm{T}}(k) + L(k)\bar{Q}_vL^{\mathrm{T}}(k))) \\
&= \partial_{L(k)}tr(ML(k)\hat{\bar{f}}_a^{-1}(k)\bar{C}\tilde{P}(k)\bar{C}^{\mathrm{T}}\hat{\bar{f}}_a^{-1}(k)L^{\mathrm{T}}(k) + L(k)\bar{Q}_vL^{\mathrm{T}}(k)) \\
&- 2\partial_{L(k)}tr(MA\tilde{P}(k)\bar{C}^{\mathrm{T}}\hat{\bar{f}}_a^{-1}(k)L^{\mathrm{T}}(k)).
\end{aligned} \tag{6.80}$$

Let $\partial_{L(k)}J_M(k+1) = 0$, then

$$\begin{aligned}
&\partial_{L(k)}tr(ML(k)\hat{\bar{f}}_a^{-1}(k)\bar{C}\tilde{P}(k)\bar{C}^{\mathrm{T}}\hat{\bar{f}}_a^{-1}(k)L^{\mathrm{T}}(k)) + \partial_{L(k)}tr(ML(k)\bar{Q}_vL^{\mathrm{T}}(k)) \\
&\quad = 2\partial_{L(k)}tr(MA\tilde{P}(k)\bar{C}^{\mathrm{T}}\hat{\bar{f}}_a^{-1}(k)L^{\mathrm{T}}(k)).
\end{aligned} \tag{6.81}$$

Considering $M = M^{\mathrm{T}} > 0$ and the operations of the matrix traces defined in Definition 6.2, we can get:

$$ML(k)(\hat{\bar{f}}_a^{-1}(k)\bar{C}\tilde{P}(k)\bar{C}^{\mathrm{T}}\hat{\bar{f}}_a^{-1}(k)+\bar{Q}_v) = MA\tilde{P}(k)\bar{C}^{\mathrm{T}}\hat{\bar{f}}_a^{-1}(k) \quad (6.82)$$

$$L(k)S(k) = AR(k). \quad (6.83)$$

Thus

$$L(k) = AK(k), \quad (6.84)$$

where $K(k) = R(k)S^{-1}(k)$, $R(k) = \tilde{P}(k)\bar{C}^{\mathrm{T}}\hat{\bar{f}}_a^{-1}(k)$, $S(k) = \hat{\bar{f}}_a^{-1}(k)\bar{C}\tilde{P}(k)\bar{C}^{\mathrm{T}}\hat{\bar{f}}_a^{-1}(k) + \bar{Q}_v$.

6.2.4 Simulation analysis

In order to verify the effectiveness and feasibility of the proposed sensor fault estimation based on the constrained zonotopic Kalman filter for system with constraints, two simulation examples are used for simulation verification.
Example 1: Consider the following linear discrete system:

$$\begin{cases} x(k+1) = Ax(k) + Bu(k) + Ew(k), \\ y(k) = Cx(k) + Fv(k), \end{cases} \quad (6.85)$$

where

$$A = \begin{bmatrix} 0.40 & -0.04 \\ 1.20 & 0.50 \end{bmatrix}, B = \begin{bmatrix} 5 \\ -3 \end{bmatrix}, C = \begin{bmatrix} 1 & 0 \\ 0 & 1 \end{bmatrix}, E = \begin{bmatrix} 1 & 0 \\ 0 & 1 \end{bmatrix}, F = \begin{bmatrix} 1 & 0 \\ 0 & 1 \end{bmatrix}.$$

The known input $u(k) = 6$, $w(k)$ and $v(k)$ satisfy $\|w(k)\|_\infty \leq 0.15$ and $\|v(k)\|_\infty \leq 0.2$, the initial state, process interference, measurement noise meet:

$$\hat{x}(0) \in X = \left\langle \begin{bmatrix} 0 \\ 0 \end{bmatrix}, \begin{bmatrix} 10 & 0 \\ 0 & 10 \end{bmatrix} \right\rangle,$$

$$w(k) \in W = \left\langle \begin{bmatrix} 0 \\ 0 \end{bmatrix}, \begin{bmatrix} 0.5 & 0 \\ 0 & 0.5 \end{bmatrix} \right\rangle,$$

$$v(k) \in V = \left\langle \begin{bmatrix} 0 \\ 0 \end{bmatrix}, \begin{bmatrix} 1 & 0 \\ 0 & 1 \end{bmatrix} \right\rangle.$$

It is known that the constraint of the system satisfies $|h(x)| = |\gamma(k) - H(k)x(k)| \leq \tilde{h}$ in the fault-free state after the system reaches stability, where

$$\gamma(k) = \begin{bmatrix} 45.2 \\ 72.4 \end{bmatrix}, H(k) = \begin{bmatrix} 1 & 0 \\ 0 & 1 \end{bmatrix}, \tilde{h} = \begin{bmatrix} 0.1 \\ 0.1 \end{bmatrix}.$$

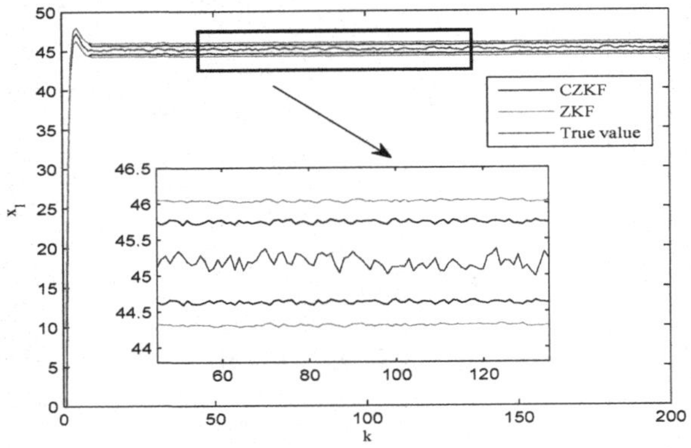

Figure 6.6: Estimation result of state x_1 in the fault-free state.

6.2.4.1 Fault-free

For systems with constraints, in order to comprehensively analyze the effectiveness and feasibility of the proposed CZKF algorithm, the CZKF algorithm is compared with the ZKF algorithm in the state estimation.

It can be seen from Figs. 6.6–6.8 that before the system's state reaches stability, the system is not subject to constraints. In this case, the CZKF algorithm is equal to the ZKF algorithm. At this time, the upper and lower bound curves of the estimated state set obtained by the two algorithms are similar, and the zonotopic sets of CZKF algorithm (black line) and ZKF algorithm (magenta line) are coincident at the time $k = 5$. For this system, it can be obtained from the curves that the system quickly reaches stability, and the constraints of the system are added at the time $k = 9$. After

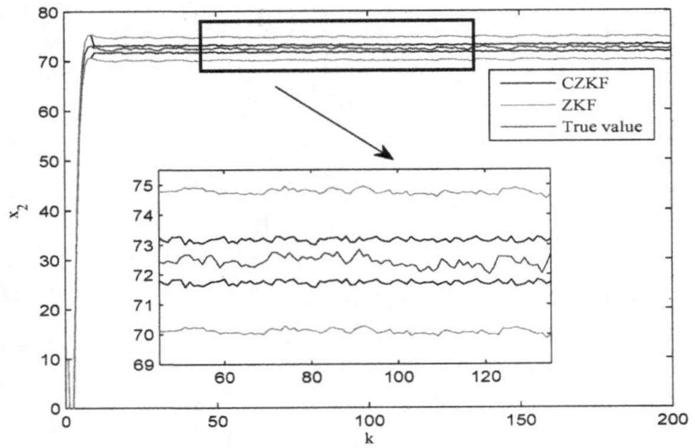

Figure 6.7: Estimation result of state x_2 in the fault-free state.

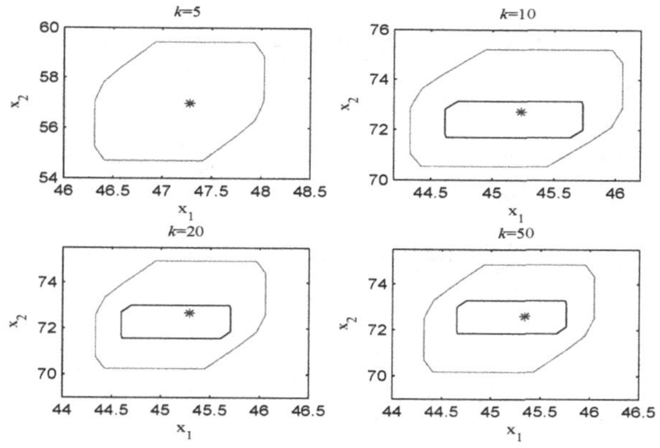

Figure 6.8: State estimation results of the two algorithms at different times in the fault-free state.

this time, the interval between the upper and lower bounds of the estimated state set corresponding to the CZKF algorithm is obviously smaller than that corresponding to the ZKF algorithm, and the zonotopic sets corresponding to the CZKF algorithm are always contained in the zonotopic sets corresponding to the ZKF algorithm, indicating that the system's constraints play an effective role in the system state estimation, making the zonotopic set of the state estimation corresponding to the CZKF algorithm smaller than that corresponding to the ZKF algorithm.

Thus, under the comparative analyses, the CZKF algorithm is less conservative owing to the constraints. In order to highlight the superiority of the proposed CZKF algorithm and the effect of constraints on the algorithm, the following analyses are focused on the comparisons between the CZKF algorithm and ZKF algorithm, and the effectiveness and advantages of the CZKF algorithm in the sensor fault estimation.

6.2.4.2 Additive sensor fault

It is assumed that the system has three different additive sensor faults at the time of $k = 100$, $k = 150$ and $k = 200$, and $f(k) = [\ 1.5 \quad 1\ ]^{\mathrm{T}}$ $(100 \leq k < 150)$, $f(k) = [\ 40 \quad 5\ ]^{\mathrm{T}}$ $(150 \leq k < 200)$ and $f(k) = [\ 5 \quad 60\ ]^{\mathrm{T}}$ $(200 \leq k \leq 250)$, respectively. It is worth noting that in order to reduce the uncertainty of $\hat{f}_a(k)$, the selection range of $\hat{f}_a(k)$ in $\hat{X}_f(k)$ can be appropriately limited in the simulation. In this simulation, $\hat{f}_a(k)$ is an arbitrary value in a zonotopic set with $\hat{p}_f(k)$ as the center, $[0.1 \cdot \hat{G}_f(k)]$ as the generator matrix. The CZKF-FE algorithm and ZKF-FE algorithm proposed in this section are used in the additive sensor fault estimation and the fault estimation results are shown in Figs. 6.9–6.12.

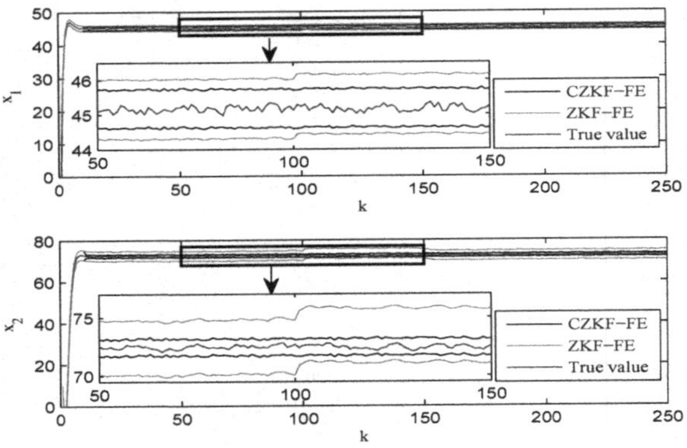

Figure 6.9: State estimation results in the additive sensor fault state.

Fig. 6.10 shows the relationship between the true output of the system and the estimated output results of the two algorithms at different time. The period during the fault-free state, the periods during the first, second and third additive sensor fault states are selected as the representative periods and four points are selected in the four representative periods as examples, respectively. In the fault-free state, the zonotopic sets of the estimated output based on the CZKF-FE algorithm (black line) and ZKF-FE algorithm (magenta line) both contain the true output value (blue star) of the system, indicating there is no fault occurred in the system. And throughout the whole process of state estimation and fault estimation, the estimated zonotopic sets based on the CZKF-FE algorithm are all included in the estimated zonotopic sets based on the ZKF-FE algorithm, indicating that compared with the ZKF-FE algorithm, the

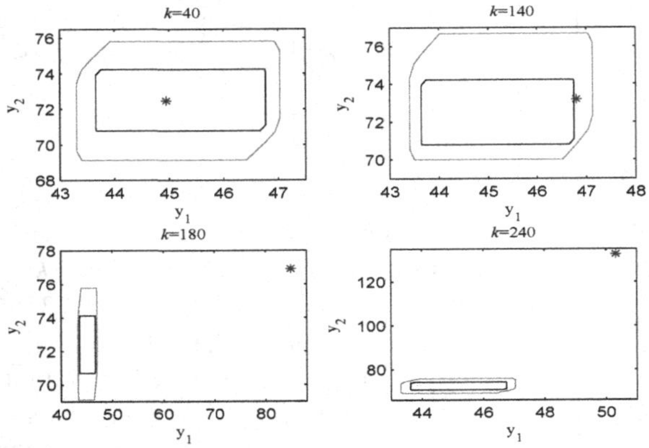

Figure 6.10: Estimated output of two algorithms and true output at different times in the additive sensor fault state.

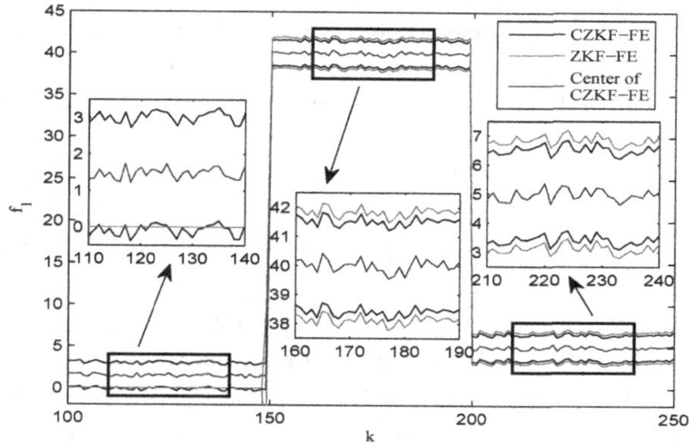

Figure 6.11: Estimated result of fault f_1 in the additive sensor fault state.

CZKF-FE algorithm is less conservative. Once an additive sensor fault occurs in the system, the relationship between the true output of the system and the zonotopic set of the estimated output will change. In the first fault state, the zonotopic set of the estimated output based on the CZKF-FE algorithm does not contain the true output of the system, indicating there is a sensor fault occurred in the system. However, since the sensor fault is relatively small at this moment, the change in the true output of the system is not obvious. At the same time, due to the great conservativeness of the ZKF-FE algorithm, the estimated output set is large. Even if the system fails, the estimated output set still contains the true output, which leads to the failed detection of the ZKF-FE algorithm at this faulty moment. As a result, there is a fault missing detection at this time. Therefore, compared with the CZKF-FE algorithm, the

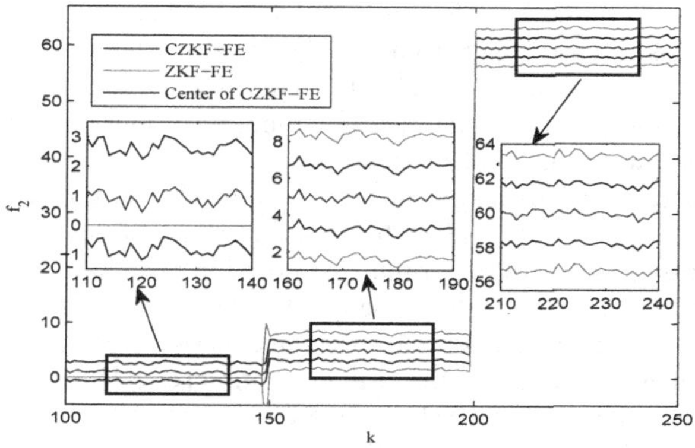

Figure 6.12: Estimated result of fault f_2 in the additive sensor fault state.

ZKF-FE algorithm has lower fault detection accuracy at this time. In the second and third fault states, the output changes are much more obvious than the first fault state. At these two periods, the estimated output sets of the two algorithms do not contain the true output, which indicates that both algorithms have detected that the system has failed during these two time periods.

It can be seen from Fig. 6.9 that the true value of the system state has been included in the upper and lower bounds of the CZKF-FE and ZKF-FE algorithms during the entire working process, but the upper and lower bounds of the ZKF-FE algorithm have changed during the time period of $k = 100 - 150$, and it can be clearly seen that the center of ZKF-FE algorithm has jumped during this period of time, which is different from the true state value, this is because that the ZKF-FE algorithm cannot detect the sensor fault due to the great conservativeness. So the estimated state of the algorithm is affected by the faulty output and changed. However, during the entire fault working process, the upper and lower bounds obtained based on the CZKF-FE algorithm basically unchanged compared with the normal working state, indicating that the algorithm can detect the faults in time in these three fault time periods, and the additive sensor fault has no effect on the system's state estimation result and an accurate state estimation value can still be obtained based on the fault estimation algorithm. In addition, compared with the ZKF-FE algorithm, the interval between the upper and lower bounds of the estimated state obtained by the CZKF-FE algorithm is smaller, indicating that the algorithm is less conservative owing to the constraints, and it can contain the true state of the system more accurately.

According to the analysis of Figs. 6.11 and 6.12, the fault estimation curve of the ZKF-FE algorithm is equal to zero at time $k = 100 - 150$, indicating this algorithm cannot estimate the system's fault at this time period because of the missing fault detection. But the curves of the CZKF-FE algorithm is not zero at this time period, and is close to the true fault value of the additive sensor fault, which indicates that the fault detection and fault estimation have been successfully realized at this moment. Besides, the fault estimation curves of the CZKF-FE and ZKF-FE algorithms are not equal to zero from time $k = 150$, indicating that both algorithms can determine that the system has a sensor fault from this time, and the fault estimation is made subsequently. Therefore, it can be concluded that both algorithms can realize fault detection quickly and accurately. At the same time, according to the change curves of the upper and lower bounds of the fault estimation sets obtained by the two algorithms and the center of the fault estimation obtained by the CZKF-FE algorithm, both algorithms can accurately realize the fault estimation of the system, indicating that the CZKF-FE and ZKF-FE algorithms proposed in this section are feasible and effective in the additive sensor fault estimation. In addition, compared with the ZKF-FE algorithm, the interval between the upper and lower bounds of the CZKF-FE algorithm is smaller, which means the fault estimation set calculated by the CZKF-FE algorithm can contain the true value of the fault more tightly, and contributing a less conservative performance in the CZKF-FE algorithm.

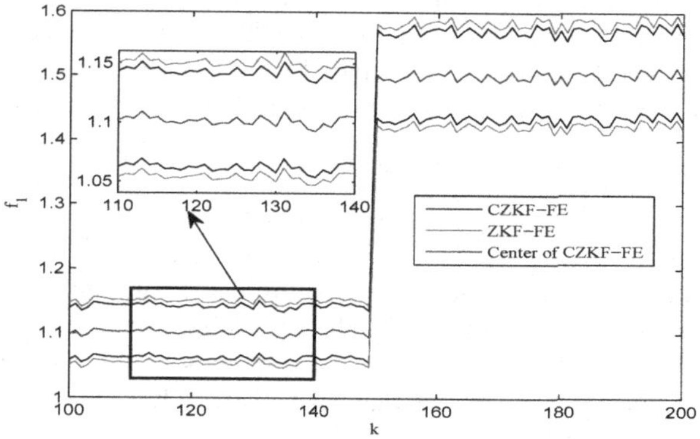

Figure 6.13: Estimated result of fault f_1 in the multiplicative sensor fault state.

6.2.4.3 Multiplicative sensor fault

It is assumed that the system has two different multiplicative sensor faults at the time of $k = 100$ and $k = 150$, and $f(k) = \begin{bmatrix} 1.1 & 0 \\ 0 & 1.2 \end{bmatrix}$ $(100 \leq k < 150)$ and $f(k) = \begin{bmatrix} 1.5 & 0 \\ 0 & 1.8 \end{bmatrix}$ $(150 \leq k \leq 200)$, respectively. Similar to the additive sensor fault, in order to reduce the uncertainty of the simulation, $\hat{f}_{ai}(k)(1,\ldots,n_y)$ is selected as an arbitrary value in the interval with $\frac{\hat{f}_i^+(k)+\hat{f}_i^-(k)}{2}$ as the center and $0.1 \cdot \frac{\hat{f}_i^+(k)-\hat{f}_i^-(k)}{2}$ as the distance. The CZKF-FE and ZKF-FE algorithms proposed in this section are applied to the fault estimation of the multiplicative sensor fault and the fault estimation results are shown in Figs. 6.13 and 6.14. The analysis shows that both

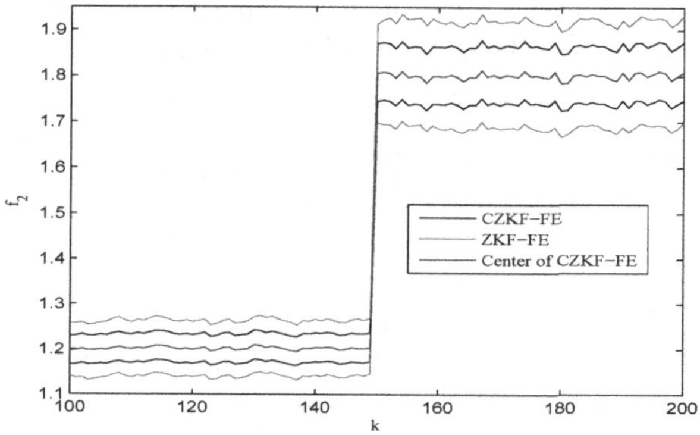

Figure 6.14: Estimated result of fault f_2 in the multiplicative sensor fault state.

algorithms can detect that the system has a fault at time $k = 100$, and there are faults with different fault values occurred at time $k = 100$ and $k = 150$. According to the center of the estimated zonotopic set obtained by the CZKF-FE algorithm, it can be obtained that the system has a multiplicative sensor fault with a fault value around $f(k) = \begin{bmatrix} 1.1 & 0 \\ 0 & 1.2 \end{bmatrix}$ within the time period of $k = 100 - 150$, and a multiplicative sensor fault with a fault value around $f(k) = \begin{bmatrix} 1.5 & 0 \\ 0 & 1.8 \end{bmatrix}$ within the time period of $k = 150 - 200$, which shows that the algorithm can accurately estimate the sensor fault value and has the advantage of high fault estimation accuracy. At the same time, the interval between the upper and lower bounds of the fault estimation set of the CZKF-FE algorithm is smaller than that of the ZKF-FE algorithm, indicating that the CZKF-FE algorithm has great advantage in the conservativeness of fault estimation.

Example 2: In the lithium battery, Thevenin model is often used in the study of the equivalent circuit model of the lithium battery because of its characteristics of simple structure and completeness. At the same time, a RC circuit can be added in the Thevenin model to form the second-order RC circuit model as shown in Fig. 6.15 [113, 123].

In the lithium battery circuit, R_1 and R_2 are the polarization internal resistances, and C_1 and C_2 are the polarization capacitances. U_1 and U_2 denote the voltages generated by the polarization phenomenon. U_{oc} represents the open circuit voltage of the battery, U is the terminal voltage of the battery, I stands the battery current and R_0 is the ohmic internal resistance. According to Kirchhoff's law of voltage, the terminal voltage $U = U_{oc} - R_0 I - U_1 - U_2$. At the same time, according to the working principle of the circuit, the following voltage dynamic characteristics can be obtained:

$$\begin{cases} \dot{U}_1 = -\frac{1}{R_1 C_1} U_1 + \frac{1}{C_1} I, \\ \dot{U}_2 = -\frac{1}{R_2 C_2} U_2 + \frac{1}{C_2} I. \end{cases} \tag{6.86}$$

When the battery works at high power, it will release a lot of heat, which will cause the temperature inside the battery to rise, making the temperature of the battery core higher than the surface. Therefore, in order to ensure the battery performance, it

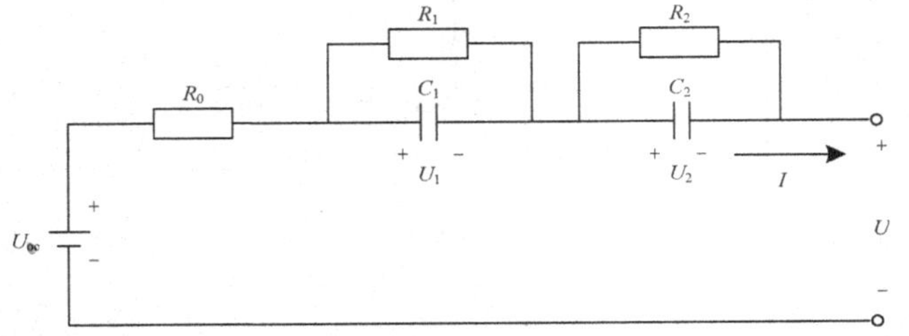

Figure 6.15: Schematic diagram of second-order RC circuit model.

is of great practical significance to estimate the battery temperature and the possible temperature sensor fault to ensure the normal operation of the battery. According to the principles of heat generation and heat conduction, a two-state thermal sub-model can be established:

$$\begin{cases} C_c\dot{T}_c = \frac{T_s-T_c}{R_c} + Q_{gen}, \\ C_s\dot{T}_s = -\frac{T_s-T_c}{R_c} + \frac{T_e-T_s}{R_u}, \end{cases} \tag{6.87}$$

where C_s is the heat capacity coefficient of the internal material of the battery, C_c is the heat capacity coefficient of the battery surface, R_c is the thermal resistance between the core and the surface of the battery and R_u is the convection resistance between the battery surface and the cooling air, $Q_{gen} = I(U_{oc} - U) = I(R_0I + U_1 + U_2)$ is the heating power of the battery core. Take heating power Q_{gen} and ambient temperature T_e as inputs, core temperature T_c and battery surface temperature T_s are states. After discretization, the state space equation of the thermal model can be established as:

$$\begin{cases} x(k+1) = Ax(k) + Bu(k) + w(k), \\ y(k) = Cx(k) + v(k), \end{cases} \tag{6.88}$$

where $w(k)$ and $v(k)$ are the unknown but bounded process disturbance and measurement noise of the system, $x = [T_c\ T_s]^{\mathrm{T}}$, $u = [Q_{gen}\ T_e]^{\mathrm{T}}$, $y = [T_c\ T_s]^{\mathrm{T}}$, Δt represents the sampling time, and

$$A = \begin{bmatrix} 1 - \frac{\Delta t}{R_cC_c} & \frac{\Delta t}{R_cC_c} \\ \frac{\Delta t}{R_cC_s} & 1 - \frac{\Delta t}{R_cC_s} - \frac{\Delta t}{R_uC_s} \end{bmatrix}, B = \begin{bmatrix} \frac{\Delta t}{C_c} & 0 \\ 0 & \frac{\Delta t}{R_uC_s} \end{bmatrix}, C = \begin{bmatrix} 1 & 0 \\ 0 & 1 \end{bmatrix}.$$

When the lithium battery is working in the fault-free state, set the parameters $R_1 = 0.0298\,\Omega$, $C_1 = 1787.7\,\mathrm{F}$, $R_2 = 0.03819\,\Omega$, $C_2 = 5.26\,\mathrm{F}$, $I = 5\,\mathrm{A}$, $R_0 = 0.0501\,\Omega$, $R_c = 1.98°\mathrm{C/W}$, $R_u = 1.718°\mathrm{C/W}$, $C_c = 63.5\,\mathrm{J/°C}$, $C_s = 4.5\,\mathrm{J/°C}$, $T_e = 18°\mathrm{C}$, $\Delta t = 1\,\mathrm{s}$, $w(k)$ and $v(k)$ satisfy $\|w(k)\|_\infty \le 0.1$ and $\|v(k)\|_\infty \le 0.1$. The initial state of the system, process noise and measurement noise meet

$$\begin{aligned} \hat{x}(0) \in X &= \left\langle \begin{bmatrix} 18 \\ 18 \end{bmatrix}, \begin{bmatrix} 4 & 0 \\ 0 & 4 \end{bmatrix} \right\rangle, \\ w(k) \in W &= \left\langle \begin{bmatrix} 0 \\ 0 \end{bmatrix}, \begin{bmatrix} 0.5 & 0 \\ 0 & 0.5 \end{bmatrix} \right\rangle, \\ v(k) \in V &= \left\langle \begin{bmatrix} 0 \\ 0 \end{bmatrix}, \begin{bmatrix} 1 & 0 \\ 0 & 1 \end{bmatrix} \right\rangle. \end{aligned}$$

For lithium battery operating systems with higher precision, the core temperature is generally restricted to remain within a certain range and the core temperature and surface temperature meet specific differential conditions in the normal operating state. For this system, the core temperature $T_c \in [28.6°\mathrm{C}, 29.4°\mathrm{C}]$, and the difference between the core temperature and the surface temperature is $5.85°\mathrm{C} \pm 0.1°\mathrm{C}$ when lithium battery system is stable.

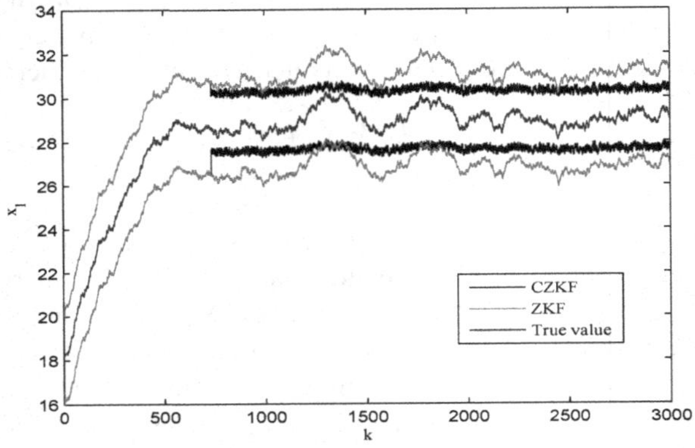

Figure 6.16: Estimation result of state x_1 in the fault-free state.

6.2.4.4 *Fault-free state*

For the thermal model in the lithium battery system, as shown in Figs. 6.16–6.18, the CZKF algorithm and ZKF algorithm can be used in the system's state estimation. The figures show that both CZKF algorithm and ZKF algorithm can realize the state estimation of the system. When the system is stable, the system's constraints can be added in the CZKF algorithm, making the true value of the system contain in the estimated zonotopic set more tightly. Thus, the CZKF algorithm has a less conservative performance.

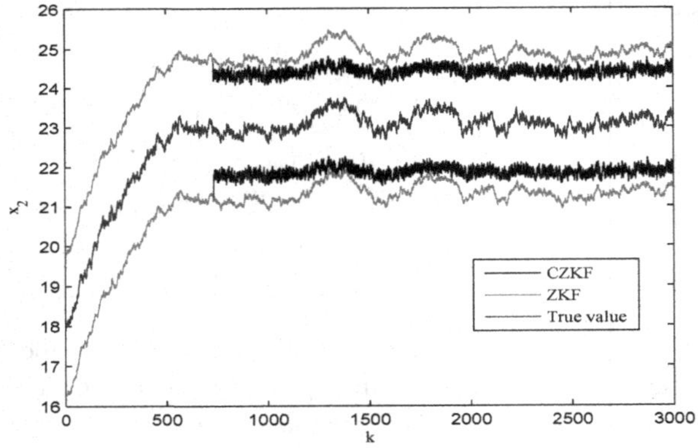

Figure 6.17: Estimation result of state x_2 in the fault-free state.

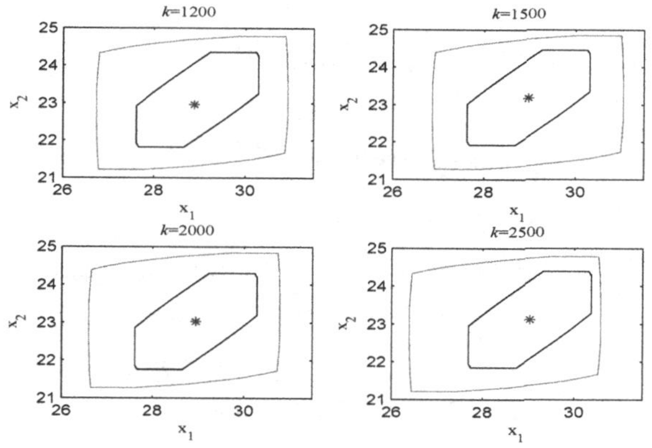

Figure 6.18: State estimation results of the two algorithms at different times in the fault-free state.

6.2.4.5 Additive sensor fault

Assuming that the lithium battery system has three different additive sensor faults at the time of $k = 2000$, $k = 3000$ and $k = 4000$, and the specific values are $f(k) = [\ 10 \quad 0\ ]^{\mathrm{T}}$ $(2000 \leq k < 3000)$, $f(k) = [\ 10 \quad 5\ ]^{\mathrm{T}}$ $(3000 \leq k < 4000)$, $f(k) = [\ 5 \quad -5\ ]^{\mathrm{T}}$ $(4000 \leq k \leq 5000)$. $\hat{f}_a(k)$ is selected as an arbitrary value in a zonotopic set $\langle \hat{p}_f(k), [0.1 \cdot \hat{G}_f(k)] \rangle$.

The CZKF-FE algorithm and the ZKF-FE algorithm are applied to the fault estimation of the additive sensor fault for the thermal model in the lithium battery system. From Figs. 6.19 and 6.20, the fault estimation curves show that both algorithms can detect the fault in time and track the true value of the fault quickly. It can be

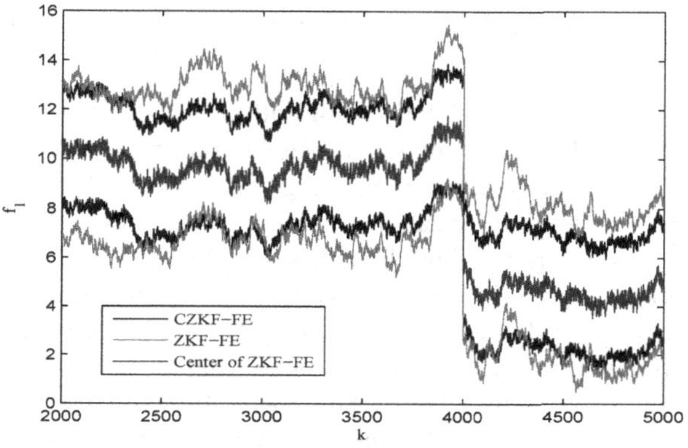

Figure 6.19: Estimated result of fault f_1 in the additive sensor fault state.

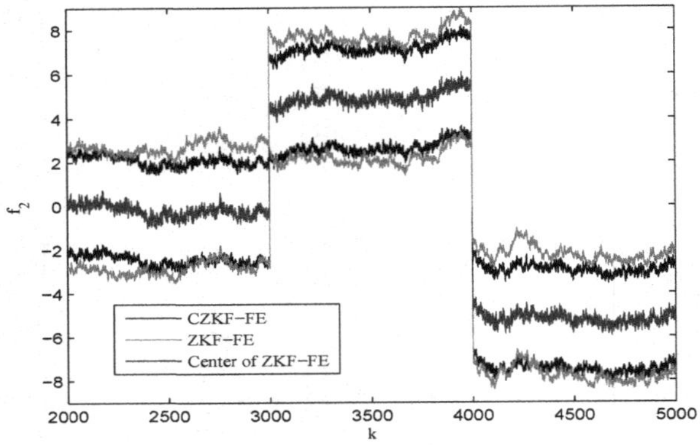

Figure 6.20: Estimated result of fault f_2 in the additive sensor fault state.

concluded from the curves that three different additive sensor faults occurred in the system at time $k = 2000$, $k = 3000$ and $k = 4000$. At the same time, due to the constraints, the interval between the upper and lower bounds of the fault estimation set obtained by the CZKF-FE algorithm are smaller than that obtained by the ZKF-FE algorithm, so the CZKF-FE algorithm can contain the true fault value of the system more tightly.

6.2.4.6 Multiplicative sensor fault

Assuming that the lithium battery system has two different multiplicative sensor faults at the time of $k = 2000$ and $k = 3000$, and the specific values are $f(k) = \begin{bmatrix} 2 & 0 \\ 0 & 1.5 \end{bmatrix}$ $(2000 \leq k < 3000)$, $f(k) = \begin{bmatrix} 1.5 & 0 \\ 0 & 2.2 \end{bmatrix}$ $(3000 \leq k \leq 4000)$. $\hat{f}_{ai}(k), (i = 1, \ldots, n_y)$ is selected as an arbitrary value in the interval with $\frac{\hat{f}_i^+(k)+\hat{f}_i^-(k)}{2}$ as the center and $0.1 \cdot \frac{\hat{f}_i^+(k)-\hat{f}_i^-(k)}{2}$ as the distance.

The CZKF-FE algorithm and ZKF-FE algorithm are used to estimate the multiplicative sensor fault of for the thermal model in the lithium battery system. Through the analyses of the fault estimation curves in Figs. 6.21 and 6.22, it can be concluded that there are two different multiplicative sensor faults occurred in the system at time $k = 2000$ and $k = 3000$. Similarly, the fault estimation result obtained by the CZKF-FE algorithm has a tighter interval. It is worth mentioning that the division in interval operation is introduced in the fault estimation of the multiplicative sensor. According to the system equation (6.39), system's output has different values because of the different multiplicative sensor faults, which will affect the interval calculation of the fault estimation, and the size of the fault estimation interval is inconsistent in the different multiplicative sensor fault states.

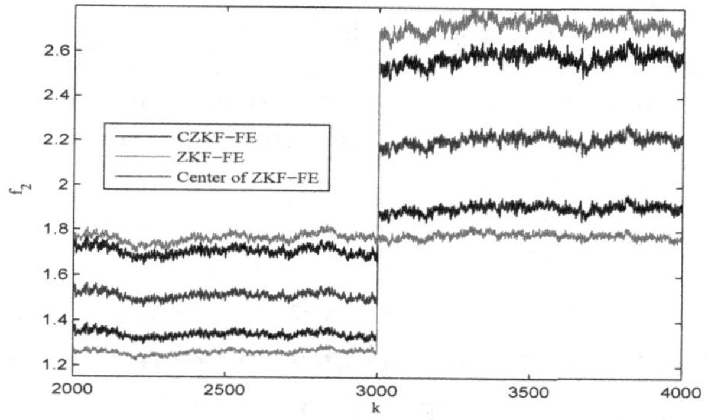

Figure 6.21: Estimated result of fault f_1 in the multiplicative sensor fault state.

6.2.5 Conclusion

Two constrained zonotopic Kalman filter based sensor fault estimators are designed in this section for the additive sensor fault and multiplicative sensor fault in the constrained system with unknown but bounded noises. The proposed fault estimation method uses the constrained zonotopic Kalman filter to estimate the system's state in the normal state. The fault is detected by checking whether the true value of the system's output is within the upper and lower bounds of the corresponding estimated zonotopic set. Sensor fault estimation is completed through the designed constrained zonotopic Kalman filter based fault estimator. The sensor fault estimation of the numerical model and the thermal model in the lithium battery are used as examples to verify the effectiveness and feasibility of the method, and the simulation results show that the method has the advantages of fast fault detection speed and high fault

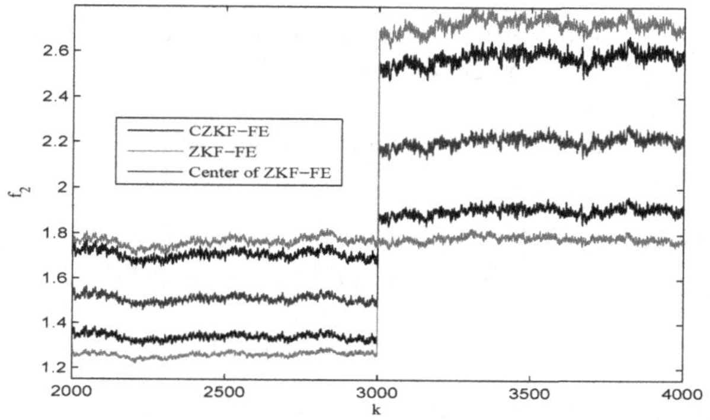

Figure 6.22: Estimated result of fault f_2 in the multiplicative sensor fault state.

estimation accuracy. Comparative analyses indicate that the CZKF-FE algorithm is less conservative, compared with the ZKF-FE algorithm.

The fault estimation method proposed in this section can also be applicable to the sensor fault estimation of other constrained systems with unknown but bounded noise. Moreover, it can further extend the constrained fault estimation method to the systems with other faulty types.

6.3 Optimal zonotopic Kalman filter-based state estimation and fault diagnosis algorithm for linear discrete-time system with time delay

6.3.1 Problem formulation and preliminaries

Considering the following linear discrete uncertain system with time delay:

$$\begin{cases} x(k+1) = Ax(k) + A_h x(k-h) + Bu(k) + Dw(k), \\ y(k) = Cx(k) + Fv(k), \end{cases} \tag{6.89}$$

where $x(k) \in \mathbb{R}^{n_x}$ $u(k) \in \mathbb{R}^{n_u}$ and $y(k) \in \mathbb{R}^{n_y}$ are the state, input and measurement output vectors of the system at time k, respectively; A, A_h, B, D, C and F are parameter matrices of the appropriate dimensions, $h \in \mathbb{Z}_+$ denotes a known constant time delay and $w(k) \in \mathbb{R}^{n_w}$ and $v(k) \in \mathbb{R}^{n_v}$ are the unknown but bounded process interference and measurement noise of the system, respectively.

The essential definitions and properties presented in this study are as follows:

Definition 6.7 The r-order zonotope $\mathcal{Z} \subset \mathbb{R}^n, (n \leq r)$ is defined as follows [124]:

$$\mathcal{Z} = p \oplus G\mathbf{B}^r = \{z \in \mathbb{R}^n : z = p + Gb, b \in \mathbf{B}^r\} = \langle p, G \rangle, \tag{6.90}$$

where $\mathbf{B}^r = \begin{bmatrix} -1, 1 \end{bmatrix}^r$ is a hypercube, $p \in \mathbb{R}^n$ is the center of $\mathcal{Z}$ and $G \in \mathbb{R}^{n \times r}$ is the generator matrix that is used to define the shape of $\mathcal{Z}$.

Zonotope is a special class of the geometrical sets and has the following certain computational properties:

Property 11 *For two zonotopes $\mathcal{Z}_1 = \langle p_1, G_1 \rangle \subset \mathbb{R}^n$ and $\mathcal{Z}_2 = \langle p_2, G_2 \rangle \subset \mathbb{R}^n$, the Minkowski sum is also a zonotope, and can be defined as follows [124]:*

$$\langle p_1, G_1 \rangle \oplus \langle p_2, G_2 \rangle = \langle p_1 + p_2, [G_1 \; G_2] \rangle. \tag{6.91}$$

Property 12 *The linear image of the zonotope $\mathcal{Z} = \langle p, G \rangle \subset \mathbb{R}^n$ by $L \in \mathbb{R}^{l \times n}$ is computed using the following matrix product [124]:*

$$L \odot \langle p, G \rangle = \langle Lp, LG \rangle. \tag{6.92}$$

Property 13 *The smallest interval hull* $\mathrm{Box}(\mathcal{Z}) = [\mathrm{z}^-,\mathrm{z}^+] = \{\mathrm{z} : \mathrm{z} \in \mathcal{Z}, \mathrm{z}^- \leq \mathrm{z} \leq \mathrm{z}^+\}$ *containing the zonotope* $\mathcal{Z} = \langle p, G\rangle \subset \mathbb{R}^n$ *can be described as follows [124]:*

$$\begin{cases} z_i^- = p_i - \sum_{j=1}^{r} |G_{i,j}|, i = 1, \ldots, n, \\ z_i^+ = p_i + \sum_{j=1}^{r} |G_{i,j}|, i = 1, \ldots, n, \end{cases} \tag{6.93}$$

where z_i^-, z_i^+ *and* p_i *are the* ith *components of* z^-, z^+ *and* p*, respectively. According to the aforementioned definitions, the interval hull of* $\mathcal{Z}$ *can also be expressed as* $\langle p, rs(G)\rangle$*, where* $rs(G)$ *is a diagonal matrix and* $rs(G)_{ij} = \sum_{j=1}^{r} |G_{i,j}|$, $i = 1, \ldots, n$.

Lemma 6.5

The reduction operator $\downarrow_s$ *can be used to simplify the zonotopes. For a zonotope* $\mathcal{Z} = p \oplus G^r \subset \mathbb{R}^n$, $(n \leq s \leq r)$*, it can be bounded by a more conservative zonotope with lower dimensions as follows [27]:*

$$\mathcal{Z} = \langle p, G\rangle \subseteq \langle p, \downarrow_s G\rangle, \tag{6.94}$$

Thus, the implement procedure of $\downarrow_s G$ *can be summarized as follows:*

- *Obtain the norm decreasing matrix* $\downarrow G$ *by rearranging the columns in the matrix G in decreasing Euclidean norm order as follows:*

  $$\downarrow G = [g_1, g_2, \ldots, g_r],\ \|g_j\|_2 \geq \|g_{j+1}\|_2,\ j = 1, \ldots, r-1.$$

- *Divide* $\downarrow G$ *into two parts,* $G_>$ *and* $G_<$*, and replace* $G_<$ *by a diagonal matrix* $rs(G_<) \in \mathbb{R}^{n\times n}$ *as follows:*

  If $r \leq s$ ***then*** $\downarrow_s G = \downarrow G$.
  Else $\downarrow_s G = [G_>, rs(G_<)] \in \mathbb{R}^{n\times s}$, $G_> = [g_1, \ldots, g_{s-n}]$ *and* $G_< = [g_{s-n+1}, \ldots, g_r]$.

Definition 6.8 The Frobenius radius of the zonotope $\mathcal{Z} = \langle p, G\rangle$ can be regarded as a size criterion of the zonotope, and can be defined as the Frobenius norm of G [27]:

$$\|\langle p, G\rangle\|_F = \|G\|_F = \sqrt{tr(G^{\mathrm{T}}G)} = \sqrt{tr(GG^{\mathrm{T}})}. \tag{6.95}$$

Similarly, in the weighted case, $\|G\|_{F,M} = \sqrt{tr(G^{\mathrm{T}}MG)}$, where $M \in \mathbb{R}^{n\times n}$ is a weighted symmetric positive definite matrix, $tr(\cdot)$ is the trace of a matrix.

Definition 6.9 The covariation of a zonotope $\mathcal{Z} = \langle p, G\rangle$ is defined as $cov(\langle p, G\rangle) = GG^{\mathrm{T}}$ [27]. The Frobenius radius minimization is equivalent to its covariation trace minimization, so $J = tr(GG^{\mathrm{T}}) = \|G\|_F^2$ can also be the size criterion of the zonotope $\mathcal{Z} = \langle p, G\rangle$, and $J_M = tr(MGG^{\mathrm{T}}) = tr(G^{\mathrm{T}}MG) = \|G\|_{F,M}^2$ in the weighted case.

Definition 6.10 Certain correlation operations about the matrix traces are as follows [27, 103]:

$$tr(A) = tr(A^{\mathrm{T}}), \tag{6.96}$$

$$tr(AB) = tr(BA), \tag{6.97}$$

$$\frac{\partial tr(AX^{\mathrm{T}}B)}{\partial X} = A^{\mathrm{T}}B^{\mathrm{T}}, \tag{6.98}$$

$$\frac{\partial tr(AXBX^{\mathrm{T}}C)}{\partial X} = BX^{\mathrm{T}}CA + B^{\mathrm{T}}X^{\mathrm{T}}A^{\mathrm{T}}C^{\mathrm{T}}, \tag{6.99}$$

where A, B, C and X are matrices with appropriate dimensions.

Assumption 5 *The initial state, process interference and measurement noise of the system (6.89) are confined to the zonotopic sets as follows:*

$$x(k-h) \in X(k-h) = \langle p(k-h), G(k-h)\rangle,\ 0 \leqslant k \leqslant h, \tag{6.100}$$

$$w(k) \in W = \langle 0, G_w\rangle,\ k \geqslant 0, \tag{6.101}$$

$$v(k) \in V = \langle 0, G_v\rangle,\ k \geqslant 0, \tag{6.102}$$

where $p(k-h) \in \mathbb{R}^{n_x}$, $G(k-h) \in \mathbb{R}^{n_x \times n_x}$ $G_w \in \mathbb{R}^{n_w \times n_w}$ and $G_v \in \mathbb{R}^{n_v \times n_v}$ are known constant vector and matrices.

6.3.2 Design of the optimal ZKF for the system with time delay

The fault diagnosis for the system with time delay is based on the estimated states, and the states contained in the zonotopic sets are estimated by designing an optimal ZKF.

For system (6.89), the design of the optimal ZKF is as follows:

$$\hat{x}(k+1) = A\hat{x}(k) + A_h\hat{x}(k-h) + Bu(k) + Dw(k) + L(k)(y(k) - C\hat{x}(k) - Fv(k)), \tag{6.103}$$

where $L(k)$ is the estimated optimal gain matrix of the optimal ZKF, and $\hat{x}(k)$ represents the estimated state at time k.

Theorem 6.8

The estimated state at time $k+1$ can be contained in the zonotopic set as follows when $0 \leqslant k \leqslant h$:

$$\hat{x}(k+1) \in \hat{X}(k+1) = \langle \hat{p}(k+1), \hat{G}(k+1)\rangle, \tag{6.104}$$

where

$$\hat{p}(k+1) = (A - L(k)C)\hat{p}(k) + A_h\hat{p}(k-h) + Bu(k) + L(k)y(k), \tag{6.105}$$

$$\hat{G}(k+1) = [(A - L(k)C)\tilde{\hat{G}}(k) \quad A_h\tilde{\hat{G}}(k-h) \quad DG_w \quad -L(k)FG_v], \tag{6.106}$$

$$\tilde{\hat{G}}(k) = \downarrow_s \hat{G}(k),\ \tilde{\hat{G}}(k-h) = \downarrow_s \hat{G}(k-h), \tag{6.107}$$

$$\hat{p}(k-h) = p(k-h),\ 0 \leqslant k \leqslant h, \tag{6.108}$$

$$\hat{G}(k-h) = G(k-h),\ 0 \leqslant k \leqslant h. \tag{6.109}$$

Proof 6.8 According to the description of system (6.89), the time delay only depends on the initial condition when $0 \leqslant k \leqslant h$. Based on Eq. (6.103), and the properties of the zonotope, we can obtain the following:

$$\begin{aligned}\hat{x}(k+1) &= A\hat{x}(k) + A_h\hat{x}(k-h) + Bu(k) + Dw(k) + L(k)(y(k) - C\hat{x}(k) - Fv(k))\\ &\in \big((A - L(k)C) \odot \langle \hat{p}(k), \hat{G}(k)\rangle\big) \oplus (A_h \odot \langle \hat{p}(k-h), \hat{G}(k-h)\rangle)\\ &\quad \oplus Bu(k) \oplus (D \odot \langle 0, G_w\rangle) \oplus L(k)y(k) \oplus (-L(k)F \odot \langle 0, G_v\rangle)\\ &\subseteq \langle (A - L(k)C)\hat{p}(k) + A_h\hat{p}(k-h) + Bu(k) + L(k)y(k),\\ &\quad [(A - L(k)C)\tilde{\hat{G}}(k)\ A_h\tilde{\hat{G}}(k-h)\ DG_w\ -L(k)FG_v]\rangle,\end{aligned} \tag{6.110}$$

where $\tilde{\hat{G}}(k) = \downarrow_s \hat{G}(k)$ and $\tilde{\hat{G}}(k-h) = \downarrow_s \hat{G}(k-h)$ are the complexity reduction matrices of $\hat{G}(k)$ and $\hat{G}(k-h)$, respectively.

Theorem 6.9
For the zonotopic set in Eq. (6.104), we define $\tilde{P}(k) = \tilde{\hat{G}}(k)\tilde{\hat{G}}^{\mathrm{T}}(k)$, $Q_v = FG_vG_v^{\mathrm{T}}F^{\mathrm{T}}$. *Then, the optimal gain matrix* $L(k)$ *can be designed as follows:*

$$L(k) = AK(k), \tag{6.111}$$

$$K(k) = R(k)S^{-1}(k), \tag{6.112}$$

$$R(k) = \tilde{P}(k)C^{\mathrm{T}}, \tag{6.113}$$

$$S(k) = C\tilde{P}(k)C^{\mathrm{T}} + Q_v. \tag{6.114}$$

Proof 6.9 $J_M(k+1) = tr(\hat{G}^{\mathrm{T}}(k+1)M\hat{G}(k+1))$ is the index that determines the size of the zonotope $\hat{X}(k+1)$, and the optimal value of $L(k)$ can be obtained such that $\partial_{L(k)}J_M(k+1) = 0$.

$$\begin{aligned}&\partial_{L(k)}J_M(k+1) = \partial_{L(k)}tr(\hat{G}^{\mathrm{T}}(k+1)M\hat{G}(k+1))\\ &= \partial_{L(k)}tr\big(M(A - L(k)C)\tilde{P}(k)(A - L(k)C)^{\mathrm{T}} + M\tilde{P}_h(k-h) + MQ_w + ML(k)Q_vL^{\mathrm{T}}(k)\big)\\ &= \partial_{L(k)}tr\big(M(L(k)C\tilde{P}(k)C^{\mathrm{T}}L^{\mathrm{T}}(k) + L(k)Q_vL^{\mathrm{T}}(k))\big) - 2\partial_{L(k)}tr(MA\tilde{P}(k)C^{\mathrm{T}}L^{\mathrm{T}}(k)).\end{aligned} \tag{6.115}$$

Let $\partial_{L(k)} J_M(k+1) = 0$, then

$$\partial_{L(k)} tr\big(M(L(k)C\tilde{P}(k)C^{\mathrm{T}}L^{\mathrm{T}}(k) + L(k)Q_v L^{\mathrm{T}}(k))\big) = 2\partial_{L(k)} tr(MA\tilde{P}(k)C^{\mathrm{T}}L^{\mathrm{T}}(k)). \tag{6.116}$$

Considering the operations of the matrix traces and $M = M^{\mathrm{T}} > 0$, we can obtain the following:

$$L(k)(C\tilde{P}(k)C^{\mathrm{T}} + Q_v) = A\tilde{P}(k)C^{\mathrm{T}}, \tag{6.117}$$

$$L(k)S(k) = AR(k), \tag{6.118}$$

$$L(k) = AK(k), \tag{6.119}$$

where $K(k) = R(k)S^{-1}(k)$, $R(k) = \tilde{P}(k)C^{\mathrm{T}}$, $S(k) = C\tilde{P}(k)C^{\mathrm{T}} + Q_v$, $\tilde{P}(k) = \tilde{\hat{G}}(k)\tilde{\hat{G}}^{\mathrm{T}}(k)$, $\tilde{P}_h(k-h) = A_h\tilde{\hat{G}}(k-h)\tilde{\hat{G}}^{\mathrm{T}}(k-h)A_h^{\mathrm{T}}$, $Q_w = DG_wG_w^{\mathrm{T}}D^{\mathrm{T}}$, $Q_v = FG_vG_v^{\mathrm{T}}F^{\mathrm{T}}$.

Theorem 6.10

The estimated state at time $k+1$ can be contained in the zonotopic set as follows when $k > h$:

$$\hat{x}(k+1) \in \hat{X}(k+1) = \langle \hat{p}(k+1), \hat{G}(k+1) \rangle, \tag{6.120}$$

where

$$\begin{aligned}
\hat{p}(k+1) = {} & \Big((A - L(k)C)\prod_{i=1}^{h}(A - L(k-i)C) + A_h\Big)\hat{p}(k-h) \\
& + (A - L(k)C)\sum_{i=1}^{h}\left(\prod_{j=1}^{i-1}(A - L(k-j)C)A_h\hat{p}(k-h-i)\right) \\
& + (A - L(k)C)\sum_{i=1}^{h}\left(\prod_{j=1}^{i-1}(A - L(k-j)C)(Bu(k-i)\right. \\
& \left. + L(k-i)y(k-i))\vphantom{\prod_{j=1}^{i-1}}\right) + Bu(k) + L(k)y(k),
\end{aligned} \tag{6.121}$$

$$\hat{G}(k+1) = \left[\left((A - L(k)C)\prod_{i=1}^{h}(A - L(k-i)C) + A_h\right)\tilde{\hat{G}}(k-h)\ H_1(k)\ H_2(k)\ H_3(k)\right], \tag{6.122}$$

$$H_1(k) = [H_{11}(k) \quad H_{12}(k) \quad \cdots \quad H_{1h}(k)], \tag{6.123}$$

$$H_{1i}(k) = (A - L(k)C)\prod_{j=1}^{i-1}(A - L(k-j)C)A_h\tilde{\hat{G}}(k-h-i), \tag{6.124}$$

$$H_2(k) = [H_{21}(k)\, H_{22}(k) \cdots H_{2h}(k)\, DG_w], \tag{6.125}$$

$$H_{2i}(k) = (A - L(k)C)\prod_{j=1}^{i-1}(A - L(k-j)C)DG_w, \tag{6.126}$$

$$H_3(k) = [H_{31}(k)\ H_{32}(k)\ \cdots\ H_{3h}(k)\ \ -L(k)FG_v], \tag{6.127}$$

$$H_{3i}(k) = -(A - L(k)C)\prod_{j=1}^{i-1}(A - L(k-j)C)L(k-i)FG_v, \tag{6.128}$$

and $\tilde{\hat{G}}(k-h) = \downarrow_s \hat{G}(k-h)$, $\tilde{\hat{G}}(k-h-i) = \downarrow_s \hat{G}(k-h-i)$, $\hat{G}(k+1) \in \mathbb{R}^{n_x \times (r+n_{H_1}+n_{H_2}+n_{H_3})}$, $H_1(k) \in \mathbb{R}^{n_x \times n_{H_1}}$, $H_2(k) \in \mathbb{R}^{n_x \times n_{H_2}}$, $H_3(k) \in \mathbb{R}^{n_x \times n_{H_3}}$.

Proof 6.10 The time delay of the system (6.89) is related to the operation of the system when $k > h$, thus the influence of the time delay on the state estimation of the system at time $k+1$ should be considered. The relationship between $x(k)$ and $x(k-h)$ is considered, and the state of the system is iterated from $x(k-h)$ to $x(k+1)$.

When $k > h$, we can obtain the following from Eq. (6.103):

$$\hat{x}(k+1) = (A - L(k)C)\hat{x}(k) + A_h\hat{x}(k-h) + Bu(k) + Dw(k) + L(k)y(k) - L(k)Fv(k),$$

$$\vdots \qquad\qquad \vdots$$

$$\begin{aligned}\hat{x}(k-i) = {} & (A - L(k-1-i)C)\hat{x}(k-1-i) + A_h\hat{x}(k-h-1-i) + Bu(k-1-i) \\ & + Dw(k-1-i) + L(k-1-i)y(k-1-i) - L(k-1-i)Fv(k-1-i),\end{aligned}$$

$$\vdots \qquad\qquad \vdots$$

$$\begin{aligned}\hat{x}(k-h) = {} & (A - L(k-1-h)C)\hat{x}(k-1-h) + A_h\hat{x}(k-h-1-h) + Bu(k-1-h) \\ & + Dw(k-1-h) + L(k-1-h)y(k-1-h) - L(k-1-h)Fv(k-1-h).\end{aligned}$$

Thus,

$$\begin{aligned}\hat{x}(k+1) = {} & (A - L(k)C)\hat{x}(k) + A_h\hat{x}(k-h) + Bu(k) + Dw(k) + L(k)y(k) - L(k)Fv(k) \\ = {} & (A - L(k)C)(A - L(k-1)C)\hat{x}(k-1) + A_h\hat{x}(k-h) + (A - L(k)C)A_h\hat{x}(k-h-1) \\ & + Bu(k) + (A - L(k)C)Bu(k-1) + Dw(k) + (A - L(k)C)Dw(k-1) + L(k)y(k) \\ & + (A - L(k)C)L(k-1)y(k-1) - L(k)Fv(k) - (A - L(k)C)L(k-1)Fv(k-1) \\ & \vdots \qquad\qquad \vdots \\ = {} & \left((A - L(k)C)\prod_{i=1}^{h}(A - L(k-i)C) + A_h\right)\hat{x}(k-h) \\ & + (A - L(k)C)\sum_{i=1}^{h}\left(\prod_{j=1}^{i-1}(A - L(k-j)C)A_h\hat{x}(k-h-i)\right) \\ & + (A - L(k)C)\sum_{i=1}^{h}\left(\prod_{j=1}^{i-1}(A - L(k-j)C)(Bu(k-i)\right.\end{aligned}$$

$$
\begin{aligned}
&+Dw(k-i)+L(k-i)y(k-i)-L(k-i)Fv(k-i))\Bigg) \\
&+Bu(k)+Dw(k)+L(k)y(k)-L(k)Fv(k) \\
\subseteq&\left\langle\left(\left((A-L(k)C)\prod_{i=1}^{h}(A-L(k-i)C)+A_h\right)\hat{p}(k-h)\right.\right. \\
&+(A-L(k)C)\sum_{i=1}^{h}\left(\prod_{j=1}^{i-1}(A-L(k-j)C)A_h\hat{p}(k-h-i)\right) \\
&+(A-L(k)C)\sum_{i=1}^{h}\left(\prod_{j=1}^{i-1}(A-L(k-j)C)\right. \\
&\left.\left.\cdot(Bu(k-i)+L(k-i)y(k-i))\right)+Bu(k)+L(k)y(k), \\
&\left.\left[\left((A-L(k)C)\prod_{i=1}^{h}(A-L(k-i)C)+A_h\right)\tilde{\hat{G}}(k-h)\ H_1(k)\ H_2(k)\ H_3(k)\right]\right\rangle \\
=&\langle\hat{p}(k+1),\hat{G}(k+1)\rangle,
\end{aligned}
$$

where

$$
\begin{aligned}
H_1(k)&=[H_{11}(k)\quad H_{12}(k)\quad\cdots\quad H_{1h}(k)], \\
H_{1i}(k)&=(A-L(k)C)\prod_{j=1}^{i-1}(A-L(k-j)C)A_h\tilde{\hat{G}}(k-h-i), \\
H_2(k)&=[H_{21}(k)\ H_{22}(k)\ \cdots\ H_{2h}(k)\ DG_w], \\
H_{2i}(k)&=(A-L(k)C)\prod_{j=1}^{i-1}(A-L(k-j)C)DG_w, \\
H_3(k)&=[H_{31}(k)\ H_{32}(k)\ \cdots\ H_{3h}(k)\ \ -L(k)FG_v], \\
H_{3i}(k)&=-(A-L(k)C)\prod_{j=1}^{i-1}(A-L(k-j)C)L(k-i)FG_v,
\end{aligned}
$$

and $\tilde{\hat{G}}(k-h)=\downarrow_s\hat{G}(k-h)$ and $\tilde{\hat{G}}(k-h-i)=\downarrow_s\hat{G}(k-h-i)$ are the complexity reduction matrices of $\hat{G}(k-h)$ and $\hat{G}(k-h-i)$, respectively, $\hat{G}(k+1)\in\mathbb{R}^{n_x\times(r+n_{H_1}+n_{H_2}+n_{H_3})}$, $H_1(k)\in\mathbb{R}^{n_x\times n_{H_1}}$, $H_2(k)\in\mathbb{R}^{n_x\times n_{H_2}}$, $H_3(k)\in\mathbb{R}^{n_x\times n_{H_3}}$.

Theorem 6.11
For the zonotopic set of Eq. (6.120), define

$$
\tilde{P}(k)=\left(\prod_{i=1}^{h}(A-L(k-i)C)\right)\tilde{\hat{G}}(k-h)\tilde{\hat{G}}^{\mathrm{T}}(k-h)\left(\prod_{i=1}^{h}(A-L(k-i)C)\right)^{\mathrm{T}}, \tag{6.129}
$$

$$\tilde{P}_{1i}(k) = \left(\prod_{j=1}^{i-1}(A-L(k-j)C)\right) A_h\tilde{\hat{G}}(k-h-i)\tilde{\hat{G}}^{\mathrm{T}}(k-h-i)A_h^{\mathrm{T}}\left(\prod_{j=1}^{i-1}(A-L(k-j)C)\right)^{\mathrm{T}}, \tag{6.130}$$

$$\tilde{P}_{2i}(k) = \left(\prod_{j=1}^{i-1}(A-L(k-j)C)\right) DG_wG_w^{\mathrm{T}}D^{\mathrm{T}}\left(\prod_{j=1}^{i-1}(A-L(k-j)C)\right)^{\mathrm{T}}, \tag{6.131}$$

$$\tilde{P}_{3i}(k) = \left(\prod_{j=1}^{i-1}(A-L(k-j)C)\right) L(k-i)FG_vG_v^{\mathrm{T}}F^{\mathrm{T}}L^{\mathrm{T}}(k-i)\left(\prod_{j=1}^{i-1}(A-L(k-j)C)\right)^{\mathrm{T}}, \tag{6.132}$$

$$Q_v = FG_vG_v^{\mathrm{T}}F^{\mathrm{T}}. \tag{6.133}$$

The optimal gain matrix $L(k)$ can be designed as follows:

$$L(k) = AK_1(k) + A_hK_2(k), \tag{6.134}$$

$$K_1(k) = R_1(k)S^{-1}(k), \tag{6.135}$$

$$K_2(k) = R_2(k)S^{-1}(k), \tag{6.136}$$

$$R_1(k) = \left(\tilde{P}(k) + \sum_{i=1}^{h}(\tilde{P}_{1i}(k)+\tilde{P}_{2i}(k)+\tilde{P}_{3i}(k))\right)C^{\mathrm{T}}, \tag{6.137}$$

$$R_2(k) = A_h\tilde{\hat{G}}(k-h)\tilde{\hat{G}}^{\mathrm{T}}(k-h)\left(\prod_{i=1}^{h}(A-L(k-i)C)\right)^{\mathrm{T}}C^{\mathrm{T}}, \tag{6.138}$$

$$S(k) = C\left(\tilde{P}(k) + \sum_{i=1}^{h}(\tilde{P}_{1i}(k)+\tilde{P}_{2i}(k)+\tilde{P}_{3i}(k))\right)C^{\mathrm{T}} + Q_v. \tag{6.139}$$

Proof 6.11 $L(k)$ is obtained by minimizing $J_M(k+1) = tr(\hat{G}^{\mathrm{T}}(k+1)M\hat{G}(k+1))$. Thus,

$$\begin{aligned}
\partial_{L(k)}J_M(k+1) &= \partial_{L(k)}tr(\hat{G}^{\mathrm{T}}(k+1)M\hat{G}(k+1))\\
&= \partial_{L(k)}tr\big(ML(k)C\tilde{P}(k)C^{\mathrm{T}}L^{\mathrm{T}}(k) - 2MA\tilde{P}(k)C^{\mathrm{T}}L^{\mathrm{T}}(k)\big)\\
&\quad - 2\partial_{L(k)}tr\left(MA_h\tilde{\hat{G}}(k-h)\tilde{\hat{G}}^{\mathrm{T}}(k-h)\cdot\left(\prod_{i=1}^{h}(A-L(k-i)C)\right)^{\mathrm{T}}C^{\mathrm{T}}L^{\mathrm{T}}(k)\right)\\
&\quad + \partial_{L(k)}tr\left(M\sum_{i=1}^{h}(L(k)C\tilde{P}_{1i}(k)C^{\mathrm{T}}L^{\mathrm{T}}(k)) - 2M\sum_{i=1}^{h}(A\tilde{P}_{1i}(k)C^{\mathrm{T}}L^{\mathrm{T}}(k))\right)\\
&\quad + \partial_{L(k)}tr\left(M\sum_{i=1}^{h}(L(k)C\tilde{P}_{2i}(k)C^{\mathrm{T}}L^{\mathrm{T}}(k)) - 2M\sum_{i=1}^{h}(A\tilde{P}_{2i}(k)C^{\mathrm{T}}L^{\mathrm{T}}(k))\right)\\
&\quad + \partial_{L(k)}tr\left(M\sum_{i=1}^{h}(L(k)C\tilde{P}_{3i}(k)C^{\mathrm{T}}L^{\mathrm{T}}(k)) - 2M\sum_{i=1}^{h}(A\tilde{P}_{3i}(k)C^{\mathrm{T}}L^{\mathrm{T}}(k)) + ML(k)Q_vL^{\mathrm{T}}(k)\right).
\end{aligned}$$

Let $\partial_{L(k)} J_M(k+1) = 0$, then

$$
\begin{aligned}
&\partial_{L(k)} tr\left(ML(k)C\big(\tilde{P}(k)+\sum_{i=1}^{h}(\tilde{P}_{1i}(k)+\tilde{P}_{2i}(k)+\tilde{P}_{3i}(k))\big)C^{\mathrm{T}}L^{\mathrm{T}}(k)+ML(k)Q_vL^{\mathrm{T}}(k)\right)\\
&=2\partial_{L(k)} tr\left(MA\big(\tilde{P}(k)+\sum_{i=1}^{h}(\tilde{P}_{1i}(k)+\tilde{P}_{2i}(k)+\tilde{P}_{3i}(k))\big)C^{\mathrm{T}}L^{\mathrm{T}}(k)\right)\\
&+2\partial_{L(k)} tr\left(MA_h\tilde{\tilde{G}}(k-h)\tilde{\tilde{G}}^{\mathrm{T}}(k-h)\left(\prod_{i=1}^{h}(A-L(k-i)C)\right)^{\mathrm{T}} C^{\mathrm{T}}L^{\mathrm{T}}(k)\right).
\end{aligned}
$$

Considering the operations of the matrix traces and $M = M^{\mathrm{T}} > 0$, we can obtain the following:

$$
\begin{aligned}
L(k)&\left(C\big(\tilde{P}(k)+\sum_{i=1}^{h}(\tilde{P}_{1i}(k)+\tilde{P}_{2i}(k)+\tilde{P}_{3i}(k))\big)C^{\mathrm{T}}+Q_v\right)\\
&=A\left(\tilde{P}(k)+\sum_{i=1}^{h}(\tilde{P}_{1i}(k)+\tilde{P}_{2i}(k)+\tilde{P}_{3i}(k))\right)C^{\mathrm{T}}\\
&\quad+A_h\tilde{\tilde{G}}(k-h)\tilde{\tilde{G}}^{\mathrm{T}}(k-h)\left(\prod_{i=1}^{h}(A-L(k-i)C)\right)^{\mathrm{T}} C^{\mathrm{T}},\\
L(k)S(k)&=AR_1(k)+A_hR_2(k).
\end{aligned}
$$

Thus, the following can be concluded:

$$
\begin{aligned}
L(k)&=AK_1(k)+A_hK_2(k),\\
K_1(k)&=R_1(k)S^{-1}(k),\\
K_2(k)&=R_2(k)S^{-1}(k),\\
R_1(k)&=\left(\tilde{P}(k)+\sum_{i=1}^{h}(\tilde{P}_{1i}(k)+\tilde{P}_{2i}(k)+\tilde{P}_{3i}(k))\right)C^{\mathrm{T}},\\
R_2(k)&=A_h\tilde{\tilde{G}}(k-h)\tilde{\tilde{G}}^{\mathrm{T}}(k-h)\left(\prod_{i=1}^{h}(A-L(k-i)C)\right)^{\mathrm{T}} C^{\mathrm{T}},\\
S(k)&=C\left(\tilde{P}(k)+\sum_{i=1}^{h}(\tilde{P}_{1i}(k)+\tilde{P}_{2i}(k)+\tilde{P}_{3i}(k))\right)C^{\mathrm{T}}+Q_v,\\
\tilde{P}(k)&=\left(\prod_{i=1}^{h}(A-L(k-i)C)\right)\tilde{\tilde{G}}(k-h)\tilde{\tilde{G}}^{\mathrm{T}}(k-h)\left(\prod_{i=1}^{h}(A-L(k-i)C)\right)^{\mathrm{T}},\\
\tilde{P}_{1i}(k)&=\left(\prod_{j=1}^{i-1}(A-L(k-j)C)\right)A_h\tilde{\tilde{G}}(k-h-i)\tilde{\tilde{G}}^{\mathrm{T}}(k-h-i)A_h^{\mathrm{T}}\left(\prod_{j=1}^{i-1}(A-L(k-j)C)\right)^{\mathrm{T}},
\end{aligned}
$$

$$\tilde{P}_{2i}(k) = \left(\prod_{j=1}^{i-1}(A - L(k-j)C)\right) DG_wG_w^{\mathrm{T}}D^{\mathrm{T}} \left(\prod_{j=1}^{i-1}(A - L(k-j)C)\right)^{\mathrm{T}},$$

$$\tilde{P}_{3i}(k) = \left(\prod_{j=1}^{i-1}(A - L(k-j)C)\right) L(k-i)FG_vG_v^{\mathrm{T}}F^{\mathrm{T}}L^{\mathrm{T}}(k-i) \left(\prod_{j=1}^{i-1}(A - L(k-j)C)\right)^{\mathrm{T}},$$

$$Q_w = DG_wG_w^{\mathrm{T}}D^{\mathrm{T}},$$

$$Q_v = FG_vG_v^{\mathrm{T}}F^{\mathrm{T}}.$$

Remark 6.4 In the derivations of Theorems 6.9 and 6.11, $S(k)$ is assumed to be reversible. If $S(k)$ is irreversible, then the pseudo-inverse of $S(k)$ can be used to replace the inverse of $S(k)$; that is, $K(k) = R(k)S^+(k)$ in Theorem 6.9 and $K_1(k) = R_1S^+(k)$, $K_2(k) = R_2S^+(k)$ in Theorem 6.11. ▮

6.3.3 Fault diagnosis

In this section, an accurate state estimation result is obtained by applying the optimal ZKF algorithm for the system with time delay, and the estimated state is then used for fault diagnosis, which is divided into the following two steps: fault detection and fault identification.

6.3.3.1 Fault detection

When the system is fault-free, the calculation rule of the estimated output set $\hat{Y}(k) = \langle \hat{p}_y(k), \hat{G}_y(k)\rangle$ that is consistent with state $\hat{x}(k)$ is as follows:

$$\begin{aligned}\hat{y}(k) &= C\hat{x}(k) + Fv(k)\\ &\in (C \odot \langle \hat{p}(k), \hat{G}(k)\rangle) \oplus (F \odot \langle 0, G_v\rangle)\\ &= \langle C\hat{p}(k), [C\hat{G}(k)\ FG_v]\rangle\\ &= \langle \hat{p}_y(k), \hat{G}_y(k)\rangle,\end{aligned} \tag{6.140}$$

where $\hat{p}_y(k) \in \mathbb{R}^{n_y}$ and $\hat{G}_y(k) \in \mathbb{R}^{n_y \times r_{\hat{G}_y}}$. Apparently, the value of $\hat{y}(k)$ cannot be determined owing to the unknown noise $v(k)$; however, the zonotopic set $\hat{Y}(k)$ containing $\hat{y}(k)$ can be obtained according to the aforementioned calculation rule. Based on the estimated output set $\hat{Y}(k)$, the upper bound $\hat{y}_u(k)$ and lower bound $\hat{y}_l(k)$ satisfy the followings:

$$\hat{y}_u(k) = \hat{p}_y(k) + \begin{bmatrix} \sum_{j=1}^{r_{\hat{G}_y}} |\hat{G}_{y1j}(k)| \\ \vdots \\ \sum_{j=1}^{r_{\hat{G}_y}} |\hat{G}_{yn_yj}(k)| \end{bmatrix}, \tag{6.141}$$

$$\hat{y}_l(k) = \hat{p}_y(k) - \begin{bmatrix} \sum_{j=1}^{r_{\hat{G}_y}} |\hat{G}_{y1j}(k)| \\ \vdots \\ \sum_{j=1}^{r_{\hat{G}_y}} |\hat{G}_{yn_yj}(k)| \end{bmatrix}. \tag{6.142}$$

If the true output value $y_t(k)$ satisfies $\hat{y}_{l1}(k) \leq y_{t1}(k) \leq \hat{y}_{u1}(k)$, ..., $\hat{y}_{ln_y}(k) \leq y_{tn_y}(k) \leq \hat{y}_{un_y}(k)$, then it is determined that $y_t(k) \in \hat{Y}(k)$; otherwise, $y_t(k) \notin \hat{Y}(k)$. When a parameter fault occurs in the system, it leads to $y_t(k) \notin \hat{Y}(k)$, which indicates that there must be a fault in the system. $f(k)$ is defined as the fault-detection signal; it is $f(k) = 0$ when $y_t(k) \in \hat{Y}(k)$ and is $f(k) = 1$ when $y_t(k) \notin \hat{Y}(k)$, indicating the system is fault-free and faulty, respectively. To facilitate the development of this study, it is assumed that only one fault at a time occurs in the system, and the system parameters do not change over a certain period of time.

The fault-detection method based on the optimal ZKF is completed by detecting whether the true output value of the system is within the upper and lower bounds of the estimated output. The judgment criterion of fault detection can be expressed as follows: when the true output value is not within the upper and lower bounds of the estimated output, the system is determined to be faulty; otherwise, the system is determined to be fault-free. The fault-detection strategy is based on the fact that in each iteration, the upper and lower bounds of the estimated output of the system are determined by the corresponding zonotopic set $\hat{Y}(k)$. When there is no fault in the system, the true output of the system must be included in the zonotopic set $\hat{Y}(k)$; that is, it is also included in the corresponding upper and lower bounds of $\hat{Y}(k)$. If the true output value is not within the upper and lower bounds of the estimated output, the system will fail. Therefore, given the noise boundaries, the fault-detection method based on the optimal ZKF proposed in this study will not cause a false alarm, that is, the false alarm rate is zero. Namely, once a fault alarm occurs, it indicates that there must be a fault in the system.

6.3.3.2 *Fault identification*

The error between the estimated and true states of the ith fault at each time k can be calculated using Eq. (6.143) as follows:

$$e_i(k) = \|\hat{p}(k) - p_{i,0}(k)\| / \|p_{i,0}(k)\| \ (i = 1, 2, \ldots, q), \tag{6.143}$$

where q is the number of parameter fault types in the fault library, $p_{i,0}(k)$ is the fault state of the ith fault type in the fault library. The continuous-time length L is selected, and the number of estimated errors that satisfy $e_i(k-l) < \varepsilon(k)$ $(l = 1, 2, \ldots, L)$ can be counted as $l_i(k)$ $(i = 1, 2, \ldots, q)$, where $\varepsilon(k)$ is the selected threshold. The value of $\varepsilon(k)$ is relative, and it is related to the practical system and the requirements of computational load. In the practical system, the value of each fault type of each system is different, so the demand for $\varepsilon(k)$ is different when distinguishing the fault types of systems. In the aspect of the requirements of computational load, the value of $\varepsilon(k)$ directly determines the amount of calculation required to identify system faults. If it is necessary to complete the fault identification as soon as possible with a small computational load, the value of $\varepsilon(k)$ can be increased appropriately. If none of the estimated states satisfy $e_i(k-l) < \varepsilon(k)$ $(l = 1,2,\ldots,L,\ i = 1,2,\ldots,q)$, a new fault outside the fault library at time k can be considered to have occurred, which is denoted as fault f_{q+1}. The number of new faults is counted at this moment and is recorded as $l_{q+1}(k)$.

The posterior probability of each fault type in the fault library was calculated and recorded as $P_i(k)$. Considering the possible new fault types, there are $q+1$ fault types. At the initial moment, the prior probability of each fault type is the same, which is $P_i(1) = 1/(q+1)$ $(i = 1,2,\ldots,q+1)$. The posterior probability $P_i(k)$, whose fault type is fault f_i $(i = 1,2,\ldots,q+1)$, can then be calculated using Eq. (6.144) as follows:

$$P_i(k) = \frac{P_i(k-1)l_i(k)}{\sum_{j=1}^{q+1} P_j(k-1)l_j(k)}. \tag{6.144}$$

When the posterior probability satisfies Eq. (6.145), the fault diagnosis is complete; that is, if the posterior probability $P_i(k)$ satisfies Eq. (6.145), the fault is recognized as f_i:

$$P_i(k) \geq 1 - \varsigma, \tag{6.145}$$

where ς is the set recognition accuracy.

6.3.4 Simulation analysis

Example 1: Numerical simulation Consider the following linear discrete uncertain system with time delay:

$$\begin{cases} x(k+1) = Ax(k) + A_h x(k-h) + Bu(k) + Dw(k), \\ y(k) = Cx(k) + Fv(k), \end{cases} \tag{6.146}$$

where

$$A = \begin{bmatrix} 0.6 & -0.4 \\ -0.5 & 0.2 \end{bmatrix}, \quad A_h = \begin{bmatrix} -0.05 & -0.03 \\ 0.04 & -0.03 \end{bmatrix},$$
$$B = \begin{bmatrix} 0.2 \\ 0.2 \end{bmatrix}, \quad D = \begin{bmatrix} -0.1 \\ 0.1 \end{bmatrix}, \quad C = \begin{bmatrix} 1 & 0 \\ 0 & 1 \end{bmatrix},$$
$$F = \begin{bmatrix} 1 & 0 \\ 0 & 1 \end{bmatrix}, \quad h = 5.$$

$u(k) = 6$ is selected as the known input; the initial state measurement noise and process interference of the system are assumed to meet respectively as follows:

$$x(k-h) \in X(k-h) = \left\langle \begin{bmatrix} 2 \\ 3 \end{bmatrix}, \begin{bmatrix} 2 & 0 \\ 0 & 2 \end{bmatrix} \right\rangle, \quad 0 \leqslant k \leqslant h, \tag{6.147}$$

$$w(k) \in W = \langle 0, 0.1 \rangle, \quad k \geqslant 0, \tag{6.148}$$

$$v(k) \in V = \left\langle \begin{bmatrix} 0 \\ 0 \end{bmatrix}, \begin{bmatrix} 0.1 & 0 \\ 0 & 0.1 \end{bmatrix} \right\rangle, \quad k \geqslant 0. \tag{6.149}$$

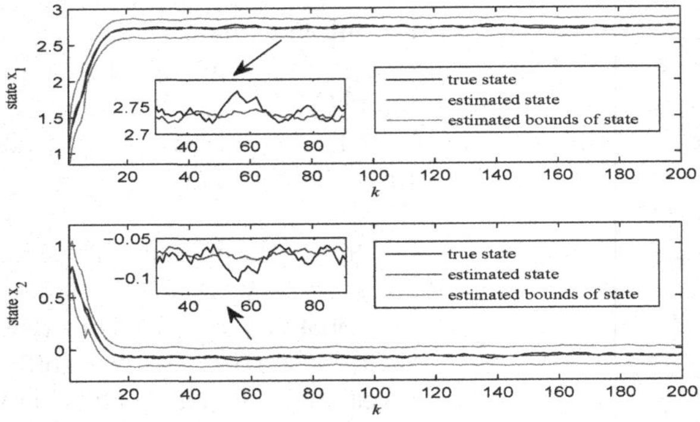

Figure 6.23: Bounds of estimated states in a fault-free state.

Next, in this numerical case, the fault-diagnosis algorithm is analyzed in the following two situations: fault-free and faulty in the fault library.

6.3.4.1 *Fault-free*

When the system is fault-free, Fig. 6.23 shows that the states can be estimated effectively by the designed optimal ZKF algorithm for the system with time delay, and the estimated state and true state are all within the upper and lower bounds of the estimated state. Fig. 6.24 illustrates that although there are small fluctuations in the true output values of the system, they are within the upper and lower bounds of the estimated output values. The upper and lower bounds of the estimated state and output fluctuate and converge to a certain extent in the initial time period and remain unchanged after the system is stable. Thus, we can conclude that the system is fault-free, and the optimal ZKF method proposed in this study is effective and feasible for the state estimation of the system with time delay.

6.3.4.2 *Fault in the fault library*

Assuming that the system fault is caused by the parameter matrix A, and there are three types of faults in the fault library of the numerical system with time delay, the specific parameter changes are as follows: $A_1 = \begin{bmatrix} 0.6 & -0.4 \\ -0.4 & 0.2 \end{bmatrix}$, $A_2 = \begin{bmatrix} 0.6 & -0.4 \\ -0.4 & 0.5 \end{bmatrix}$ and $A_3 = \begin{bmatrix} 0.4 & -0.4 \\ -0.4 & 0.2 \end{bmatrix}$, and the corresponding faults are f_1, f_2 and f_3, respectively. The filters $fl_1 \sim fl_4$ corresponding to faults $f_1 \sim f_4$ are provided to indicate the types of faults identified, and f_4 represents the new fault outside the fault library. In this faulty state, fault f_1 is added to the system during the time period $k \in \{50, 120\}$. Considering the effect of historical data during the fault period, the difference between the true and fault states in the fault library was large in the beginning, and subsequently decreased rapidly and was maintained within a certain

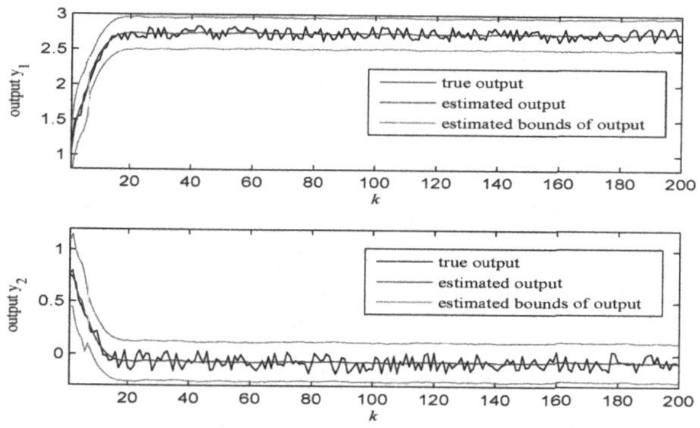

Figure 6.24: Bounds of estimated outputs in a fault-free state.

small range. Therefore, the value of $\varepsilon(k)$ should be set dynamically; specifically, the value of $\varepsilon(k)$ changes from 1.2 to 0.1 within ten-time points after the fault-detection point, and then remains unchanged at 0.1 in the subsequent fault identification process. For $\varsigma = 0.3$ and $L = 5$, the fault-diagnosis results are shown in Figs. 6.25–6.28.

As shown in Fig. 6.25, the true output values of the system are within the upper and lower bounds of the output values estimated by the optimal ZKF algorithm at the beginning, indicating that the system is in a normal state. The true output values of the system began to change and went beyond the upper and lower bounds of the output values estimated by the optimal ZKF algorithm at approximately $k = 50$, indicating that the system has a fault at that moment; hence, the zonotopic set of the output estimated by the optimal ZKF algorithm no longer contains the true output value. The

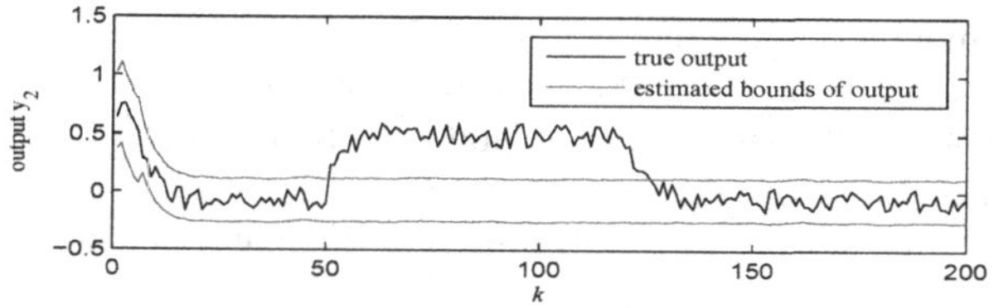

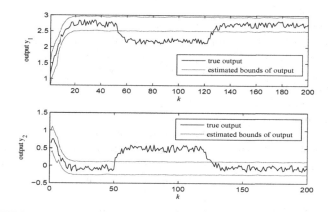

Figure 6.25: True output in the scenario of fault in the fault library and the estimated bounds of y.

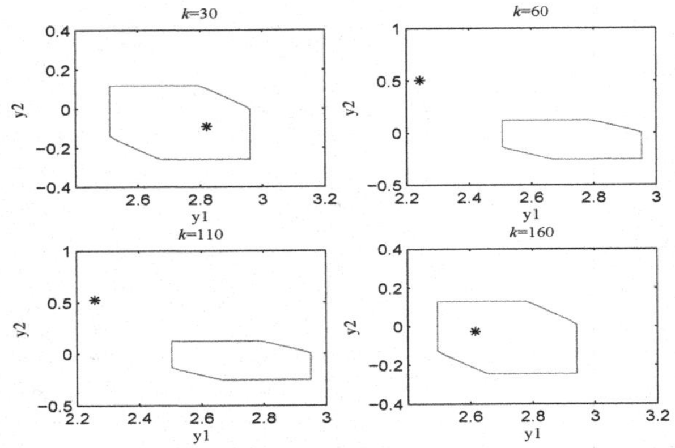

Figure 6.26: True output in the scenario of fault in fault library and the estimated output at different times.

true output values of the system return to be within the upper and lower bounds of the output values estimated by the optimal ZKF algorithm at approximately $k = 120$, indicating that the system returns to the normal from the fault state at this time, and the system only has a fault between $k = 50 - 120$ during the entire operating process.

Fig. 6.26 presents the relationship between the true output and the estimated zonotopic set more intuitively; the normal time period before the fault, the faulty time period during the fault, and the normal time period after the fault are selected as the representative time periods, considering one point, two points and one point in the three representative time periods as examples, respectively. The magenta zonotope is the approximate feasible set of the output estimated by the optimal ZKF algorithm, and the black star indicates the true output value of the system. As indicated by Fig. 6.26, when $k = 60$ and $k = 110$, the system is in a fault state, and the zonotopic sets of output y estimated by the optimal ZKF algorithm do not contain the true output values of the system with the time delay at the fault time. Moreover, when the system is in a normal state, at times $k = 30$ and $k = 160$, the zonotopic sets estimated by the optimal ZKF algorithm contain the true outputs of the system, indicating that the detection result is consistent with the real system.

To obtain a more specific fault-occurrence time, the curve of the fault-detection signal f is shown in Fig. 6.27; it indicates that the system is detected to have a fault during time $k = 50 - 127$ by the proposed fault-detection algorithm, which is close to the real fault time when slight delays are ignored. Subsequently, the algorithm enters the fault identification stage; $fp_1 \sim fp_4$ represents the fault-matching probability of $f_1 \sim f_4$.

A normal fault-matching method is compared to the proposed fault-diagnosis algorithm to further demonstrate its advantages. The normal fault-matching method separates the fault by making the error between the estimated and true states of

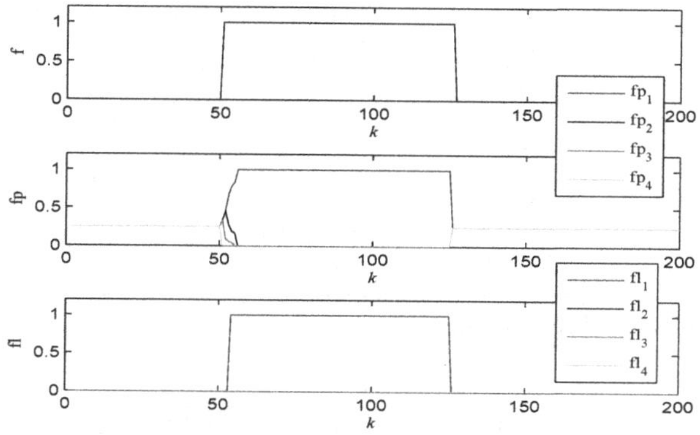

Figure 6.27: Fault diagnosis results in the scenario of fault in the fault library.

the ith fault less than a fixed threshold $\varepsilon'(k)$ in a continuous period L'. The comparison curves of the fault identification obtained by the two algorithms are shown in Fig. 6.28. In this simulation, L' and $\varepsilon'(k)$ were set to 20 and 0.1, respectively. $fl_1 \sim fl_4$ in Fig. 6.28 are the fault-diagnosis filters corresponding to $f_1 \sim f_4$ using the fault-diagnosis algorithm combining the fault-matching method and Bayesian theory, and $fl_{11} \sim fl_{41}$ indicate the fault identification results corresponding to $f_1 \sim f_4$ through the normal fault-matching algorithm. According to the figure, the normal fault-matching algorithm can identify the fault type of the pitch system; however, fault identification is achieved at $k = 72$, which is 19 time points slower than the proposed algorithm, indicating that its fault identification speed is significantly slower than that of the fault-diagnosis algorithm proposed in this study.

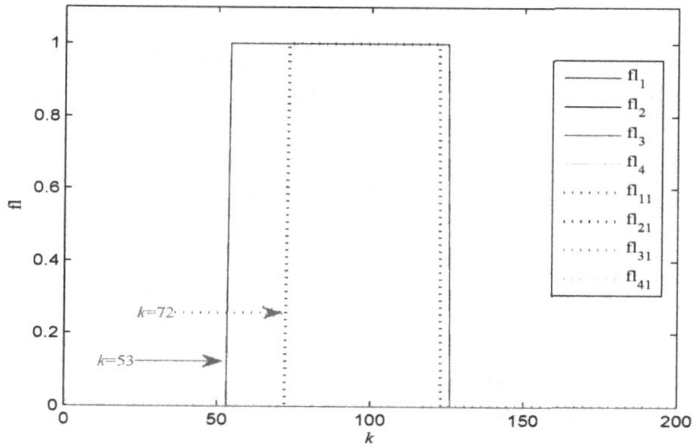

Figure 6.28: Comparison of fault identification results of the two algorithms for scenarios of fault in the fault library.

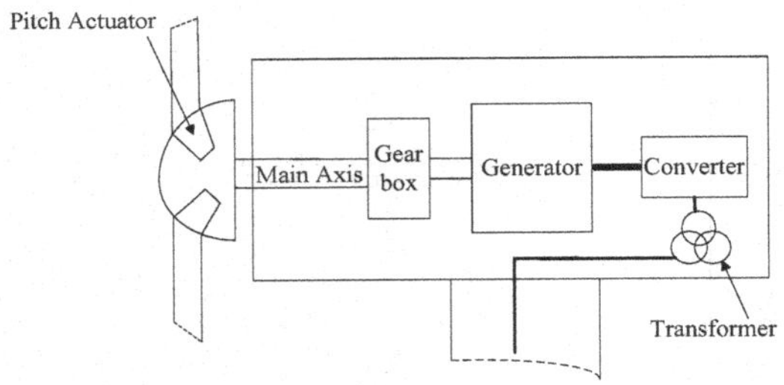

Figure 6.29: Wind turbine system.

Example 2: Pitch system of wind turbine Wind turbines are prone to failure in the process of power generation, and several scholars have studied the corresponding fault diagnosis [37]. As shown in Fig. 6.29, the pitch system is an important part of the blade control and pitch angle conversion of a wind turbine.

The mathematical model of the pitch system can be expressed as follows [72]:

$$\begin{bmatrix} \dot{\beta} \\ \dot{\beta}_a \end{bmatrix} = \begin{bmatrix} 0 & 1 \\ -\omega_n^2 & -2\zeta\omega_n \end{bmatrix} \begin{bmatrix} \beta \\ \beta_a \end{bmatrix} + \begin{bmatrix} 0 \\ \omega_n^2 \end{bmatrix} \beta_r, \tag{6.150}$$

where β and β_a are the pitch angle and angular velocity, respectively; β_r is the reference value of the pitch, and ω_n and ζ are the system parameters. Let $\omega_n = 11.11$ rad/s and $\zeta = 0.6$ be the natural frequency and damping coefficient of the system, respectively. The system can be expressed as a continuous-time state-space equation as follows:

$$\dot{x} = \begin{bmatrix} 0 & 1 \\ -\omega_n^2 & -2\zeta\omega_n \end{bmatrix} x + \begin{bmatrix} 0 \\ \omega_n^2 \end{bmatrix} u + \begin{bmatrix} 0 & 0 \\ -\frac{\omega_n^2}{2} & -\frac{\omega_n^2}{2} \end{bmatrix} w, \tag{6.151}$$

$$y = Cx + Fv, \tag{6.152}$$

where $x = [\beta\ \beta_a]^T$, $u = \beta_r$ and w and v are the interference and output noise, respectively. Let the sampling time be $T_s = 0.01$s, the system is discretized, where

$$A = \begin{bmatrix} 0.9941 & 0.0093 \\ -1.1532 & 0.8695 \end{bmatrix}, B = \begin{bmatrix} 0.0059 \\ 1.1532 \end{bmatrix},$$

$$F = \begin{bmatrix} 1 & 0 \\ 0 & 1 \end{bmatrix}, D = \begin{bmatrix} -0.0030 & -0.0030 \\ -0.5766 & -0.5766 \end{bmatrix}, C = \begin{bmatrix} 1 & 0 \\ 0.5 & 0.1 \end{bmatrix}.$$

We consider the influence of the time delay of the state on the pitch system, and select $A_h = \begin{bmatrix} -0.005 & -0.003 \\ 0.004 & -0.003 \end{bmatrix}$, $h = 6$. The sinusoidal signal $u(k) = 1.5\sin(6k) + 7$ is selected as the external excitation signal, and it is assumed that the initial state, process interference, and measurement noise of the system meet, respectively, as

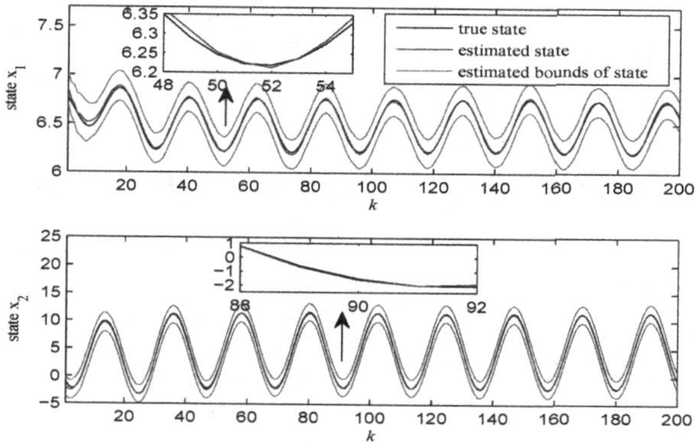

Figure 6.30: Bounds of x in a fault-free state.

follows:

$$x(k-h) \in X(k-h) = \left\langle \begin{bmatrix} 6.5 \\ 0 \end{bmatrix}, \begin{bmatrix} 5 & 0 \\ 0 & 5 \end{bmatrix} \right\rangle, \ 0 \leqslant k \leqslant h, \tag{6.153}$$

$$w(k) \in W = \left\langle \begin{bmatrix} 0 \\ 0 \end{bmatrix}, \begin{bmatrix} 0.1 & 0 \\ 0 & 0.1 \end{bmatrix} \right\rangle, \ k \geqslant 0, \tag{6.154}$$

$$v(k) \in V = \left\langle \begin{bmatrix} 0 \\ 0 \end{bmatrix}, \begin{bmatrix} 0.1 & 0 \\ 0 & 0.1 \end{bmatrix} \right\rangle, \ k \geqslant 0. \tag{6.155}$$

Subsequently, in this pitch system case, the fault-diagnosis algorithm is analyzed in the following two situations: fault-free and fault outside the fault library.

6.3.4.3 Fault-free

As shown in Fig. 6.30, when the pitch system is fault-free, the optimal ZKF algorithm proposed in this study for a linear system with time delay has a sufficient state estimation performance. The estimated state curve rapidly follows the true state of the system; the two curves are included in the upper and lower bounds of the estimated state. As shown in Fig. 6.31, the true output value of the system is within the upper and lower bounds of the output value estimated by the optimal ZKF algorithm when the pitch system is fault-free.

6.3.4.4 Fault outside the fault library

The following three faults are in the fault library of the pitch system [110]: pressure drop, pump wear and air content increase, which are defined as faults f_1, f_2 and f_3, respectively. The new fault outside the fault library is defined as fault f_4. The faulty parameter values are listed in Table 6.3.

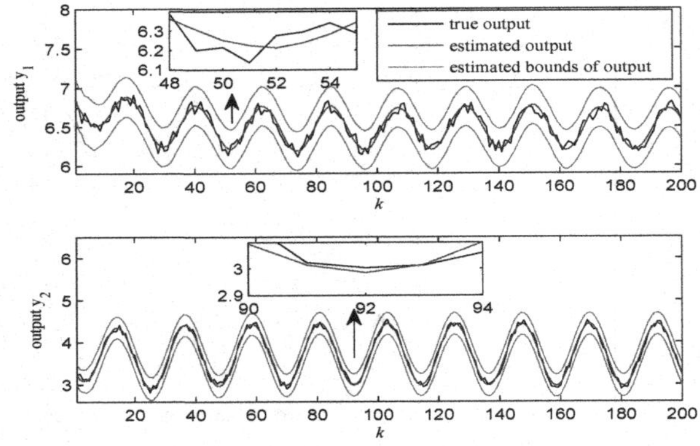

Figure 6.31: Bounds of y in a fault-free state.

Given the actual system, we assume that the types of faults that may occur in the actual system are those faults in the fault library and the new fault outside the fault library when $60 \leqslant k \leqslant 115$, which are defined as $f_1 \sim f_4$; filters corresponding to faults $f_1 \sim f_4$ are provided to indicate the types of faults identified. In this scenario, f_4 is included in Eq. (6.151) when $60 \leqslant k \leqslant 115$. Figs. 6.32–6.35 show the fault diagnosis results of the proposed fault-diagnosis algorithm, and Fig. 6.35 presents the comparison results of the fault identification using the proposed algorithm and the normal fault-matching algorithm.

Set $\varsigma = 0.3$, $L = 8$ and $\varepsilon(k)$ changes dynamically during the fault period and its value decreases from 0.2 to 0.1 within 10 time points after the fault-detection point, and then remains at 0.1 until the end of the fault. As shown in Fig. 6.34, the fault-matching probability fp_4 gradually increases to 1 during the fault period. When fp_4 is greater than $1 - \varsigma$, the corresponding fault identification filter fl_4 is set to be 1 at time $k = 64$, which indicates that during the entire working period of the pitch system, the new fault f_4 outside the fault library occurs after a period of normal state.

Fig. 6.32 indicates that the true output values of the pitch system begin to go beyond the upper and lower bounds of the output values estimated by the optimal ZKF algorithm at approximately $k = 60$, and return to be within the bounds at approximately $k = 120$, indicating that the system fault occurs and ends at approximately

Table 6.3: Parameters of pitch system in fault conditions

	Fault	Parameters
f_1	Pressure drop	$\xi_1 = 0.45$, $\omega_{n1} = 5.73$ rad/s
f_2	Pump wear	$\xi_2 = 0.75$, $\omega_{n2} = 7.27$ rad/s
f_3	Air content increase	$\xi_3 = 0.90$, $\omega_{n3} = 3.42$ rad/s
f_4	New fault	$\xi_4 = 1.60$, $\omega_{n4} = 15.00$ rad/s

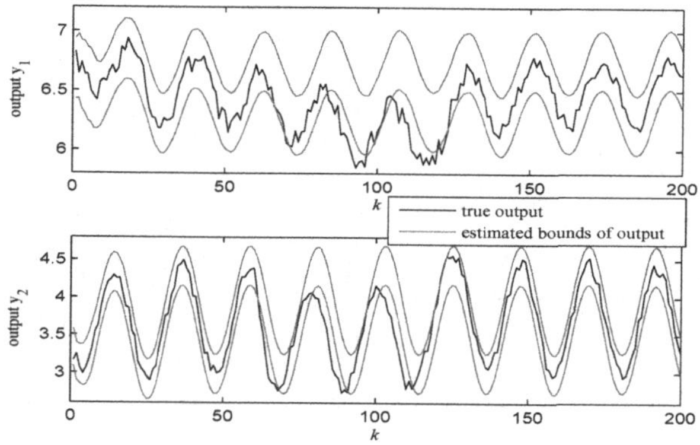

Figure 6.32: True output in the scenario of fault outside the fault library and bounds of y.

$k = 60$ and $k = 120$, respectively; thereafter, the system resumes normal operation. The output values estimated by the optimal ZKF algorithm at different time points in Fig. 6.33 verify the correctness of the proposed algorithm more specifically. When $k = 30$ and $k = 160$, the zonotopic sets of the output values estimated by the optimal ZKF algorithm (red lines) all contain the true output values (black stars), while the zonotopic sets of the output values do not contain the true output values when $k = 80$ and $k = 110$, which is consistent with the fault state of the real pitch system.

The curve of the fault-detection signal f is shown in Fig. 6.34, indicating that the specific time of the faults detected by the fault-diagnosis algorithm is $k \in \{61-124\}$,

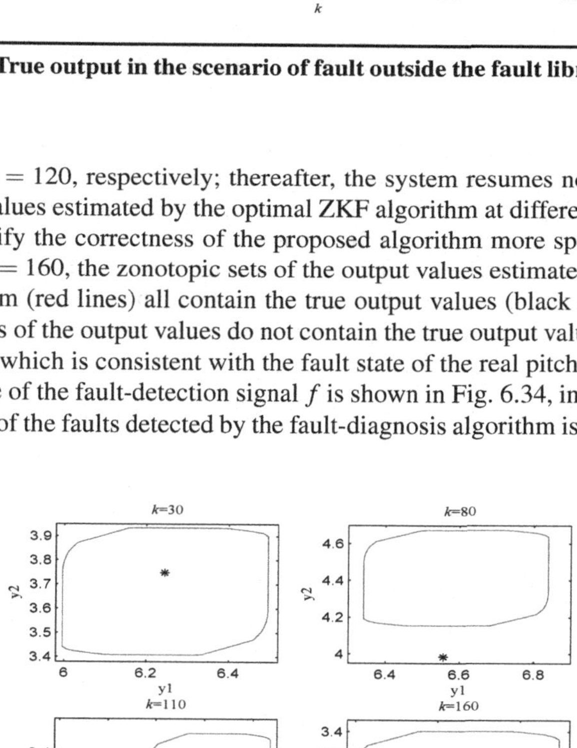

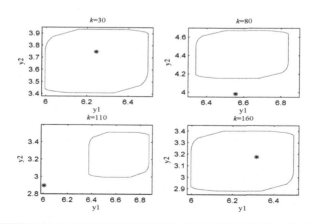

Figure 6.33: True output in the scenario of a fault outside the fault library and the estimated output at different times.

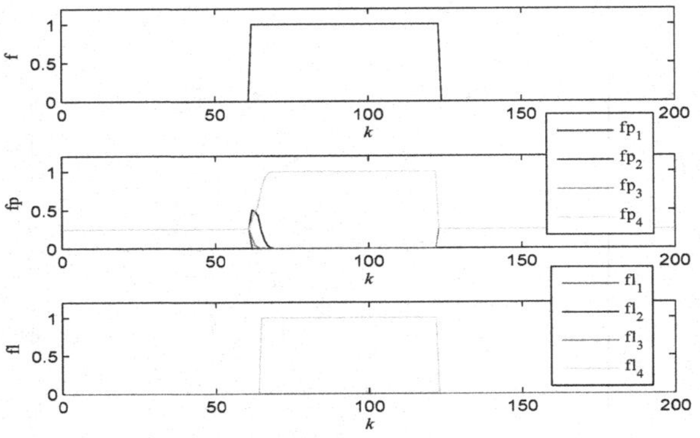

Figure 6.34: Fault diagnosis results in the scenario of fault outside the fault library.

which is close to the real time of the system fault, thus we can conclude that the proposed algorithm has a sufficient fault-detection performance.

As shown in Fig. 6.35, the comparison curves between the two algorithms show that the fault identification time of the fault-diagnosis algorithm based on the combination of the fault-matching method and Bayesian theory is 19 time points faster than that of the normal fault-matching algorithm. Thus, the proposed fault-diagnosis algorithm has an advantage in terms of the speed of fault identification.

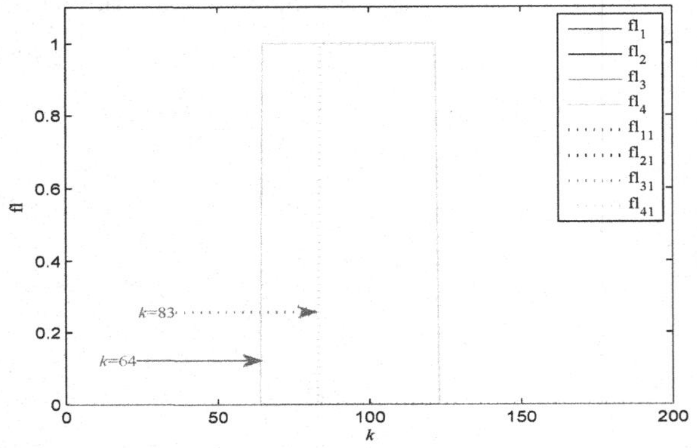

Figure 6.35: Comparison of fault identification results of the two algorithms in the scenario of a fault outside the fault library.

6.3.5 Conclusion

In this study, a state estimation and fault-diagnosis method based on the optimal zonotopic Kalman filter is presented for a linear discrete-time system with time delay. The optimal observer gain of the optimal zonotopic Kalman filter was designed by minimizing the size of the zonotopic set to estimate the state set. The parameter fault-diagnosis problems of a numerical system and the pitch system of a wind turbine are used as examples to verify the effectiveness and feasibility of the algorithm.

The fault-diagnosis method proposed in this study can also be applied to combine certain parameter estimation algorithms [105, 112] to study the fault-diagnosis problems of other time-delay industrial processes with unknown but bounded noise. In addition, based on the proposed fault-diagnosis algorithm, the study can be further extended to the fault diagnosis of other types of delay, such as unknown time delay systems, and time-varying delay systems.

Chapter 7

Summary

Nowadays, the theory and technology of automatic control are changing with each passing day. The research of automatic control system is based on mathematical model, which is the mathematical description of system dynamic characteristics, and system modeling is the basis of all subsequent design and control. State estimation is very important to understand the internal operation state of the system. Due to the complexity of modern systems, once a fault occurs, it will cause irreparable losses. Therefore, fault diagnosis technology is one of the methods to improve the reliability and safety of the system and reduce the losses.

The theory and development of system modeling, state estimation and fault diagnosis are a gradual process. When the filtering method is applied to solve these problems, it can effectively solve the low result accuracy caused by the inaccuracy of input and output data and observation data. However, the traditional method based on filter design usually requires the process interference and measurement noise of the system to meet the specific distribution requirements, which is obviously contrary to the interference and noise of the actual system. Therefore, the solution of this problem is widely concerned by scholars.

The non-probabilistic filter design method based on set-membership estimation only requires that the noise and interference are unknown but bounded. Compared with the traditional filter based design method, it has obvious advantages in universality, practicability and robustness. Therefore, the method based on set-membership filter has gradually become a research hotspot at home and abroad in recent years.

In this book, based on the research of traditional filtering method design, this book studies the set-membership filtering design method based on ellipsoid, polytope, interval, parallelotope and zonotope. Different set-membership filtering methods are developed in different spatial wrapping forms. At the same time, innovative research is carried out according to different spatial shape characteristics and specific problems at this stage. In addition, at a deeper level, this book proposes two

DOI: 10.1201/b23146-7

methods based on composite set-membership filtering, which effectively combines the advantages of two or more spatial shapes. Compared with the set-membership filtering method with single spatial shape, this method has obvious advantages in dealing with system modeling, state estimation, fault diagnosis and so on.

The main research results obtained in this paper are summarized as follows:

1. Aiming at the problem of the traditional time-varying parameter estimation method that uses ellipsoids and zonotopes as the feasible parameter set is highly conservative and complexity, this book proposes a time-varying parameter system estimation method based on zonotope-ellipsoid double filtering. Simulation results confirm that compared with the existing methods, the proposed algorithm improves the operation efficiency of the identification process; simultaneously, it effectively improves the identification accuracy and reduces the conservativeness of the identification results.

2. A variety of state estimation methods of zonotope and its derived algorithms are proposed. For bi-directional DC-DC converter of lithium battery formation, this book proposes set-membership filtering to state estimation. For the spring-mass-damper nonlinear system, a zonotopic set-valued observer based state estimation algorithm is proposed, and the unknown noise term is wrapped in a zonotope during each recursive step. The second-order polynomial Stirling interpolation improves the linearization accuracy and reduces the calculation amount. The method that combines sequence updating and tightening strips reduces the accumulation of errors and improves the estimation accuracy. For solving the linear system state estimation problem such as the temperature recognition of Lithium battery system, this book proposes a system state estimation method based on zonotopic particle filter. The particles are extracted from the given initial distribution, and a new group of particles which are closer to the real value is obtained by constructing a multi cell group according to the zonotopic particle group at the previous time and the system observation value at that time to restrict the diffusion distribution of particles, and to replace the particles outside the restricted area, so as to improve the particle diversity in the resampling process of particle filter.

3. For the state estimation problems of nonlinear systems, the state estimation method based on convex space is proposed. First, the hyperparallel space set-membership filtering state estimation method is proposed. Second this book presented a set-membership filtering algorithm for axisymmetric box space wrapping errors to solve the state estimation problem of nonlinear systems under the interference of unknown and bounded uncertain noise. The interval estimation of the noise is obtained. Then design the axisymmetric box space wrapping noise and the state feasible set, and the prediction set is updated by the measurement of the axisymmetric box space through the prediction and update steps, and the state feasible set of the nonlinear system is obtained. In the intersection operation of the axisymmetric box space, this book proposes the idea of splitting the interval box, which not only ensures the accuracy,

but also improves the boundary contraction speed of the state estimation and reduces the conservativeness of the state estimation algorithm.

4. For the fault diagnosis of for linear discrete-time systems, the set inversion interval observer filtering-based guaranteed fault estimation algorithm and a novel interval observer filtering-based fault diagnosis method are proposed in this book, the DC motor simulations show that these algorithms can not only obtain more compact estimation results, but also solve the problem of the exponential increase in the computation time of the traditional interval set inversion algorithm. Besides, a filter fault diagnosis method based on directional expansion of orthometric hyperparallel space is proposed , fault detection is completed by detecting whether the parameter feasible set is empty. When the feasible set is empty, the system fails, and the specific failed parameters are isolated from the state of the hyperparallel space test sets. When the failed parameters are isolated, After directional expansion, the orthometric hyperparallel space wrapped with fault parameters in fault state is obtained to complete fault identification. The simulation of wind turbine system verifies the effective of the proposed method.

5. This book proposes some fault diagnosis methods based on zonotopic Kalman filtering. First A novel fault diagnosis method based on zonotopic Kalman filter for linear system with state constraints is proposed. Second, two constrained zonotopic Kalman filter based sensor fault estimators are designed in this book for the additive sensor fault and multiplicative sensor fault in the constrained system with unknown but bounded noises. Finally, a state estimation and fault-diagnosis method based on the optimal zonotopic Kalman filter is presented for a linear discrete-time system with time delay.

To sum up, this book mainly studies the system modeling, state estimation and fault diagnosis methods based on set-membership filtering, and makes a supplement to these methods, but there are still many problems to be further studied and explored.

1. This book mainly studies linear systems, and the study of nonlinear systems is limited to simple linearization of nonlinear systems. In the actual production process, there are many complex nonlinear systems, such as Wiener nonlinear systems, Hammerstein-Wiener systems and Error-in-variables systems, etc. Thus these problems need to be further studied.

2. A lot of the research work in this book is carried out around the proposal and improvement of the set-membership filtering algorithm. However, in-depth analysis of the nature of the algorithm (convergence and robustness to the underestimation of the error bound, etc.) and the processing of special cases (the existence of measurement outliers and unknown error bounds) are currently rare, the above issues are worth discussing.

References

[1] A. Abur and Y. Lin. Robust state estimation against measurement and network parameter error. *IEEE Transaction on Power Systems*, 35(5):4751–4759, 2018.

[2] M. Ahwiadi and W. Wang. An adaptive particle filter technique for system state estimation and prognosis. *IEEE Transactions on Instrumentation and Measurement*, 69(9):6756–6765, 2020.

[3] T. Alamo, J. M. Bravo, and E. F. Camacho. Bounded error identification of systems with tome-varying parameters. *IEEE Transactions on Automatic Control*, 51(7):1144–1150, 2006.

[4] T. Alamo, J. M. Bravo, and E. F. Camacho. Guaranteed state estimation by zonotopes. *Automatica*, 41:1035–1043, 2014.

[5] T. Alamo, J. M. Bravo, and M. J. Redondo. A set-membership state estimation algorithm based on dc programming. *Automatica*, 44(1):216–224, 2008.

[6] A. M. Andrew. *Applied Interval Analysis with Examples in Parameter and State Estimation, Robust Control and Robotics*. 2001.

[7] V. Antonio and Z. Giovanni. Sequential approximation of feasible parameter sets for identification with set membership uncertainty. *IEEE Transactions on Automatic Control*, 41(6):774–785, 1996.

[8] B. Zhou, J. D. Han, and G. J. Liu. A UD factorization-based nonlinear adaptive set-membership filter for ellipsoidal estimation. *International Journal of Robust and Nonlinear Control*, 18(16):1513–1531, 2010.

[9] R. V. Beard. *Failure accommodation in linear systems through self-reorganization*. Massachusetts Institute of Technology, 1971.

[10] N. Bedioui, R. Houimli, and M. Besbes. Simultaneous sensor and actuator fault estimation for continuous-time polytopic LPV system. *International Journal of Systems Science*, 50(6):1290–1302, 2019.

[11] G. Belforte and B. Bona. An improved parameter identification algorithm for signals with unknown-but-bounded errors. *IFAC Proceedings Volumes*, 18(5):1507–1512, 1985.

[12] E. Bernardi and E. J. Adam. Observer-based fault detection and diagnosis strategy for industrial processes. *Journal of the Franklin Institute*, 357(14):10054–10081, 2020.

[13] D. Bertsekas and I. Rhodes. Recursive state estimation for a set-membership description of uncertainty. *IEEE Transactions on Automatic Control*, 16(2):117–128, 1971.

[14] J. Blesa, J. Saludes, and V. Puig. Robust fault detection using polytope-based set-membership consistency test. In *IET*, pages 1767–1777, 2013.

[15] J. M. Bravo, T. Alamo, and E. F. Camacho. Bounded error identification of systems with time-varying parameters. *IEEE Transactions on Automatic Control*, 51(7):1144–1150, 2006.

[16] M. Buciakowski, M. Witczak, M. Mrugalski, and D. Theilliol. A quadratic boundedness approach to robust dc motor fault estimation. *Control Engineering Practice*, 66:181–194, 2017.

[17] Z. Cao, L. Xin, and Y. Fu. Adaptive dynamic surface control for vision-based stabilization of an uncertain electrically driven nonholonomic mobile robot. *Robotica*, 34(2):449–467, 2016.

[18] E. V. Carlos and P. Radoslav. Effective recursive set-membership state estimation for robust linear MPC. *IFAC Papers Online*, 52(1):486–491, 2019.

[19] P. Casau, P. Rosa, S. M. Tabatabaeipour, C. Silvestre, and J. Stoustrup. A set-valued approach to FDI and FTC of wind turbines. *IEEE Transactions on Control Systems Technology*, 23(1):245–263, 2015.

[20] S. B. Chabane, C. S. Maniu, T. Alamo, E. F. Camacho, and D. Dumur. Sensor fault detection and diagnosis using zonotopic set-membership estimation. In *IEEE Mediterranean Conference on Control and Automation*, pages 261–266, 2014.

[21] W. Chai, X. F. Sun, and J. F. Qiao. Improved zonotopic method to set membership identification for systems with time-varying parameters. *IET Control Theory and Applications*, 5(17):2039–2044, 2011.

[22] W. Chai, X. F. Sun, and J. F. Qiao. Set membership state estimation with improved zonotopic description of feasible solution set. *International Journal of Robust and Nonlinear Control*, 23:1642–1654, 2013.

[23] X. Q. Chen, R. Sun, and M. Liu. Two-stage exogenous Kalman filter for time-varying fault estimation of satellite attitude control system. *Journal of the Franklin Institute*, 357:2354–2370, 2020.

[24] M. Cheung, S. Yurkovich, and K. Passino. An optimal volume ellipsoid algorithm for parameter set estimation. *IEEE Transactions on Automatic Control*, 38(8):1292–1296, 1992.

[25] L. Chisci, A. Garulli, and G. Zappa. Recursive state bounding by parallelotopes. *IFAC Papers Online*, 32(7):1049–1055, 1996.

[26] C. Combastel. A state bounding observer for uncertain non-linear continuous-time systems based on zonotopes. In *Decision and Control, 2005 and 2005 European Control Conference. CDC-ECC '05. 44th IEEE Conference on*, pages 7228–7234, 2005.

[27] C. Combastel. Zonotopes and Kalman observers: Gain optimality under distinct uncertainty paradigms and robust convergence. *Automatica*, 55:265–273, 2015.

[28] C. Combastel. An extended zonotopic and Gaussian Kalman filter (EZGKF) merging set-membership and stochastic paradigms: Toward non-linear filtering and fault detection. *Annual Reviews in Control*, 42:232–243, 2016.

[29] T. Cui, F. Ding, X. Li and T. Hayat. Kalman filtering based gradient estimation algorithms for observer canonical state-space systems with moving average noises. *Journal of the Franklin Institute*, 356:5485–5502, 2019.

[30] J. R. Deller. Set membership identification in digital signal processing. *IEEE ASSP Magazine*, 6(4):4–20, 1989.

[31] X. X. Duan, S. L. Feng, Q. Huang, R. S. Luo, and J. J. Liu. Modeling and controlling of bi-directional dc/dc converter based on hybrid automata. *Electrical Measurement and Instrumentation*, 57:125–131, 2020.

[32] S. Eelco and E. C. Mark. A nonlinear set-membership filter for on-line applications. *International Journal of Robust and Nonlinear Control*, 13:1337–1358, 2003.

[33] J. Feng, Y. Yao, S. X. Lu, and Y. Liu. Domain knowledge-based deep-broad learning framework for fault diagnosis. *IEEE Transactions on Industrial Electronics*, 4(68):3454–3464, 2021.

[34] E. Fogel. System identification via membership set constraints with energy constrained noise. *Automatic Control, IEEE Transactions on*, 24(5):752–758, 1979.

[35] E. Fogel and Y. F. Huang. On the value of information in system identification-bounded noise case. *Automatica*, 18(2):229–238, 1982.

[36] C. M. García, V. Puig, C. M. Astorga-Zaragoza, and G. L. Osorio-Gordillo. Robust fault estimation based on interval Takagi-Sugeno unknown input observer. *IFAC-PapersOnLine*, 51(24):508–514, 2018.

[37] L. I. Gliga, B. D. Ciubotaru, H. Chafouk, D. Popescu, and C. Lupu. Fault diagnosis of a direct drive wind turbine using a bank of Goertzel filters. In *2019 6th International Conference on Control, Decision and Information Technologies (CoDIT)*, pages 1729–1734, 2019.

[38] A. S. Hadi and M. S. Shaker A new estimation/decoupling approach for robust observer-based fault reconstruction in nonlinear systems affected by simultaneous time varying actuator and sensor faults. *Journal of the Franklin Institute*, 357(13):8956–8979, 2020.

[39] L. Y. Hao, J. C. Han, G. Huo, and L. L. Li. Robust sliding mode fault-tolerant control for dynamic positioning system of ships with thruster faults. *Control and Decision*, 35(6):1291–1296, 2020.

[40] D. M. Himmelblau. *Fault Detection and Diagnosis in Chemical and Petrochemical Process*. Amsterdam: Elsevier, 1978.

[41] A. Hy and Y. Dan. Adaptive fault-tolerant fixed-time tracking consensus control for high-order unknown nonlinear multi-agent systems with performance constraint. *Journal of the Franklin Institute*, 357(16):11448–11471, 2020.

[42] L. Jaulin. Interval constraint propagation with application to bounded-error estimation. *Automatica*, 36(10):1547–1552, 2000.

[43] L. Jaulin and E. Walter. Set inversion via interval analysis for nonlinear bounded-error estimation. *Automatica*, 29(4):1053–1064, 1993.

[44] L. Jaulin, M. Kieffer, and I. Braems. Guaranteed non-linear estimation using constraint propagation on sets. *International Journal of Control*, 74(18):1772–1782, 2001.

[45] B. Jiang, W. U. Yun-Kai, L. U. Ning-Yun, and Z. H. Mao. Review of fault diagnosis and prognosis techniques for high-speed railway traction system. *Control and Decision*, 5(33):841–855, 2018.

[46] Z. P. Jiang. Robust exponential regulation of nonholonomic systems with uncertainties. *Automatica*, 36:3520–3524, 2000.

[47] F. Joyce and S. Dmitry. Wavefront reconstruction with defocus and transverse shift estimation using Kalman filtering. *Optics and Lasers in Engineering*, 11(1):122–129, 2018.

[48] M. Khalili, X. Zhang, and Y. Cao. Distributed fault-tolerant control of multi-agent systems: An adaptive learning approach. *IEEE Transactions on Neural Networks and Learning Systems*, 31(2):420–432, 2020.

[49] M. Kieffer, L. Jaulin, and E. Walter. Guaranteed recursive nonlinear state bounding using interval analysis. *International Journal of Adaptive Control and Signal Processing*, 16(3):193–218, 2002.

[50] W. Kühn. Rigorously computed orbits of dynamical systems without the wrapping effect. *Computing*, 61(1):47–67, 1998.

[51] R. Lamouchi, T. Raissi, M. Amairi, and M. Aoun. Interval observer framework for fault-tolerant control of linear parameter-varying systems. *International Journal of Control*, 91(3):524–533, 2018.

[52] V.T.H. Le, C. Stoica, T. Alamo, E. F. Camacho, and D. Dumur. Uncertainty representation based on set theory. In F. Castanié, editor, *Zonotopes: From Guaranteed State Estimation to Control*, pages 17–18. New York: Wiley, 2013.

[53] P. Lenka and L. Jirsa. Recursive Bayesian estimation of autoregressive model with uniform noise using approximation by parallelotopes. *International Journal of Adaptive Control and Signal Processing*, 31(8):1184–1192, 2017.

[54] H. Li, Y. Hui, J. Qu, and H. Sun. Fault diagnosis using particle filter for MEA typical components. *Journal of Engineering*, (13):603–606, 2018.

[55] J. Li, Z. Wang, Y. Shen, and M. Rodrigues. Zonotopic fault detection observer for linear parameter-varying descriptor systems. *International Journal of Robust and Nonlinear Control*, 29(11):3426–3445, 2019.

[56] M. H. Li, S. F. Hu, and J. W. Xia. Dissolved oxygen model predictive control for activated sludge process model based on the fuzzy c-means cluster algorithm. *International Journal of Control, Automation, and Systems*, 18(9):2435–2444, 2020.

[57] X. Li, F. Zhu, and J. Zhang. State estimation and simultaneous unknown input and measurement noise reconstruction based on adaptive H observer. *International Journal of Control Automation and Systems*, 14(3):647–654, 2016.

[58] X. G. Lin, Y. Z. Jiao, K. Liang, and H. Li. Application of the nonlinear filtering algorithm with a correlation noise in the dynamic positioning. *Control Theory and Applications*, 33(8):1081–1088, 2016.

[59] X. Liu, B. Zhao, and D. Liu. Fault tolerant tracking control for nonlinear systems with actuator failures through particle swarm optimization-based adaptive dynamic programming. *Applied Soft Computing*, 97:106–116, 2020.

[60] Y. W. Liu. Design of fault detection system for a heavy duty gas turbine with state observer and tracking filter. In *ASME Turbo Expo: Turbomachinery Technical Conference and Exposition*, 2017.

[61] Z. Liu, Z. Wang, Y. Wang, H. P. Ju, and Z. Ji. Directional expansion-based fault diagnosis algorithm using orthotopic and ellipsoidal filtering. *IET Control Theory and Applications*, 14(18):2836–2846, 2020.

[62] N. Y. Lu, X. F. Meng, B. Jiang, and H. P. Zhao. Compound fault diagnosis method based on multi-signal model and blind source separation. *Control and Decision*, 11(31):1945–1952, 2016.

[63] D. G. Luenberger. Observing the state of a linear system. *IEEE Transactions on Military Electronics*, 8(2):74–80, 1964.

[64] M. Wang and A. Tayeli. Nonlinear state estimation for inertial navigation systems with intermittent measurements. *Automatica*, 122:109–244, 2020.

[65] N. Magnus, K. P. Niels, and R. Ole. New developments in state estimation for nonlinear systems. *Automatica*, 36(11):1627–1638, 2000.

[66] D. G. Maksarov and J. P. Norton. Computationally efficient algorithms for state estimation with ellipsoidal approximations. *International Journal of Adaptive Control and Signal Processing*, 16(6):411–434, 2002.

[67] X. Meng, D. Zhai, Z. Fu, and X. Xie. Adaptive fault tolerant control for a class of switched nonlinear systems with unknown control directions. *Applied Mathematics and Computation*, 370, 2020.

[68] D. Merhy, M. Cristina . Stoica, T. Alamo, E. F. Camacho, and C. S. Ben. Guaranteed set-membership state estimation of an octorotor's position for radar applications. *International Journal of Control*, 93(11):2760–2770, 2020.

[69] J. Meseguer, V. Puig, and T. Escobet. Approximating fault detection linear interval observers using λ-order interval predictors. *International Journal of Robust and Nonlinear Control*, 31(7):1040–1060, 2017.

[70] M. Milanese and G. Belforte. Estimation theory and uncertainty intervals evaluation in presence of unknown but bounded errors: Linear families of models and estimators. *IEEE Transactions on Automatic Control*, 27(2):408–414, 1982.

[71] M. Milanese and A. Vicino. Estimation theory for nonlinear models and set membership uncertainty. *Automatica*, 27(2):403–408, 1991.

[72] R. Mostafa, A. Moosa, H. Reza, and M. Mohammad. Finite time estimation of actuator faults, states, and aerodynamic load of a realistic wind turbine. *Renewable Energy*, 130:256–267, 2018.

[73] N. Moustakis, B. Zhou, and T. L. Quang. Fault detection and identification for a class of continuous piecewise affine systems with unknown subsystems and partitions. *International Journal of Adaptive Control and Signal Processing*, 32(7):980–993, 2018.

[74] P. Murdoch. Observer design for a linear functional of the state vector. *IEEE Transactions on Automatic Control*, 18(3):308–310, 1973.

[75] T. Raissi, N. Ramdani, and Y. Candau. Set membership state and parameter estimation for systems described by nonlinear differential equations. *Automatica*, 40(10):1771–1777, 2004.

[76] T. Rassi, N. Ramdani, and Y. Candau. Set membership parameter estimation in the frequency domain based on complex intervals. *International Journal of Control Automation and Systems*, 7(5):824–834, 2009.

[77] L. Ravanbod, C. Jauberthie, N. Verdière, and L. Travé-Massuyès. Improved solutions for ill-conditioned problems involved in set-membership estimation for fault detection and isolation. *Journal of Process Control*, 58:139–151, 2017.

[78] M. Rayyam and M. Zazi. A novel metaheuristic model-based approach for accurate online broken bar fault diagnosis in induction motor using unscented Kalman filter and ant lion optimizer. *Transactions of the Institute of Measurement and Control*, 8(42):1537–1546, 2019.

[79] V. Reppa and A. P. Tzes. Fault diagnosis based on set membership identification using output-error models. *International Journal of Adaptive Control and Signal Processing*, 30(2):224–255, 2016.

[80] S. Schoen and H. Kutterer. Using zonotopes for overestimation–free interval least-squares–some geodetic applications. *Reliable Computing*, 11(2):137–155, 2005.

[81] E. Scholte and M. E. Ca Mpbell. A nonlinear set-membership filter for online applications. *International Journal of Robust and Nonlinear Control*, 13(15):1337–1358, 2003.

[82] J. K. Scott, R. Findeisen, and R. D. Braatz. Input design for guaranteed fault diagnosis using zonotopes. *Automatica*, 50(6):1580–1589, 2014.

[83] S. A. A. Shahriari. Modelling and dynamic state estimation of a doubly fed induction generator wind turbine. *COMPEL International Journal of Computations and Mathematics in Electrical*, 39(6):1393–1409, 2020.

[84] B. Sharan and T. Jain. Spectral analysis-based fault diagnosis algorithm for 3-phase passive rectifiers in renewable energy systems. *IET Power Electronics*, 16(13):3818–3829, 2020.

[85] U. Sharma, S. Thangavel, A. Mukkula, and R. Paulen. Effective recursive parallelotopic bounding for robust output-feedback control. *IFAC-PapersOnLine*, 51(15):1032–1037, 2018.

[86] Q. Shen, J. Y. Liu, Q. Zhao, and Q. Wang. Central difference set-membership filter for nonlinear system. *Control Theory and Applications*, 36(8):1239–1249, 2019.

[87] E. Siahlooei and S. A. S. Fazeli. An application of interval arithmetic for solving fully fuzzy linear systems with trapezoidal fuzzy numbers. *Advances in Fuzzy Systems*, 2018:1–10, 2018.

[88] M. P. Spathopoulos and I. D. Grobov. A state-set estimation algorithm for linear systems in the presence of bounded disturbances. *International Journal of Control*, 63(4):799–811, 1996.

[89] J. S. Shamma and K. Y. Tu. Approximate set-valued observers for nonlinear systems. *IEEE Transactions on Automatic Control*, 42(5):648–658, 1997.

[90] G. D. Tang, X. H. Zhang, and T. Zhang. Research on double loop decoupling control scheme for bi-directional dc/dc converter. *Power Electronics*, 53(11):20–31, 2019.

[91] M. Tang, Z. Chen, and F. Yin. Robot tracking in SLAM with Masreliez-Martin unscented Kalman filter. *International Journal of Control Automation and Systems*, 18(1):2315–2325, 2020.

[92] W. Tang, Z. Wang, Y. Wang, T. Raïssi, and Y. Shen. Interval estimation methods for discrete-time linear time-invariant systems. *IEEE Transactions on Automatic Control*, 64(11):4717–4724, 2019.

[93] E. G. Tian, Z. D. Wang, and L. Zou. Probabilistic-constrained filtering for a class of nonlinear systems with improved static event-triggered communication. *International Journal of Robust and Nonlinear Control*, 29:1484–1498, 2018.

[94] S. Udit, T. Sakthi, R. G. M. Anwesh, and P. Radoslav. Effective recursive parallelotopic bounding for robust output-feedback control. *IFAC Papers Online*, 51(15):1032–1037, 2018.

[95] A. Vicino and G. Zappa. Sequential approximation of feasible parameter sets for identification with set membership uncertainty. *Automatic Control IEEE Transactions on*, 41(6):774–785, 1996.

[96] J. Wang, Y. Shi, M. Zhou, Y. Wang, and V. Puig. Active fault detection based on set-membership approach for uncertain discrete-times systems. *International Journal of Robust and Nonlinear Control*, 30(14):5322–5340, 2020.

[97] J. X. Wang, Z. W. Wang, M. A. Xiu-Zhen, and Z. G. Yuan. Decoupling and diagnosis of multi-fault of diesel engine fuel system. *Control and Decision*, 10(34):2249–2255, 2019.

[98] L. Wang, T. Miao, and L. Liu. Sliding mode control of bi-directional dc/dc converter in dc microgrid based on exact feedback linearization. *WSEAS Transactions on Circuits and Systems*, 19:206–211, 2020.

[99] S. F. Wang, Y. Luo, L. Dong, and L. I. Guo-Liang. Research on off-line identification of equivalent model parameters of second order Thevenin lithium batteries. *Electronic Design Engineering*, 2018.

[100] X. Wang, Z. X. Song, and K. Yang. State of charge estimation for lithium-bismuth liquid metal batteries. Energies, 12(1), 2019.

[101] Y. Wang and V. Puig. Zonotopic extended Kalman filter and fault detection of discrete-time nonlinear systems applied to a quadrotor helicopter. In *Control and Fault-tolerant Systems*, pages 367–372, 2016.

[102] Y. Wang, P. Vicen, and C. Gabriela. Robust fault estimation based on zonotopic Kalman filter for discrete-time descriptor systems. *International Journal of Robust and Nonlinear Control*, 28(16):5071–5086, 2018.

[103] Y. Wang, Z. H. Wang, P. Vicenç, and G. Cembrano. Zonotopic fault estimation filter design for discrete-time descriptor systems. *IFAC-Papers Online*, 50(1):5055–5060, 2017.

[104] Z. Wang, M. Rodrigues, and D. Theilliol. Actuator fault estimation observer design for discrete-time linear parameter-varying descriptor systems. *International Journal of Adaptive Control and Signal Processing*, 29(2):242–258, 2015.

[105] Z. Wang, Y. Wang, and Z. Ji. A novel two-stage estimation algorithm for nonlinear Hammerstein-Wiener systems from noisy input and output data. *Journal of the Franklin Institute*, 354(4):1937–1944, 2017.

[106] Z. Y. Wang, Z. Tang, and H. P. Ju. A novel two-stage ellipsoid filtering-based system modeling algorithm for a Hammerstein nonlinear model with an unknown noise term. *Nonlinear Dynamics*, 98(4):55–63, 2019.

[107] Z. Y. Wang, G. X. Xu, Z. X. Liu, Y. Wang, and Z. C. Ji. Orthotopic linear programming filtering based fault diagnosis method. *Control and Decision*, 35(4):807–815, 2020.

[108] J. Wei, G. Dong, and Z. Chen. Lyapunov-based thermal fault diagnosis of cylindrical lithium-ion batteries. *IEEE Transactions on Industrial Electronics*, 67(6):4670–4679, 2020.

[109] A. S. Willsky. A survey of design methods for failure detection in dynamic systems. *Automatica*, 12(6):601–611, 1975.

[110] D. H. Wu, W. Liu, S. Jin, and Y. Shen. Fault estimation and fault-tolerant control of wind turbines using the SDW-LSI algorithm. *IEEE Access*, 4:7223–7231, 2016.

[111] K. Xiong, L. Liu, and H. Zhang. Adaptive robust extended Kalman filter for nonlinear stochastic systems. *IET Control Theory and Applications*, 2(3):239–250, 2008.

[112] L. Xu, W. Xiong, A. Alsaedi, and T. Hayat. Hierarchical parameter estimation for the frequency response based on the dynamical window data. *International Journal of Control Automation and Systems*, 16(4):1756–1764, 2018.

[113] Y. D. Xu, M. Y. Hu, C. Y. Fu, K. B. Cao, and Z. Yang. State of charge estimation for lithium-ion batteries based on temperature-dependent second-order RC model. *Electronics*, 8(9):1012, 2019.

[114] M. Yang, C. Jin, and G. Dong. Weak fault feature extraction of rolling bearing based on cyclic wiener filter and envelope spectrum. *Mechanical Systems and Signal Processing*, 25(5):1773–1785, 2011.

[115] P. Yang, Z. Liu, D. Li, Z. Zhang, and Z. Wang. Sliding mode predictive active fault-tolerant control method for discrete multi-faults system. *International Journal of Control, Automation and Systems*, 19(3):1228–1240, 2021.

[116] X. Yang, T. Li, Y Wu, Y. Wang, and Y. Long. Fault estimation and fault tolerant control for discrete-time nonlinear systems with perturbation by a mixed design scheme. *Journal of the Franklin Institute*, 358(3):1860–1887, 2020.

[117] Y. Yang, S. X. Ding, and L. L. Li. Parameterization of nonlinear observer-based fault detection systems. *IEEE Transactions on Automatic Control*, 61(11):3687–3692, 2016.

[118] Z. H. Yang, S. H. Wang, C. Ma, and J. N. Huang. Development of fault diagnosis system for wind power planetary transmission based on labview. *The Journal of Engineering*, 2019(23):9170–9172, 2019.

[119] J. H. Yook, I. H. Kim, M. S. Han, and Y. I. Son. Robustness improvement of dc motor speed control using communication disturbance observer under uncertain time delay. *Electronics Letters*, 53(6):389–391, 2017.

[120] F. Q. You, H. L. Zhang, and F. L. Wang. A new set-membership estimation method based on zonotopes and ellipsoids. *Transactions of The Institute of Measurement and Control*, 40(7):2091–2099, 2018.

[121] K. K. Zhang, B. Jiang, X. G. Yan, and Z. H. Mao. Incipient fault detection for traction motors of high-speed railways using an interval sliding mode observer. *IEEE Transactions on Intelligent Transportation Systems*, 20(7):2703–2714, 2019.

[122] K. K. Zhang, B. Jiang, X. G. Yan, and C. Edwards. Interval sliding mode observer-based fault accommodation for non-minimum phase LPV systems with online control allocation. *International Journal of Control*, 93(11):2675–2689, 2020.

[123] L. J. Zhang, H. Peng, Z. S. Ning, Z. Q. Mu, and C. Y. Sun. Comparative research on RC equivalent circuit models for lithium-ion batteries of electric vehicles. *Applied Sciences*, 7(10):1002, 2017.

[124] W. H. Zhang, Z. H. Wang, S. Y. Guo, and Y Shen. Interval estimation of sensor fault based on zonotopic Kalman filter. *International Journal of Control*, 94(6):1641–1650, 2019.

[125] L. Zhao, J. Wang, T. Yu, H. Jian, and T. Liu. Design of adaptive robust square-root cubature Kalman filter with noise statistic estimator. *Applied Mathematics and Computation*, 256:352–367, 2015.

[126] B. Zhou, K. Qian, and X. D. Ma. Ellipsoidal bounding set-membership identification approach for robust fault diagnosis with application to mobile robots. *Journal of Systems Engineering and Electronics*, 28(5):986–995, 2017.

[127] B. Zhou, K. Qian, X. D. Ma, and X. Z. Dai. A new nonlinear set membership filter based on guaranteed bounding ellipsoid algorithm. *Acta Automatica Sinica*, 39(2):146–154, 2013.

[128] S. Y. Zhou, Y. D. Song, and X. S. Luo. Fault-tolerant tracking control with guaranteed performance for nonlinearly parameterized systems under uncertain initial conditions. *Journal of the Franklin Institute*, 357(11):6805–6823, 2020.

[129] F. L. Zhu, W. Zhang, J. C. Zhang, and S. H. Guo. Unknown input reconstruction via interval observer and state and unknown input compensation feedback controller designs. *International Journal of Control Automation and Systems*, 19(1):145–157, 2021.

Printed in the United States
by Baker & Taylor Publisher Services